£13-45

HETEROCYCLES IN
ORGANIC SYNTHESIS

GENERAL HETEROCYCLIC CHEMISTRY SERIES

Edward C. Taylor and Arnold Weissberger, Editors

MASS SPECTROMETRY OF HETEROCYCLIC COMPOUNDS
by Q. N. Porter and J. Baldas

NMR SPECTRA OF SIMPLE HETEROCYCLES
by T. J. Batterham

HETEROCYCLES IN ORGANIC SYNTHESIS
by A. I. Meyers

HETEROCYCLES IN ORGANIC SYNTHESIS

A. I. MEYERS

Colorado State University

A Wiley-Interscience Publication

JOHN WILEY & SONS, New York . London . Sydney . Toronto

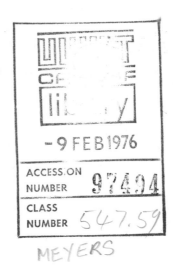
Copyright © 1974, by John Wiley & Sons, Inc.

All rights reserved. Published simultaneously in Canada.

No part of this book may be reproduced by any means, nor transmitted, nor translated into a machine language without the written permission of the publisher.

Library of Congress Cataloging in Publication Data:
Meyers, Albert Irving, 1932-
Heterocycles in organic synthesis.

(General heterocyclic chemistry series)
"A Wiley-Interscience publication."
1. Chemistry, Organic—Synthesis.
2. Heterocyclic compounds. I. Title.
QD262.M56   547'.59   73-17355
ISBN 0-471-60065-2

Printed in the United States of America

10 9 8 7 6 5 4 3 2 1

To the *other* Meyers' Group —

Joan, Harold, Jill, and Lisa

INTRODUCTION TO THE SERIES

General Heterocyclic Chemistry

The series, "The Chemistry of Heterocyclic Compounds," published since 1950 by Wiley-Interscience, is organized according to compound classes. Each volume deals with syntheses, reactions, properties, structure, physical chemistry, etc., of compounds belonging to a specific class, such as pyridines, thiophenes, pyrimidines, three-membered ring systems. This series has become the basic reference collection for information on heterocyclic compounds.

Many aspects of heterocyclic chemistry have been established as disciplines of *general* significance and application. Furthermore, many reactions, transformations, and uses of heterocyclic compounds have specific significance. We plan, therefore, to publish monographs that will treat such topics as nuclear magnetic resonance of heterocyclic compounds, mass spectra of heterocyclic compounds, photochemistry of heterocyclic compounds, X-ray structure determination of heterocyclic compounds, UV and IR spectroscopy of heterocyclic compounds, and the utility of heterocyclic compounds in organic synthesis. These treatises should be of interest to *all* organic chemists as well as to those whose particular concern is heterocyclic chemistry. The new series, organized as described above, will survey under each title *the whole field of heterocyclic chemistry* and is entitled "General Heterocyclic Chemistry," The editors express their profound gratitude to Dr. D. J. Brown of Canberra for his invaluable help in establishing the new series.

Department of Chemistry   Edward C. Taylor
Princeton University
Princeton, New Jersey

Research Laboratories   Arnold Weissberger
Eastman Kodak Company
Rochester, New York

PREFACE

For too many years organic chemistry has witnessed a trend toward over-specialization within the discipline. We have jokingly referred to specialists in the 17-position of steroids, 3-position in pyridines, and so on. Although much of this specialization is necessary for maximum competence in many complex areas of organic chemistry and although it has undoubtedly led to many exciting developments, the students in both introductory and advanced courses are easily frustrated by attempts to absorb the results of this information explosion. In particular, heterocyclic chemistry has become what many students feel is the sole domain of the dye and medicinal chemist and has little pedogogical value as far as principles and utility in the broad area of organic synthesis are concerned. It is somewhat disappointing to find that this stigma attached to heterocyclic chemistry which was recognized more than 25 years ago in Professor A. A. Morton's monograph on *The Chemistry of Heterocyclic Compounds*[1] remains essentially unchanged. In the preface to this book, which was the first attempt to integrate heterocyclic chemistry into the broad areas of organic chemistry, he wrote:

A generous third of the compounds listed in Beilstein have heterocyclic nuclei; over half of the types of compounds produced by nature have heterocyclic systems; the greater share of the drugs and medicinals and nearly all the alkaloids contain such structures; the large majority of the colors produced by nature and by man have these nuclei as essential components. Yet this wealth of chemical interest, and much more, is scarcely touched in the average textbook. Such compounds as are mentioned are usually treated as individual substances, not as members of a great class of compounds. And if classes of compounds are mentioned, the treatment is too sketchy to be worthwhile or to provide a working knowledge of the subject. Accordingly, the average student of organic chemistry comes to regard heterocyclic compounds as a very special field, to be entered if he desires to be a specialist of a kind, but certainly not to be eagerly studied as the fruition of his efforts to understand the principles of organic chemistry.

Most chemists will agree that investigators in heterocyclic chemistry spend the majority of their effort studying routes to more elaborate systems, improved methods to increase the versatility of heterocyclic syntheses, or the transposition of substituents. On the other hand, chemists who work with nonheterocyclic systems rarely turn to the heterocyclic molecule as a possible solution to their synthetic problems. It is significant that Professor Morton, in his effort to find acceptance of heterocyclic chemistry in general synthesis, actually devoted a small section of his chapter on furans to the "Preparation of Nonfuran Products from Furan Compounds" and began by stating that "Heterocyclic compounds are frequently very useful as sources of non-heterocyclic materials." This was followed by citing the formation of a-(t-alkyl) carboxylic acids from 5-(t-alkyl) furoate esters.

This philosophy, briefly touched upon in 1946, has nevertheless been widely overlooked both by the teacher and by the researcher in organic chemistry. For several years I have been primarily concerned with establishing this concept in synthesis, which has resulted in an accumulation of data from both the literature and my own laboratory.

It is, therefore, the main goal of this work to demonstrate that many useful (as well as superior) methods for obtaining functionalized organic compounds and structures of diverse architecture can be found by employing a heterocycle either as a precursor, reagent, or vehicle for formation. This approach would serve two valuable purposes. First, it would illustrate to the beginning student that a place in the repertoire of organic compounds should be reserved for heterocycles. Second, it would stimulate the imagination of investigators into looking at a heterocyclic system with the renewed interest in its potential synthetic utility.

The book is divided into chapters devoted to functional groups or classes of organic compounds, in much the same manner in which an introductory organic chemistry text is presented. This means that a variety of different heterocycles appear in each chapter, and many heterocycles reappear in other chapters. The sole requirement for various heterocycles discussed in a given section is that they, in some way, assist or be responsible for the formation of the same class of organic compound. For those interested in the behavior of a particular heterocycle, the index contains references to various classes of compounds derived from each heterocyclic system.

To appreciate the synthetic utility of a heterocycle, some knowledge is required regarding its chemical properties. To present, in a single volume, the synthetic routes and properties of heterocyclic compounds would be undoubtedly presumptuous and incomplete. Only those synthetic schemes and chemical properties that have a bearing on the philosophy already set forth are

[1]A. A. Morton, *The Chemistry of Heterocyclic Compounds*, McGraw-Hill, New York, 1946.

mentioned. For the reader concerned with detailed description of synthesis and chemical properties of heterocyclic molecules, the traditional source material is recommended.[2,3]

Searching the literature for the information that constitutes the body of this monograph proved to be a rather difficult task primarily because the chemical literature is not aligned to this particular subject. For this reason, many interesting and useful synthetic methods derived from heterocycles are certain to have been overlooked. Significant contributions were made by many researchers who replied to a form letter requesting reprints, unpublished materials, and examples known to them that would add to this effort. I am grateful for their time and concern.

Acknowledgment is warmly extended to Mrs. Linda Benedict who typed the manuscript with care and concern. Finally, a debt of gratitude is rendered to Professor E. C. Taylor for his encouragement, enthusiasm, and unending patience in reading the manuscript.

A. I. Meyers

Fort Collins, Colorado
September 1973

[2] The most extensive works dealing specifically with heterocyclic compounds are in the following series of volumes:
(a) *The Chemistry of Heterocyclic Compounds,* A. Weissberger and E. C. Taylor, eds., Interscience, New York.
(b) *Heterocyclic Compounds,* R. C. Elderfield, ed., Wiley, New York.
(c) *Advances in Heterocyclic Chemistry,* A. R. Katritzky, ed., Academic Press, New York. In Volume 7 (p. 225) of this series, source material in heterocyclic chemistry has been summarized.

[3] Monographs of recent vintage that incorporate many of today's modern mechanistic concepts related to heterocycles should also be consulted:
(a) L. A. Paquette, *Modern Heterocyclic Chemistry,* Benjamin, New York, 1968.
(b) A. Albert, *Heterocyclic Chemistry,* Oxford University Press, New York, 1968.
(c) M. H. Palmer, *The Structure and Reactions of Heterocyclic Compounds,* Arnold, London, 1967.

CONTENTS

1 GENERAL CONSIDERATIONS CONCERNING SYNTHETICALLY USEFUL HETEROCYCLIC COMPOUNDS

In the usual planning of a synthetic program, the target molecule is mentally dissected into small and recognizable fragments, which are then considered individually with regard to their accessibility. These "accessible fragments" then become the subject of an assembling process such that they can be physically reconstructed to achieve the synthetic goal. The assembling process—or organic synthesis as chemists prefer to call it—depends invariably on appropriate carbon skeletons and introduction and manipulation of functional groups that serve as "links" to allow all the fragments to be incorporated into the desired product. For the large part, "accessible fragments" have always been nonheterocyclic compounds since heterocycles are not normally considered to be qualified candidates for service in a synthetic program. To blame their lack of utility on inaccessibility would be a gross distortion of fact, as evidenced by the enormous literature pertaining to heterocyclic chemistry.

By contemplating various heterocyclic compounds as an aid in synthesis one may be surprised to learn that considerable functionality, appropriate carbon skeletons, and needed driving forces to achieve a chemical transformation are already "built in" and await the command to perform. To be sure, the chemical behavior of heterocyclic compounds is governed by the same factors that govern *all* chemical behavior, and they respond alike to electronic and solvent effects, as well as to orbital and strain theories. Hence no revolutionary concepts of synthesis need be invoked—merely recognition of their inherent abilities to carry out the desired task. To focus attention by recognition of the unique properties of heterocycles that result in useful synthetic reactions is undoubtedly the *raison d'être* of this monograph.

1

I. REACTION TYPES

In order that the reader may gain some preliminary insight into the properties of heterocycles that cause them to be synthetically useful, some reaction types well known to both the novice and the master in organic synthesis are summarized here, with appropriate illustrations.

The four following categories appear to be the most common processes by which heterocyclic compounds have demonstrated their synthetic prowess:

1. Destruction of aromaticity.
2. Drive toward aromaticity.
3. Release of ring strain.
4. Temporary formation of a heterocyclic intermediate.

A. Destruction of Aromaticity

1

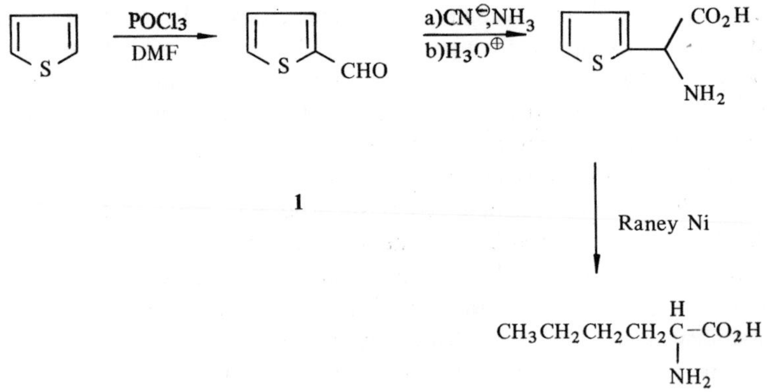

Thiophene, a highly common substance, serves as a versatile precursor to a variety of amino acids by allowing ready functionalization to the a-formyl derivative **1**. The latter may then be converted to the thiophene amino acid by the Strecker synthesis and reductive cleavage to a-aminohexanoic acid.[1] There are many variations of this route when applied to amino acids, carboxylic acids, hydrocarbons, ketones, and functionalized derivatives. These are surveyed in detail in the appropriate chapters later in the book.

Many useful synthetic applications of the isoxazole ring have been demonstrated by Stork and his students.[2] For example, the elegant annelation of the bicyclic ketone (**3**) with the appropriate isoxazole produced initially the alkylated octalindione (**4**). This was converted, via catalytic reduction of the isoxazole ring, into the tricyclic ketone, which was then transformed into

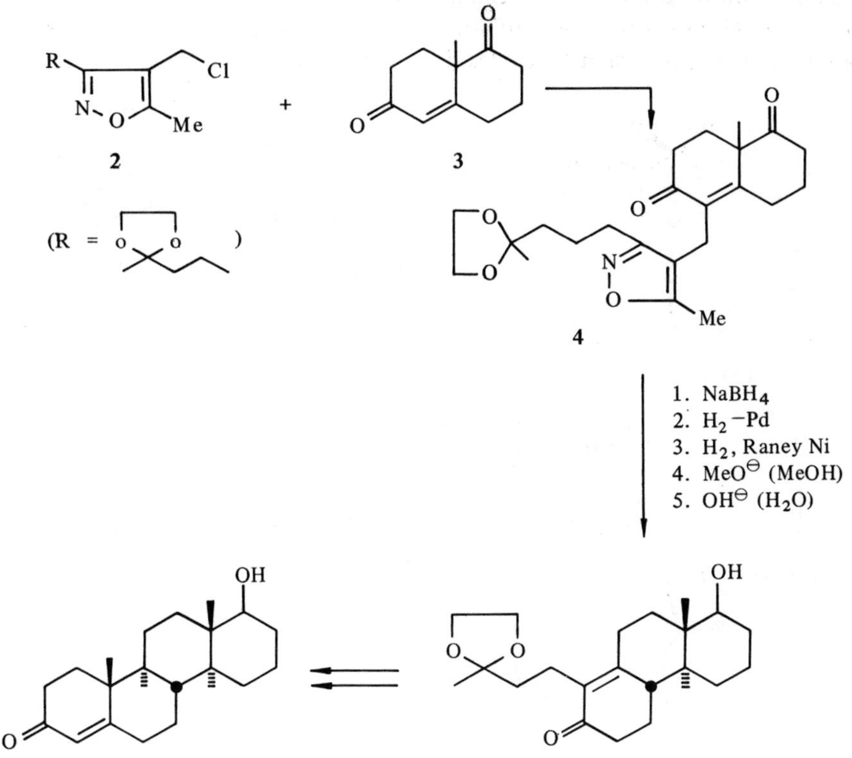

1. NaBH$_4$
2. H$_2$ –Pd
3. H$_2$, Raney Ni
4. MeO$^{\ominus}$ (MeOH)
5. OH$^{\ominus}$ (H$_2$O)

D-homotesterone **(5)** by the usual means. Thus the heterocycle in **(4)** is a *masked fused ring* held in abeyance until the appropriate period in the synthetic sequence.

We have, therefore, in both of the examples above, heterocyclic compounds with "built-in" or latent functionality[3] as well as the appropriate carbon skeletons ready to spring into action upon destruction of their aromatic or delocalized character–a rather pleasant set of circumstances.

B. Drive toward Aromaticity

Even though heterocycles have been passed over in large part by chemists confronted with synthetic problems, nature has made good use of their unique properties. In a classic example, the mild stereospecific reductions using dihydropyridines rushing toward aromaticity illustrate the point. The stereospecific reduction of pyruvate salts to chiral a-hydroxy acids[4] by diphosphopyridine nucleotide (DPN) **6** is typical of the types of important transformations that can occur when aromaticity provides the driving force.

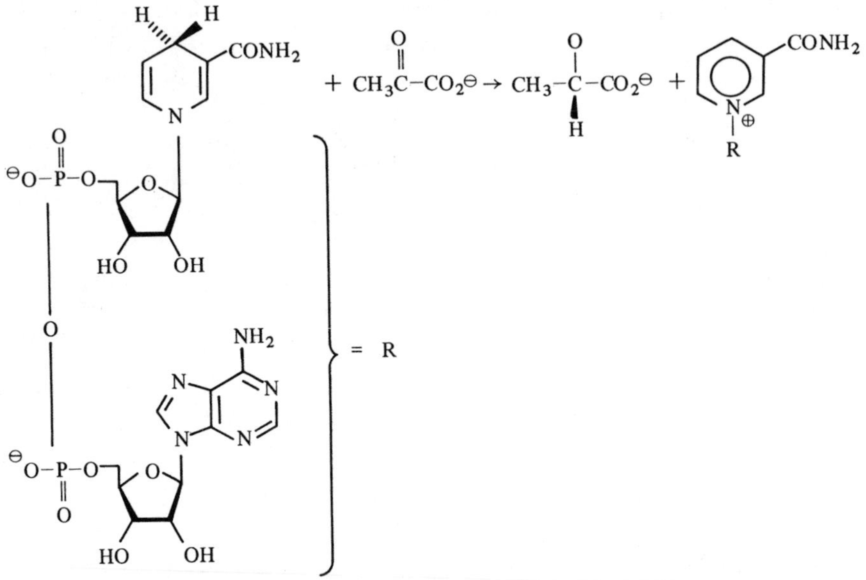

6

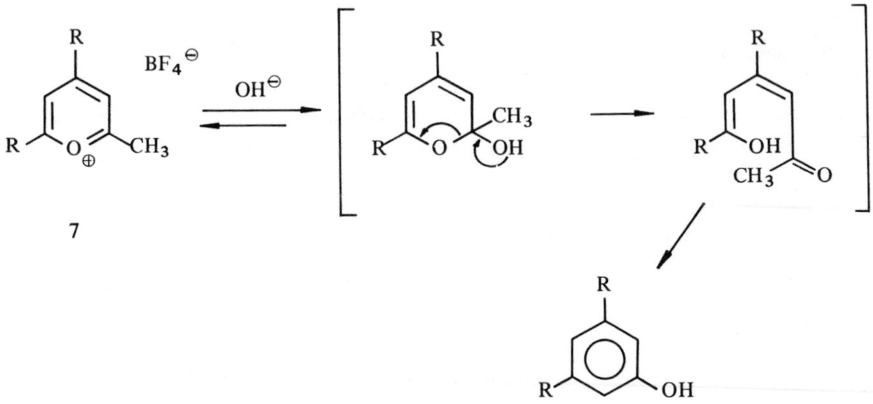

It should be obvious that homocyclic aromatic rings would be a prime target for heterocycles, which are accessible and possess the potential for isomerization in a thermodynamically "downhill" process. Such a heterocycle is found in pyrylium salts (**7**) which serve as precursors to a variety of phenols. The significant property of the pyrylium salt seen here is the electrophilic nature of the 2-position. This enhances addition by the hydroxyl ion (or other nucleophiles) producing a ring opened intermediate that undergoes a simple intramolecular aldol condensation to the phenol. In an analogous fashion a diverse host of aromatic compounds may be prepared.[5]

An example that utilizes a *heterocyclic synthesis* as the driving force to accomplish a useful transformation has been described.[6] In a study to find selective peptide cleaving reagents, the *o*-fluoroaniline derivative (8) was treated in aqueous dioxane at pH 9 giving rise to the adduct (9). An intramolecular acylation followed, releasing the amino acid fragment and the dihydroquinazolone (10). The lactam functionality in 10 is sufficiently stable to provide the necessary driving force, despite the lack of aromaticity in the resulting heterocycle.

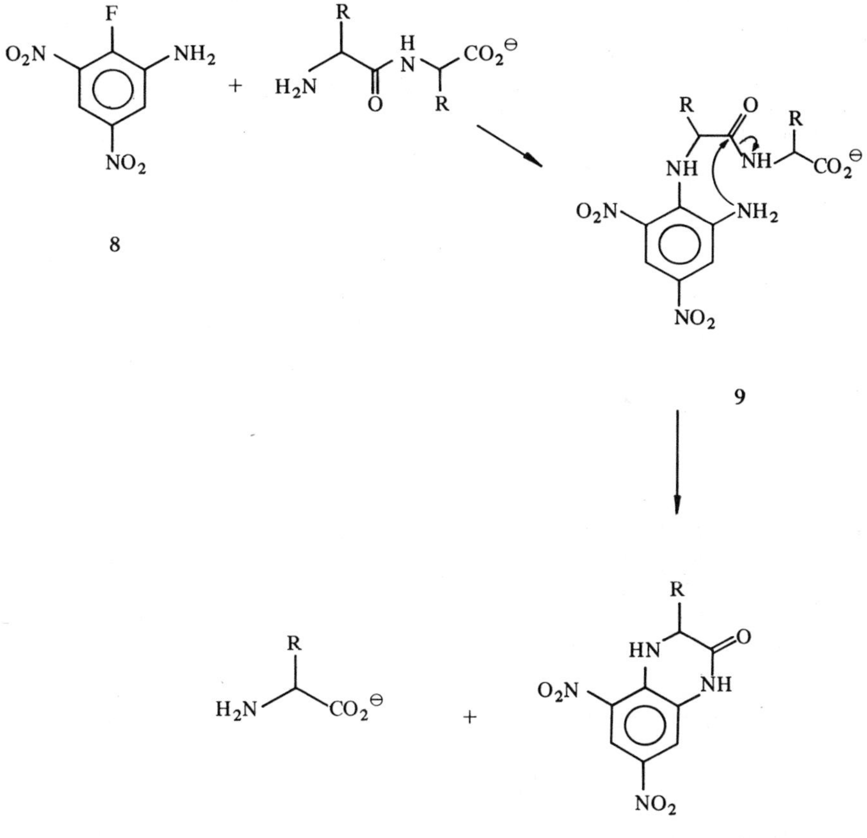

C. Release of Ring Strain

The preferred mode of reaction in small ring compounds is usually characterized by ring opening in order to relieve orbital distortions. Heteroatoms present in small rings provide no exception to this behavior.

The elegant studies of Griffin[7a] on photochemical cleavage of oxiranes to carbenes represent an excellent example of heterocycle utility originating from ring strain release. Irradiation of oxiranes in the presence of olefins produces the oxonium ylid (11), which fragments to the carbene and is subsequently trapped by a variety of olefins in a $(3 \rightarrow 2 + 1)$ process. A variety of cyclopropanes are thus produced in good yield.

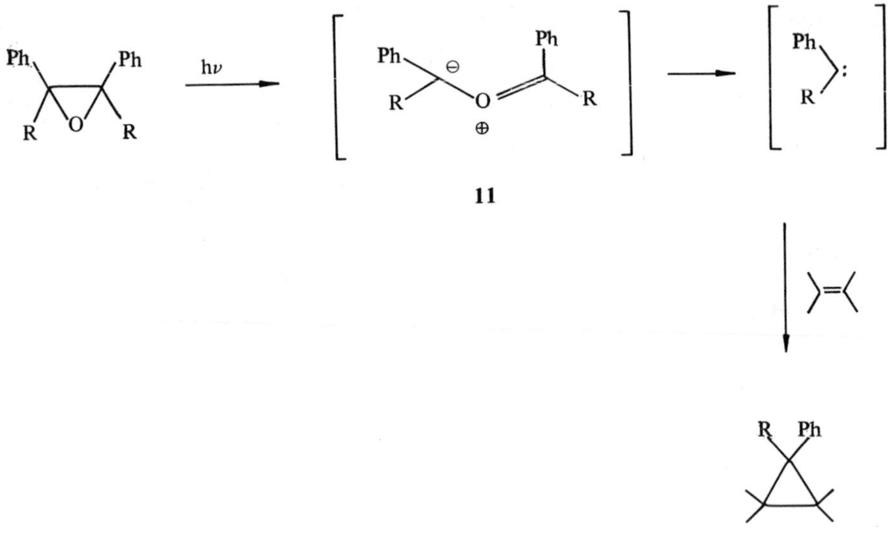

11

Strain release provides the driving force in the transformation of a,β-epoxy diazirines 12 to acetylenic ketones in the steroid series.[7b]

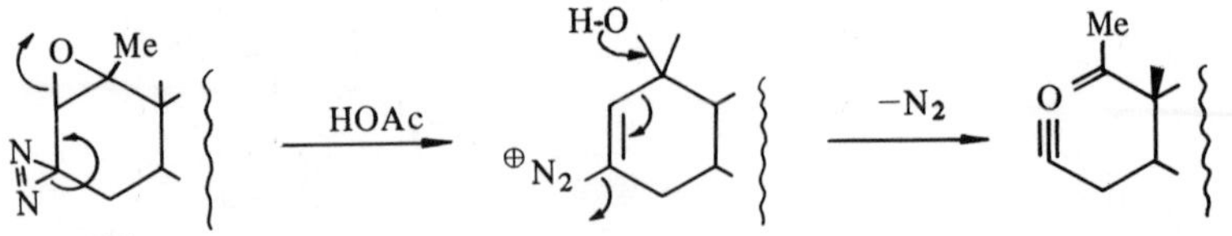

12

The diazirines are interesting, highly strained ring systems which undergo a variety of ring opening reactions and provide carbenes when pyrolyzed.[7c,7d]

An interesting and useful carbon insertion reaction (13 → 14) has been described by Sheehan[8a] which stems from the strain release of the three-membered heterocycle (15). Conversion of the α-bromoamide to the aziridone (15), followed by addition of methyl magnesium bromide gave the cyclopropyl imine (16), presumed to arise from rearrangement of the unsaturated aziridine after dehydration. The unstable nature of 16 resulted in hydration to the carbinol amine and subsequent ring cleavage to form the homologated amide 14. Although the precise nature of the intermediates has been questioned,[8b] this does not detract from the utility of the process, which illustrates the feasibility

of homologating an open chain compound by temporarily constructing a heterocycle that possesses the features necessary to perform the task. This concept represents some of the more ingenious uses of heterocyclic compounds and is discussed further in the next section.

D. Temporary Formation of a Heterocyclic Intermediate

When a heterocycle is employed merely to alter the reactivity of a molecule, that is, to make it more prone to reaction usually under mild conditions, it is undoubtedly chosen because of its unique properties. The extensive research by Staab[9] on the use of imidazoles as reagents in organic synthesis provides an excellent example. The formation of active esters (17) utilizing diimidazole carbonyl (18) allows, under mild conditions, the conversion of carboxylic acids to amides (19). This technique is particularly suitable for the preparation of peptide linkages which, under certain preparative conditions, are very susceptible to racemization. The reactive intermediate 17 is also capable of reaction with a variety of other nucleophiles.

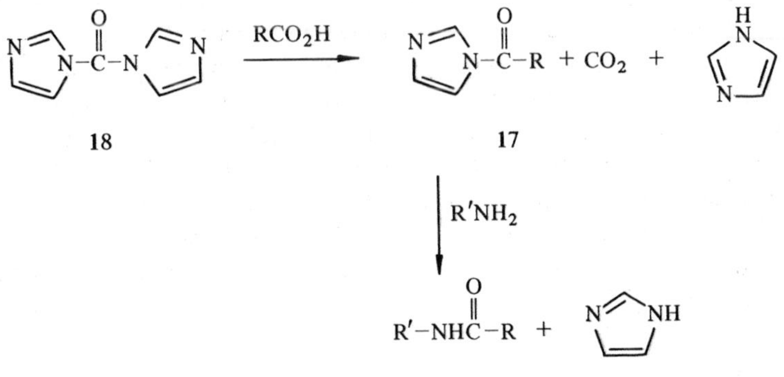

The conversion of a *trans*-amino alcohol (20) to its *cis*-isomer (21) is another process that invokes the temporary formation of a heterocyclic intermediate, namely, the 2-oxazoline system (22).[10] Since *trans*-amino alcohols are routinely prepared by nucleophilic ring opening of epoxides, this technique provides a facile entry into the *cis*-series.

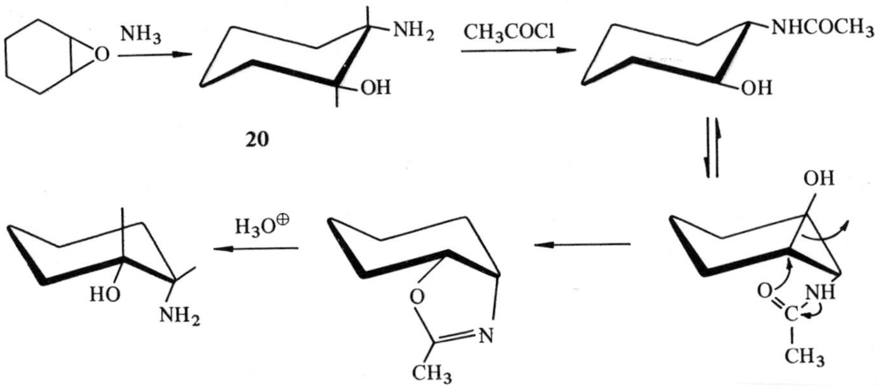

Extrusion reactions which involve loss of simple molecules (N_2, SO_2, CO_2), are quite interesting and synthetically valuable.[11] However, in most cases the process represents merely interconversion of starting materials and products.

For example, dienes react with azodicarboxylate esters to produce the tetra-hydropyridazines **(23)** in highly stereoselective fashion. Hydrolysis results in decarboxylation to the unsubstituted derivative **(24)** which may be readily oxidized to the cyclic azo compound **(25)**. This unstable product, upon warming, extrudes nitrogen to return, *in a stereospecific,* manner to the geometrically pure diene.[12] Thus a mixture of isomeric dienes may be separated and recovered pure by temporarily making use of the heterocyclic intermediate **(23)**.

The more significant synthetic dividend of this simple process is reaped when one considers the inherent properties of the temporarily constructed tetra-hydropyridazine. This is considered in the next section.

In the synthetic approach to corrins, Eschenmoser[13] described an elegant and useful transformation that originates from the well-known desulfurization of thiiranes **(26)** to olefins. The ability of a thiirane to serve as a vehicle to olefins

was dramatically demonstrated in complex molecules by the coupling of the two corrinoid precursors **27** and **28** to the advanced intermediate **30**. By planning the synthesis so that the thiirane moiety would be temporarily constructed as a labile intermediate in the valence tautomerism of this system **(29a-c)**, he was able to extrude the sulfur atom using triethylphosphite, thus coupling the two fragments.

Here, then, is an example of a heterocycle that need only be present in a fleeting moment in order to serve a synthetically useful purpose.

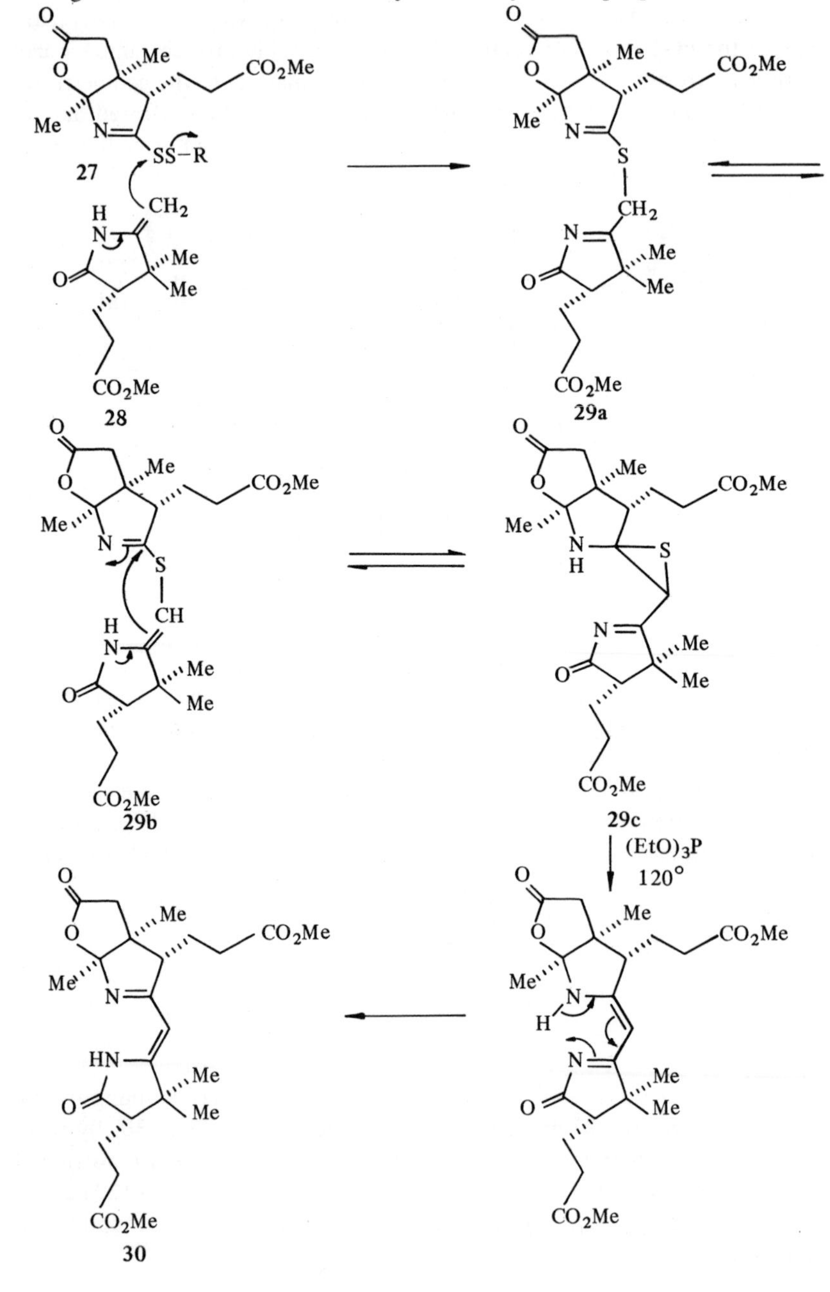

II. MODIFICATION AND ELABORATION OF SYNTHETICAL- LY USEFUL HETEROCYCLES

When a heterocyclic system possesses one or more of the characteristics just described, the chemist must next find the means (using old established methods or developing new ones) to modify or elaborate the molecule to allow for versatility in synthesis. In other words, if the heterocycle falls into the categories given above and also allows, by virtue of its chemistry, introduction of a variety of functional groups or carbon skeletons, then the heterocycle indeed serves as a *precursor* or *vehicle* for useful chemical transformations. The following are offered to illustrate this point.

To return to the previous example describing the tetrahydropyridazines, Berson[12] showed that addition of diazomethane to **23** led to a mixture of two

cyclopropane derivatives **31** (88%) and **32** (12%). By separation of the isomers, removal of the carboxyl function, and oxidation to the unstable azo compound, both heterocycles served as vehicles to deliver pure isomeric dienes **33** and **34** respectively. In the overall process just described, a conjugated diene has been homologated to a nonconjugated derivative utilizing a temporarily constructed heterocycle.

In an example given earlier, the thiophene molecule which is well known to cleave to hydrocarbons can also be readily elaborated. The Friedel-Crafts acylation[14] of thiophene leads to good yields of the diacyl derivatives (35). These lend themselves nicely as precursors to 1,6-diketones (36) after Raney nickel desulfurization.

Substituted dihydro-1,3-oxazines (37) or thiazines (38) were reported to reduce under controlled conditions using sodium borohydride to the tetrahydro derivations (39).[15,16] This, in effect, represents a reduction of a carboxylic acid derivative (37, 38) to an aldehyde derivative (39), since mild acid hydrolysis produces the aldehyde. This observation represents a potentially useful synthetic procedure if simple elaboration of these heterocycles with a variety of substituents were feasible. By developing means to alkylate the

37   X = 0
38   X = S
39

40   41   42

methyl group in 40, a series of elaborated oxazines (41) was obtained and, by reduction to 42 and destruction of the heterocycle, a versatile aldehyde synthesis was indeed in hand.[17]

It is, therefore, important that the heterocycle be sufficiently stable to withstand modification or elaboration of its nucleus and then be called upon to "do its thing." In the pages to follow, a survey of synthetically useful reactions involving heterocycles as precursors, vehicles, or reagents are discussed. For the sake of coherency and to remove the stigma regarding the "specialists domain," the chapters are divided into synthesis of classes of organic compounds—something less than an innovative format.

The author had originally intended to include a chapter entitled "Reactive Intermediates" from heterocyclic precursors. This has not been done for two reasons. *(a)* It would not conform with the traditional classes of organic compounds presented, since they would by definition be carbenes, nitrenes, arynes, carbonium ions, and carbanions. Since many heterocyclic compounds do, indeed, act as precursors to these species and traditional classes of compounds ultimately must be formed, the subject has been diffused into the appropriate chapters. *(b)* A recent paperback[18] has done an excellent job in surveying this topic, although the reactive intermediates presented are not solely limited to heterocyclic origins.

REFERENCES

1. Y. L. Goldfarb, B. P. Fabrichnyi, and I. F. Shalavina, *Tetrahedron,* **18**, 21 (1962).

2. G. Stork and J. E. McMurry, *J. Am. Chem. Soc.,* **89**, 5464 (1967).

3. An excellent account of this general concept has recently appeared: D. Lednicer, "Latent Functionality in Organic Synthesis," *Advances in Organic Chemistry,* Vol. 8, E. C. Taylor, ed., Wiley-Interscience, 1972, p. 179.

4. G. J. Karabatsos, J. S. Fleming, N. Hsi, and R. H. Abeles, *J. Am. Chem.,* **88**, 849 (1966); F. A. Loewis, F. H. Westheimer, and B. Vennesland, *ibid.,* **75**, 5018 (1953).

5. K. Dimroth and K. H. Wolf, *Newer Methods of Preparative Organic Chemistry,* Vol. 3, Academic, New York, 1964, p. 357; A. T. Balaban, W. Schroth, and G. Fischer, *Advances in Heterocyclic Chemistry,* Vol. 10, Academic Press, New York, 1969, p. 241.

6. K. L. Kirk and L. A. Cohen, *J. Am. Chem. Soc.,* **94**, 8142 (1972).

7a. G. W. Griffin, *Angew. Chem. Int. Ed.,* **10**, 540 (1971).

7b. P. Borrevang, J. Hhort, R. T. Rapala, and R. Edie, *Tetrahedron Lett.,* 4905 (1968).

7c. E. Schmitz, *Advances in Heterocyclic Chemistry,* Vol. 2, A. R. Katritzky, ed., Academic Press, New York, 1963, p. 83.

7d. H. M. Frey, *Adv. Photochem.,* **4**, 225 (1966).

8a. J. C. Sheehan and M. M. Nafissi, *J. Am. Chem. Soc.,* **91**, 4596 (1969).

8b. E. R. Talaty, A. E. Dupuy, Jr., C. K. Johnson, T. P. Pirotte, W. A. Fletcher, and R. E. Thompson, *Tetrahedron Lett.,* 4435 (1970).

9. H. A. Staab, *Angew. Chem. Int. Ed.,* **1**, 351 (1962); *ibid.,* **3**, (1964).

10. G. E. McCasland and D. A. Smith, *J. Am. Chem. Soc.,* **72**, 2194 (1950); J. W. Huffman and J. E. Engle, *J. Org. Chem.,* **24**, 1844 (1959).

11. B. P. Stark and A. J. Duke, *Extrusion Reactions,* Pergamon Press, Elmsford, N. Y., 1967.

12. J. A. Berson and S. S. Olin, *J. Am. Chem. Soc.,* **91**, 777 (1969).

13. A. Eschenmoser, *Quart. Rev.,* **24**, 366 (1970).

14. S. Gronowitz, *Advances in Heterocyclic Chemistry,* Vol. 1, A. R. Katritzky, ed., Academic Press, New York, 1963, p. 1.

15. A. I. Meyers and A. Nabeya, *Chem. Commun.,* 1163 (1967).

16.   J. C. Getson, J. M. Greene, and A. I. Meyers, *J. Heterocycl. Chem.*, **1**, 300 (1964).

17.   A. I. Meyers, A. Nabeya, H. W. Adickes, I. R. Politzer, G. R. Malone, A. C. Kovelesky, R. L. Nolen, and R. C. Portnoy, *J. Org. Chem.*, **38**, 36 (1973).

18.   T. L. Gilchrist and C. W. Rees, *Carbenes, Nitrenes, and Arynes*, Appleton-Century-Crofts, New York, 1969.

2 ALKANES AND CYCLOALKANES

Saturated aliphatic hydrocarbons are among the most fundamental of organic compounds and yet have resisted sophisticated direct methods of synthesis. Survey courses in organic chemistry still describe the formation of hydrocarbons by the Wurtz reaction or some related process[1,2] that leave much to be desired.

$$RM + R'X \longrightarrow R-R' + MX$$

$$(M = Na, MgX, K, Cd, Cu)$$

The current methods of choice for producing saturated hydrocarbons usually involve more indirect routes such as olefin reduction since these substances are more accessible. It is, however, rather unfortunate that hydrocarbon syntheses from heterocycles are rarely discussed although this route is superior to many others. The relative complexity of certain heterocycles, in most cases, precludes their introduction in elementary organic courses at the stage where hydrocarbons are discussed. Perhaps structural organic chemistry, in which chemical bonding and atomic array are presented first, would then prepare the student for the best synthetic methods rather than the simplest ones and thereby enhance his excitement for the subject.[3]

I.   FROM THIOPHENES

The thiophene molecule presents a highly useful template on which to construct a variety of saturated hydrocarbons. The reductive desulfurization of thiophene gives rise to a four-carbon unit (e.g., **1** and **2**) which carries the original thiophene substituents, R, as part of the hydrocarbon skeleton.

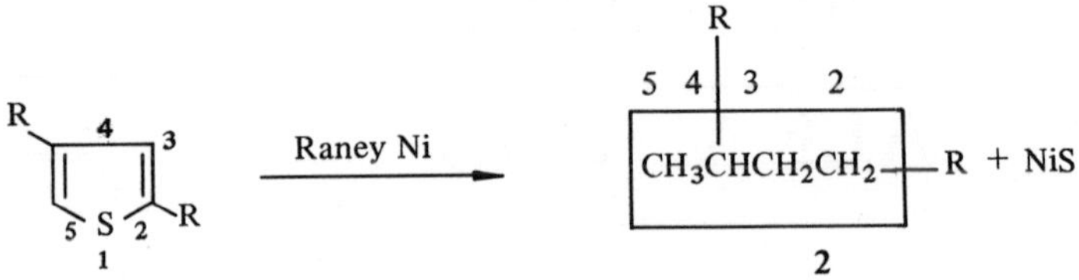

This convenient behavior of substituted thiophenes has led to a host of diverse hydrocarbon structures. The condensation of dimethylphenyl carbinol with thiophene in 70% sulfuric acid gave the dialkylated thiophene (3), which

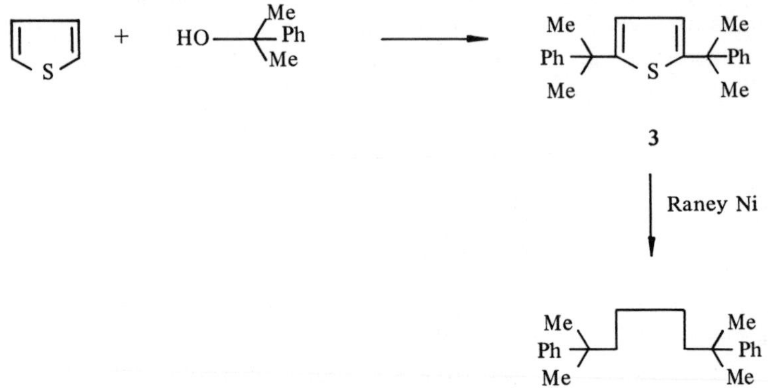

produced, after Raney nickel desulfurization, the symmetrically branched octane (4).[4] Extended chain hydrocarbons may also be prepared by condensation of more than a single molecule of thiophene. For example, reaction of acetone (2 moles) and thiophene (3 moles) in the presence of sulfuric acid produces the *tris*-(thienyl)propane (5), an excellent precursor to the highly branched hydrocarbon (6).[5] In the synthesis of 3,7,11-trimethylhentriacontane

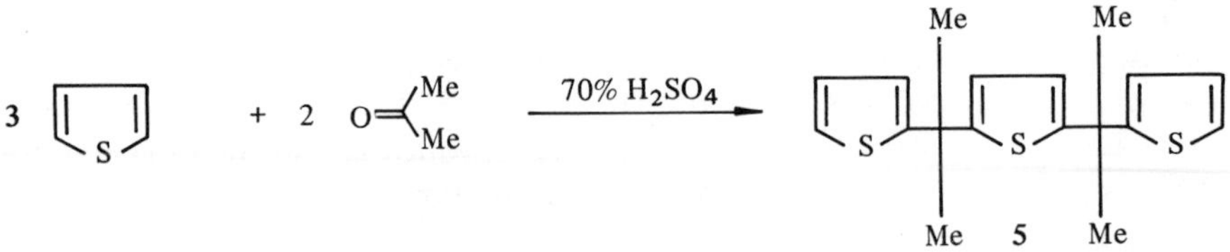

5 → Raney Ni → 6

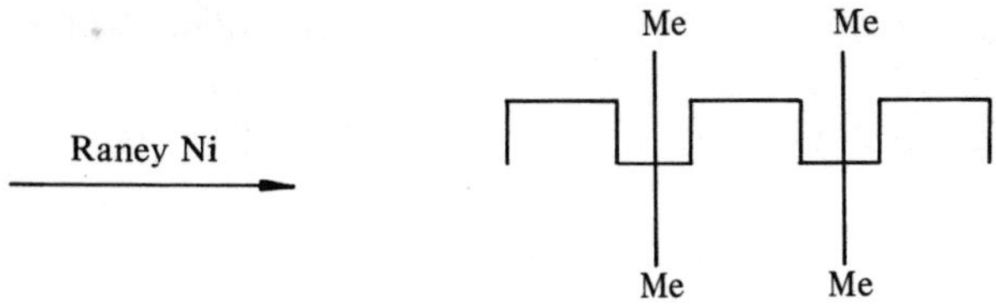

(7), a major constituent fo the Panamanian ant, *Atta columbica,* the thiophene nucleus was called upon to aid in the construction of this unusual hydrocarbon.[6]

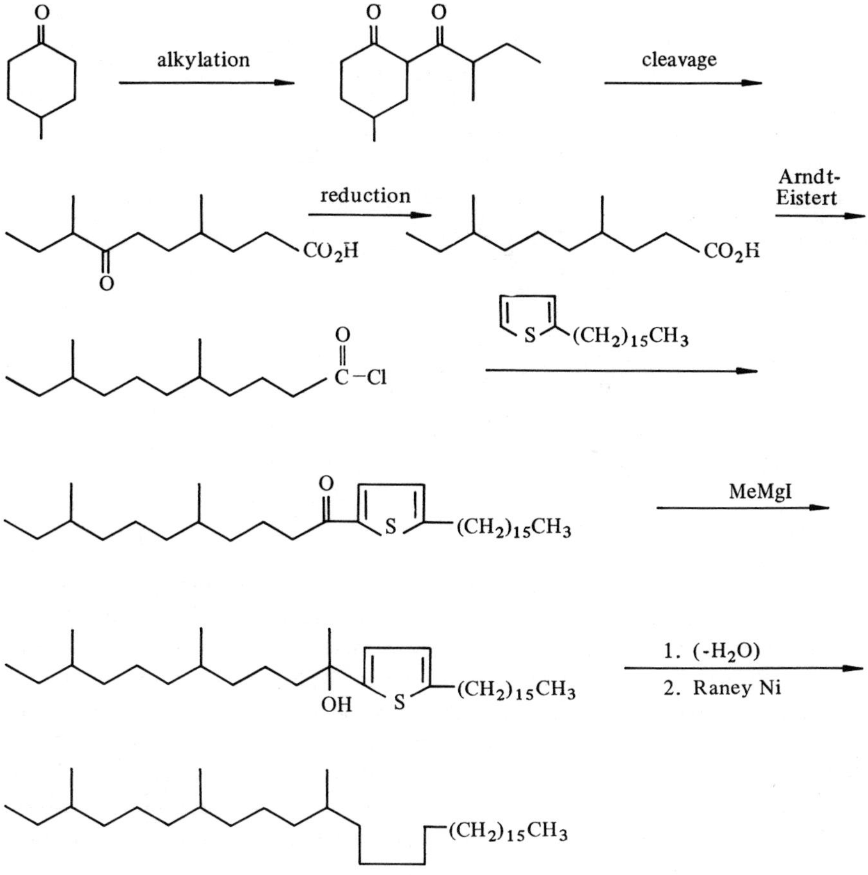

An ingenious method to test the optical activity of an asymmetric hydrocarbon, butylethylhexylpropylmethane (8), was described by Wynberg[7] and relied heavily on the ability of thiophenes to produce hydrocarbons, as well as allowing resolution of this material. The scheme employed is shown below.

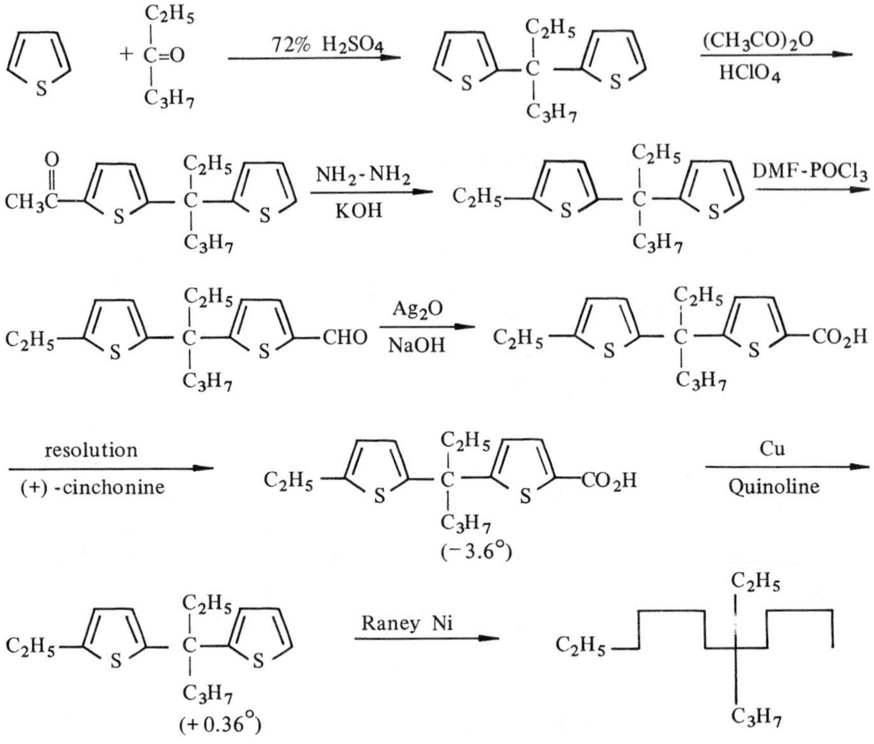

8

Of interest was the finding that, within the limits of experimental error, both enantiomeric hydrocarbons obtained by this route failed to exhibit any optical rotation. This supported the conclusion that mere differences in hydrocarbon chain length are not a sufficient condition for a quaternary carbon to rotate plane polarized light. Of interest to the subject at hand is the variety of manipulations performed on the thiophene ring, thus allowing considerable versatility in hydrocarbon syntheses—a requirement for any heterocycle that is to possess synthetic utility. Alicyclic hydrocarbons are also synthesized when thiophenes, appropriately substituted, are subjected to desulfurization. For example, when benzo[b]thiophene (9) was transformed into the norcarane system (10) and followed by a series of standard manipulations, the thieno-

tropylium ion (11) was obtained. Although the authors[8] performed the borohydride reduction (to 12) and Raney nickel reduction to ethylcycloheptane

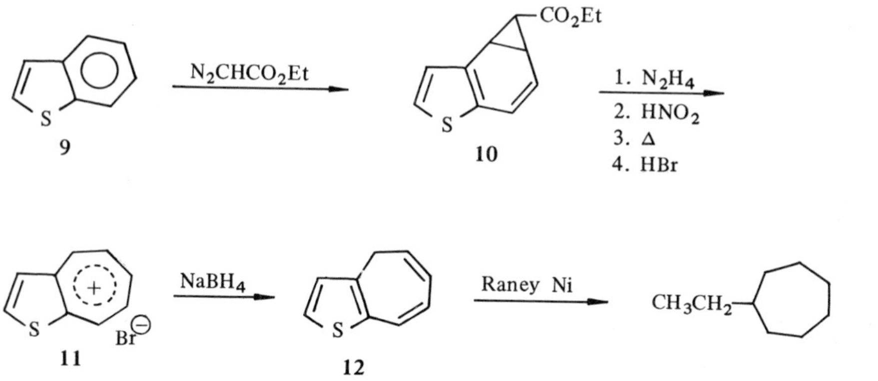

primarily as a structure proof, the method is of synthetic value. The readily
available benzo[b]thiophenes could serve as useful sources of many substituted
cycloheptane derivatives.

2,5-Disubstituted thiophenes (13) are capable of being converted into acyloins
(14) and ultimately to macrocyclic hydrocarbons by the route shown below[9]:

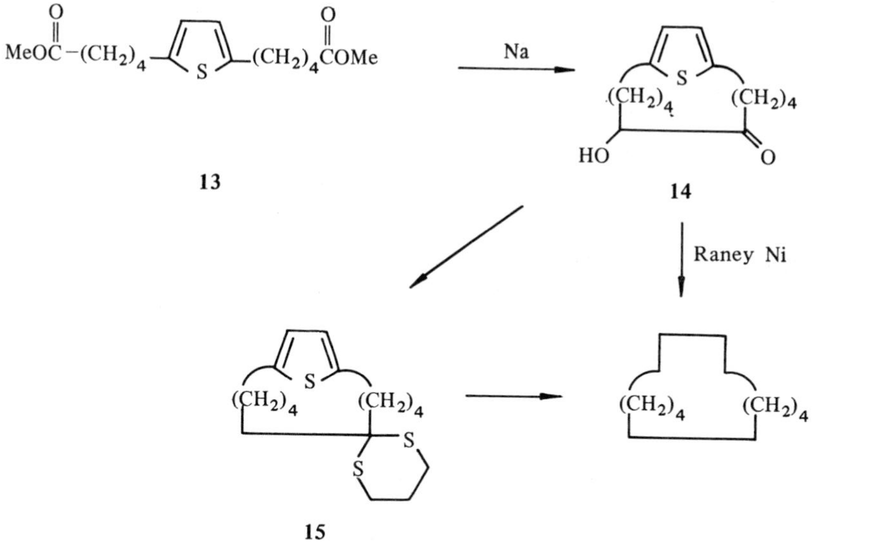

The acyloin may be reduced to the 1,3-dithane (15) by the method of Cram[23]

and also leads to the cyclic hydrocarbon.

Intramolecular alkylation of the enolate of the β-ketoester (16), followed by Raney nickel desulfurization of the resulting macrocyclic thiophene (17), gave the C_{15} cyclic hydrocarbon (18).[10]

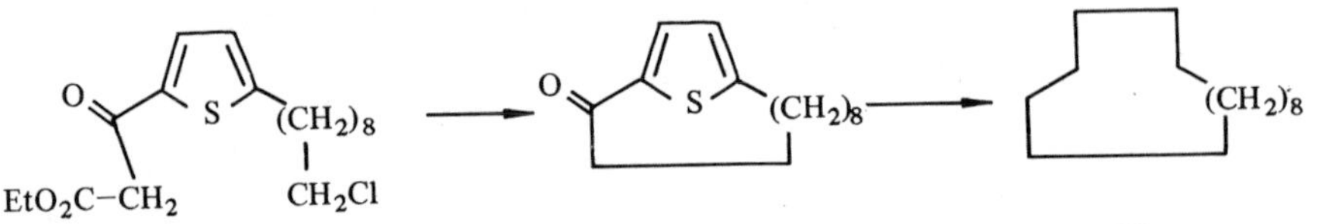

In order for the reader to rationally plan a synthesis utilizing the thiophene nucleus, it is important that he become familiar with the chemistry of thiophenes. Many excellent sources are available for this purpose,[11-15] and they shall not be reiterated here.

II. FROM 1,3-DITHIANES

The reaction of carbonyl compounds with thiols (or dithiols) under dehydrative conditions gives rise to dithioketals (19) which are well known to cleave to the methylene derivative (20) upon treatment with Raney nickel.[11]

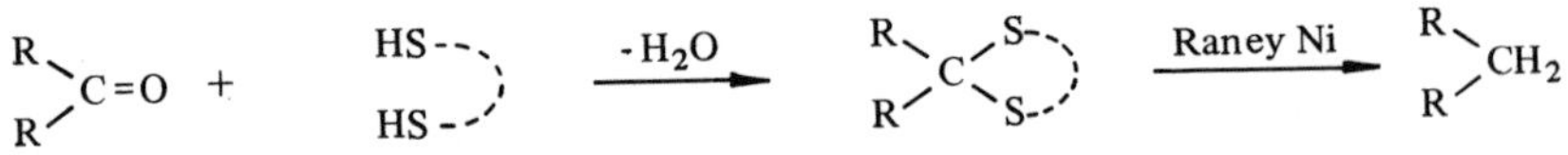

A few examples that illustrate the conversion of a carbonyl group to a methylene group using this procedure are the following.

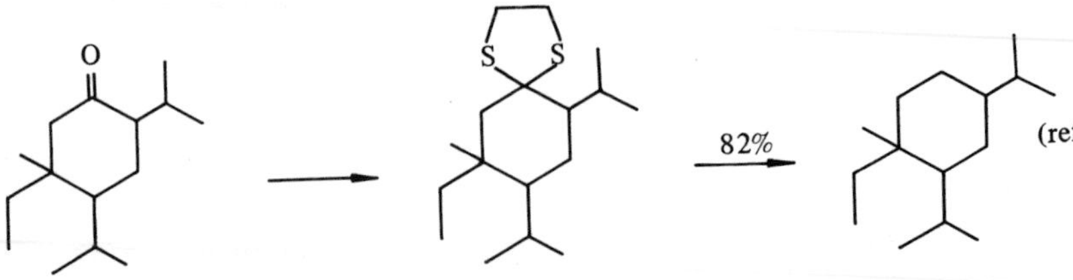

(ref.16)

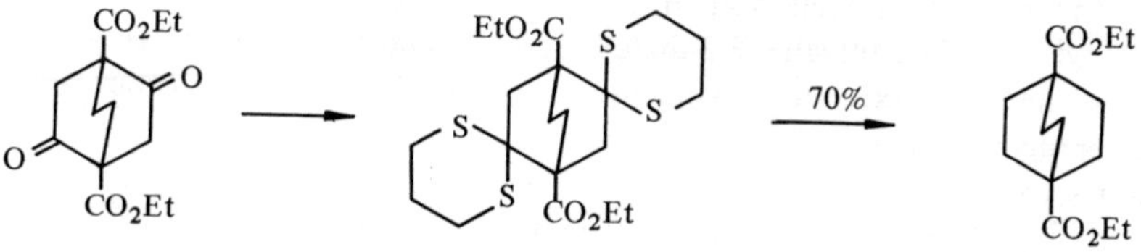

(ref.17)

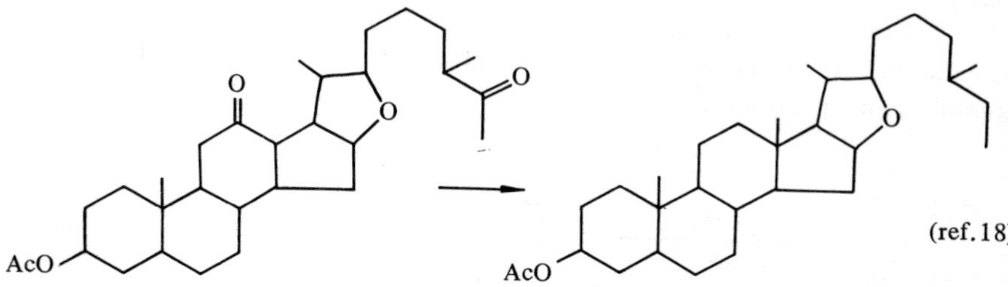

(ref.18)

This method is, therefore, a useful alternative to the Wolff-Kishner and Clemmenson reductions, since it is performed under essentially neutral conditions. Although a variety of thiols and dithiols have been employed in this fashion, the use of 1,3-dithiane (21) has been of particular significance. This readily accessible heterocycle was reported by Corey and Seebach[19] to metallate smoothly with butyllithium at low temperatures producing the stable lithio salt (22). Addition of a variety of electrophiles (e.g., alkyl halides) resulted in efficient alkylation to the 2-substituted dithiane (23). If this product is treated,

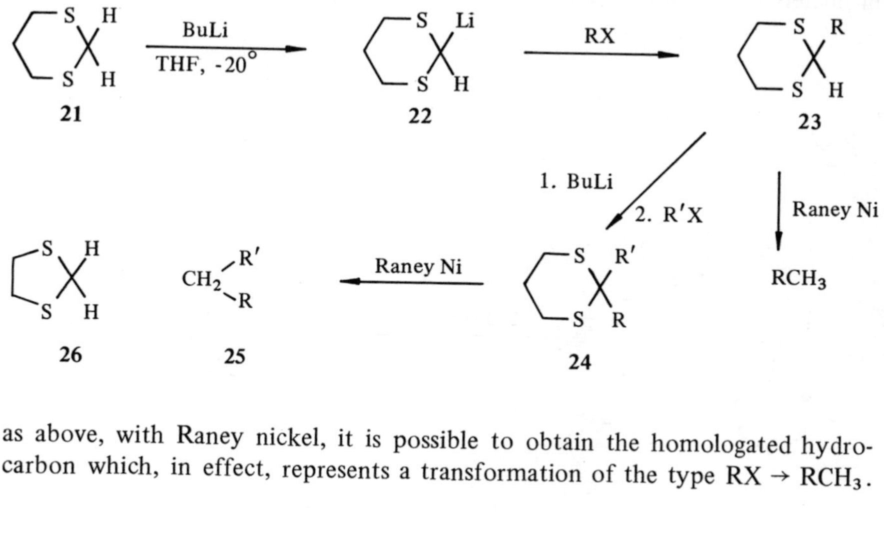

as above, with Raney nickel, it is possible to obtain the homologated hydrocarbon which, in effect, represents a transformation of the type RX → RCH$_3$.

However, the repeated metallation of **23** also proceeds readily as did the subsequent alkylation forming the 2,2-dialkyl-1,3-dithiane **(24)**. The latter is now a useful precursor to extended chain hydrocarbons **(25)** as well as other functionalized molecules. It is of interest to note that this scheme of "riveting" two alkyl halides to a methylene carbon [RX + "(-CH$_2$-)" + RX] is a process that appears to be unique only to the six-membered heterocyclic system, since attempts to duplicate this reaction with 1,3-dithiolane **(26)** have failed, primarily because of the instability of its lithio salt.

The synthesis of undecane[19] **(27)** in 80% yield from n-amyl bromide and dithiane in two successive operations followed by Raney nickel cleavage is exemplary of the potential of this route to hydrocarbons.

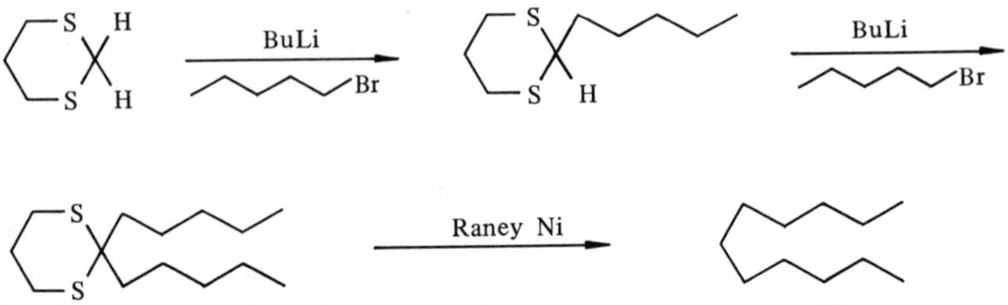

The construction of the cyclophane **(29)** was accomplished by treating the bis-dithiane derivative **(28)** via its dilithio salt with p-xylylene dibromide which gave the intramolecularly alkylated dithiane. Raney nickel desulfurization ultimately led to the cyclophane in 25% overall yield.[20]

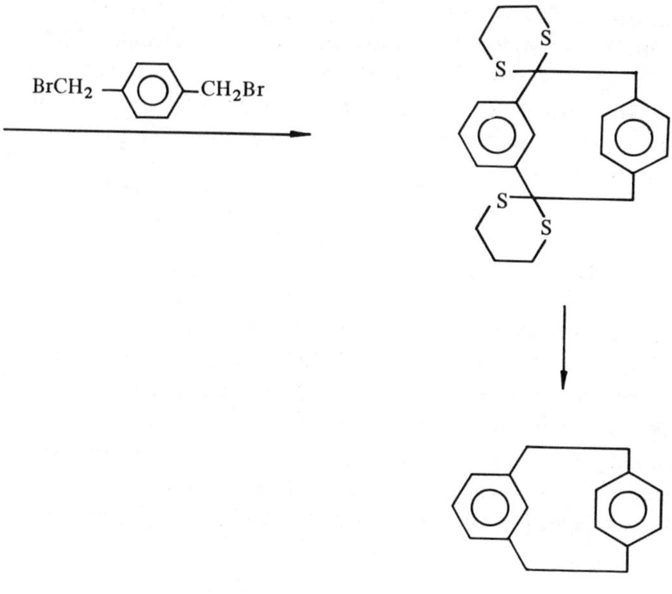

BrCH$_2$ —〈 〉— CH$_2$Br

29

While, to date, few examples of hydrocarbon syntheses have been reported, there is no doubt that this versatile method of elaborating the dithiane molecule followed by the time tested Raney nickel desulfurization will find extensive application in the future.

1,3-Dithianes have been prepared by a variety of methods involving 1,3-propanedithiol and carbonyl compounds. The parent molecule is formed in good yield by condensing methylal and 1,3-propanedithiol with boron trifluoride etherate in chloroform.[21] Monosubstituted-1,3-dithianes can be prepared directly from the carbonyl compound in good yields (80-90%) by treatment with 1,3-propanedithiol in the presence of zinc chloride, hydrogen chloride, or boron trifluoride in inert solvents.[19,22] An interesting synthesis of 1,3-dithianes was described by Cram[23] which involved the direct conversion of acyloins to the disubstituted dithiane (30). This represents a rather unique

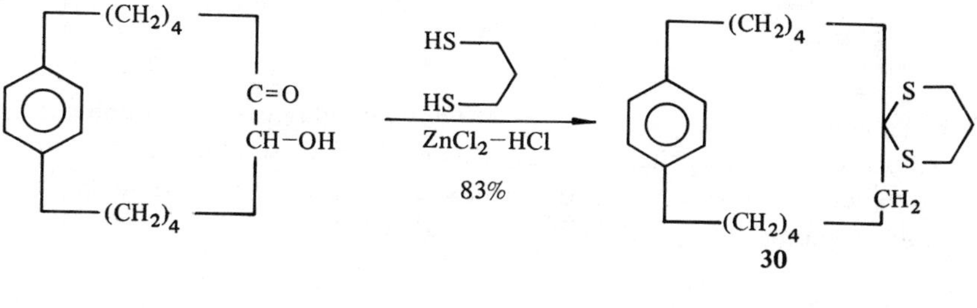

30

method for forming paracyclophanes after Raney nickel reduction.

The introduction of substituents into the dithiane molecule which serves as a useful vehicle to a variety of functionalized compounds has recently been reviewed.[19] Although there is considerable literature[22] on reactions involving other positions in dithianes, they are of no immediate value to the utility of this heterocycle in synthesis.

III. FROM THIETANES

An unusual reaction involving thietanes (31) in the form of their S-methyl salts (32) has recently been described by Trost.[24] Treatment with butyl-lithium gave rise to cyclopropanes (33) in 23-29% yield. The reaction was found to be highly stereoselective if geometrically pure thietanes are employed. Thus *trans*-2,4-dimethylthietane (34) gave 89% *cis*-1,2-dimethylcyclopropane, whereas the *cis*-dimethylthietane (35) led to 98% *trans*-1,2-dimethylcyclopropane. Although the exact mechanism is still an open question, the results showed that a

stepwise nucleophilic displacement involving 36 was highly unlikely. The formation of the cyclopropanes was considered to be the result of simultaneous

36

bond breaking of both carbon-sulfur bonds and subsequent predominant conrotatory ring closure to the products. This could be a useful method for preparing a variety of alkyl substituted cyclopropanes, and the simplicity of the reagents overcomes the less than desirable yields.

34

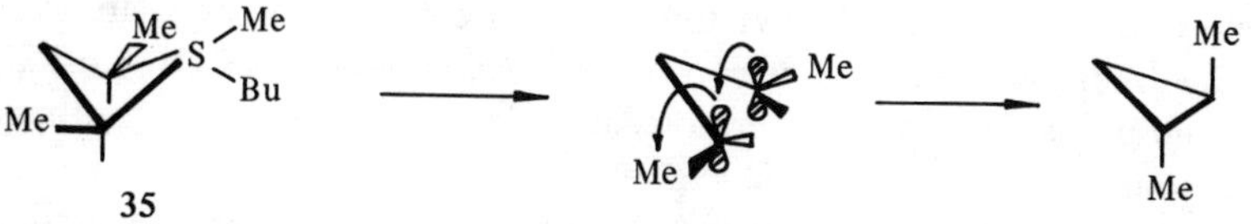

35

A variety of synthetic procedures is available for the preparation of thietanes.[24,25] These may, in turn, be treated as above to produce the corresponding cyclopropane.

IV. FROM TETRAHYDROTHIOPHENES

A rather novel synthesis of cyclopentane derivatives (40) has been developed[26] based on the readily available tetrahydrothiophene (37) molecule. Formation of the S-methyl salt was accomplished by treating tetrahydrothiophene with methyl iodide. The latter is now a highly reactive electrophile and it demonstrates its ability by ring opening to the 5,5-disubstituted-*n*-pentylmethyl sulfide 38. The reported efficiency of this ring opening is rather poor (12-43%) and could probably be improved by utilizing the fluoroborate

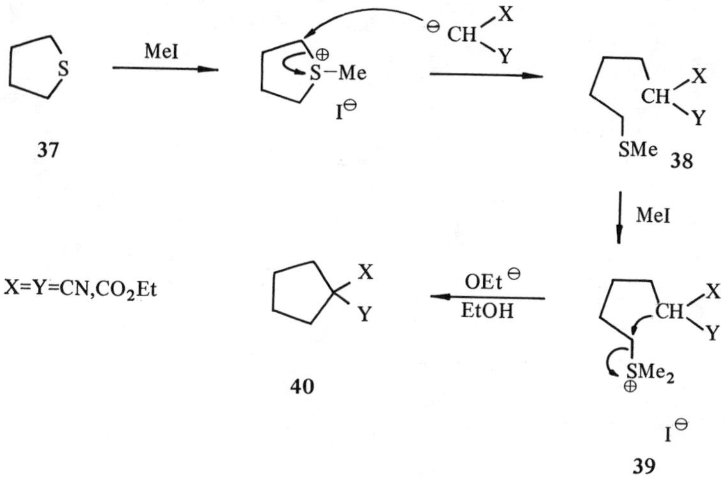

X=Y=CN,CO$_2$Et

37

38

39

40

salt rather than the iodide salt. Since the reaction leading to **38** is performed in DMSO, the iodide ion is undoubtedly a competing nucleophile which interferes with the process. The addition of a second equivalent of methyl iodide gives quantitative yields of the sulfonium salt **(39)** and subsequent treatment with base leads to the cyclopentane systems in good yields. The advantage of using tetrahydrothiophene as the four-carbon unit to construct the cyclopentanes lies mainly in the fact that dialkylated cyano esters, or malonates, are avoided. When the traditional malonic ester synthesis ($X = Y = CO_2C_2H_5$) using 1,4-dibromobutane is employed, intermolecular alkylation occurs as a bothersome side reaction.

The parent molecule is commercially available or may be prepared from 1,4-dibromobutane and potassium sulfide in aqueous or ethanolic solution.[27] The dihalide may contain alkyl or aryl substituents which would then give substituted tetrahydrothiophenes. Using the scheme above, substituted cyclopentanes would be produced.

V.   FROM 2-MERCAPTOTHIAZOLINES

The well-known S-alkylation of 2-mercaptothiazoline and the expected acidity of protons alpha to sulphur together provided an interesting and new coupling reaction leading to hydrocarbons.[28] Thus the lithio salt of 2-alkylthiothiazoline **(41)** upon treatment with alkyl halides gave the dialkylthiazoline derivative

($R'=R=$alkyl,allyl,propargyl)

(42) which was reductively cleaved to the hydrocarbon using Raney nickel. In this fashion, a new synthesis of squalene was achieved by coupling two molecules of farnesyl bromide (44) via the farnesyl thiothiazoline (43). Once again, the temporary use of a heterocycle as a "riveting agent" served a useful synthetic purpose (see p. 22 for a similar technique involving 1,3-dithianes).

VI. FROM 2-MERCAPTOPYRIMIDINE

The perennial problem of manipulating polyhydroxy compounds so that specific transformations may be carried out on, for example, a primary alcohol function in the presence of a secondary alcohol has constantly plagued organic chemists. A recent finding that may alleviate this problem involves the highly

nucleophilic 2-mercaptopyrimidine.[29] By employing the *di*-neopentyl acetal of dimethylformamide (46), a substance incapable of alkylation by nucleophiles is nevertheless able to undergo *trans*-acetalization with primary alcohols (45) by simple heating in acetonitrile, benzene, or DMF. If a slight excess of 2-mercapto-pyrimidine is present in the solution, a good yield of the thioalkylpyrimidine (47) is obtained, which is derived only from the primary alcohol function. Treatment with Raney nickel in methanol affords the deoxy compound 48 as the sole product. This potentially useful and very mild technique that employs 2-mercaptopyrimidine (undoubtedly other heterocyclic mercapto derivatives could be employed as well) should also find considerable utility in more general synthetic problems.[30,31]

VII. FROM PYRAZOLINES

The decomposition of Δ^{-1}-pyrazolines (49) has been recognized for many years as a route to cyclopropanes.[32] Although the pyrazolines exist as tautomers (49 ⇌ 50), either form is suitable as a precursor to cyclopropanes because the acid or base present during the thermolysis will maintain the mobility between both forms. In the absence of an isomerization catalyst, 50 is stable both thermally and photochemically.

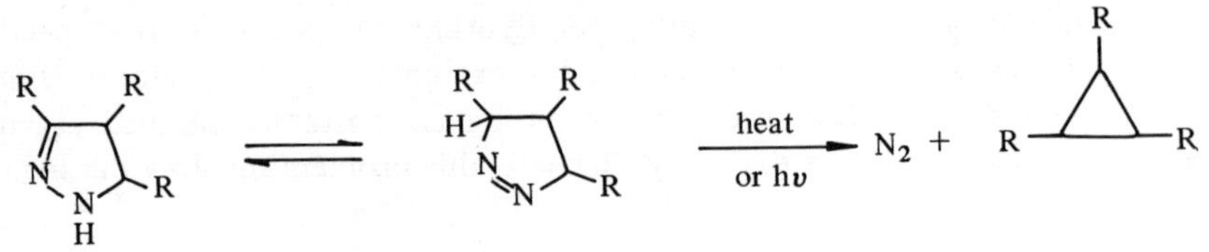

50   49

Many different cyclopropane derivatives are formed from this versatile heterocycle. Thermal decomposition, usually above 200°, extrudes nitrogen and leads to the cyclopropane (or cycloalkane) and minor amounts of olefinic products.

(Ref. 33)

(Ref. 34)

(Ref. 35)

(Ref. 36)

(Ref. 37a)

(Ref. 37b)

The stereospecificity of the thermal cleavage of pyrazolines is lacking although stereoselectivity does exist. There seems to be a preponderance of inversion of stereochemistry by a mechanism that is not yet clear. For example, when the *cis*-dimethyl pyrazoline **(51)**[33] is heated, the ratio of *cis* to *trans* dimethyl cyclopropanes is approximately 1:2. On the other hand, when the *trans*-

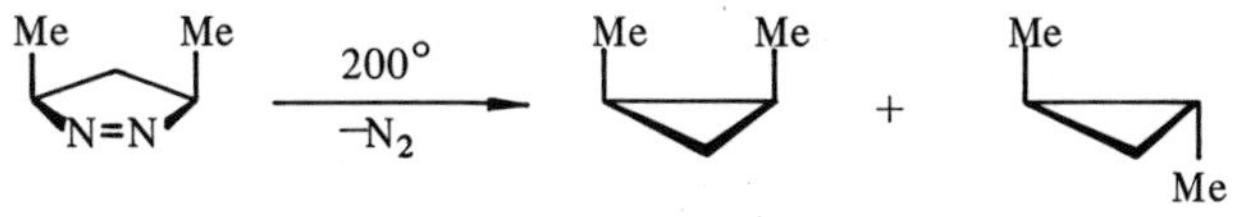

dimethyl pyrazoline **(52)** is similarly treated, the *cis*-dimethylcyclopropane is the

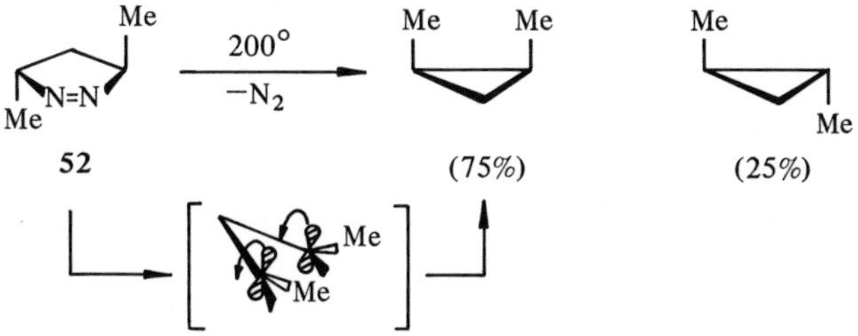

predominant product. This has been explained by invoking the π-cyclopropane intermediate **(53)** formed by "least motion" of the generated triplet species, which then cyclizes in a conrotatory fashion to the inverted *cis*-cyclopropane (from the *trans*-pyrazoline). The same argument was put forth to explain the cyclopropane products obtained by decomposition of the thietonium salts **(32)** with butyllithium. This rationale has been challenged,[34] since a π-cyclopropane **(54)** with "built-in" strain (a three-carbon bridge), and therefore a less stable species, should not lead to inverted products. However, when **55** was pyrolyzed, the major product (73%) was, in fact, the inverted cyclopropane, **56**.

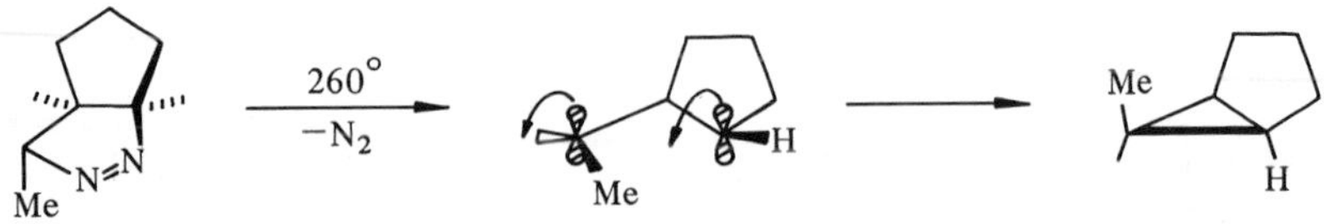

55   54   56

Recently, an even more strained "π-cyclopropane" intermediate (57) has been generated[36] by thermal pyrazoline decomposition, which proceeded with predominant inversion to the bicyclic system (58). Further study and consideration of a stepwise cleavage of both C–N bonds have been suggested to comprehend this process fully. Nevertheless, the chemist concerned primarily with synthesis in the cyclopropane series need not necessarily await the resolution of this problem to fulfill his needs.

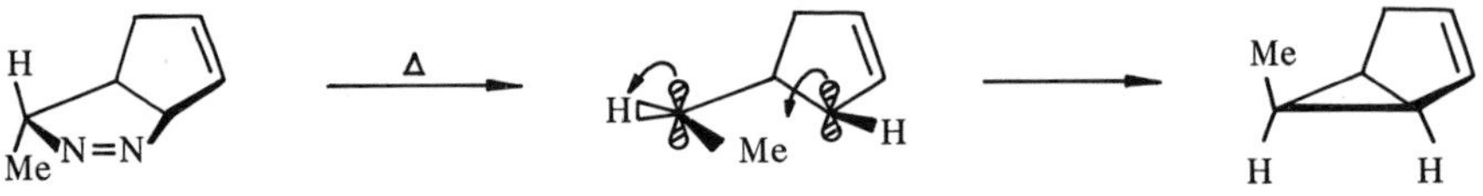

57   58

Many cyclopropanes as well as cycloalkanes have been produced in good yields using photolytic techniques to effect the decomposition of pyrazolines. In this manner greater stereoselectivity and even stereospecificity have been accomplished. The light induced decomposition of the cis-pyrazoline (59) gave the cis-dimethylcyclopropane carboxylate (60) in high yield with no detectable

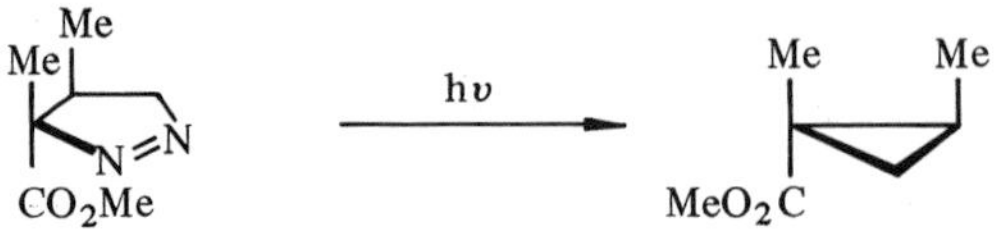

59   60

trace of isomeric products.[37] In the solid state, photolysis of the 1,2-diazacycloheptene (61) produced cis 1,2-diphenylcyclopentane stereospecifically, whereas in solution the products were isomeric mixtures.[38]

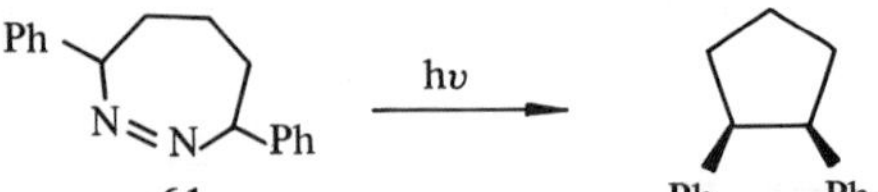

61   Ph   Ph

Many other pyrazolines may serve as useful precursors to cycloalkanes.

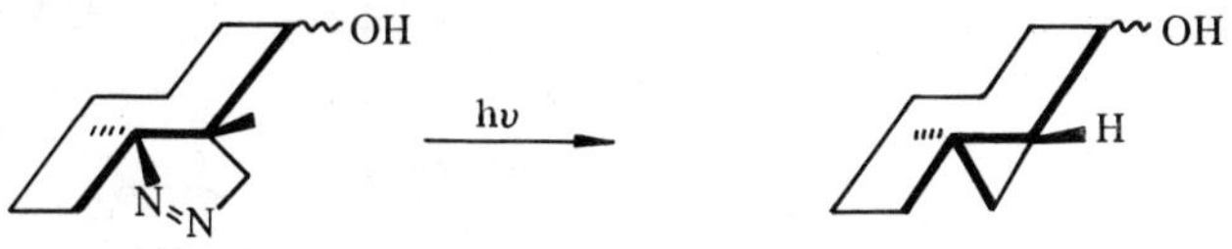

(Ref. 39)

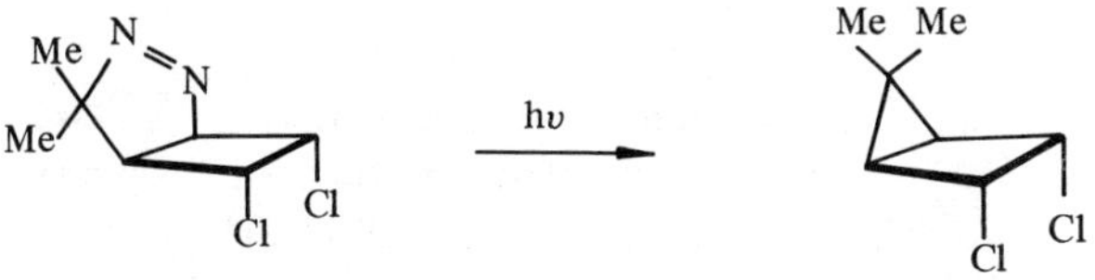

(Ref. 40)

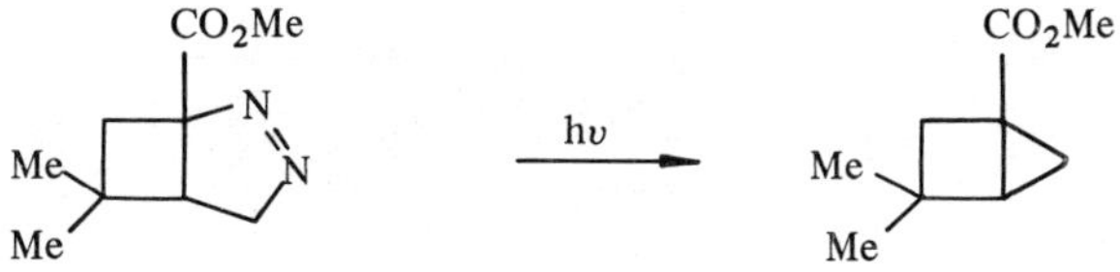

(Ref. 41)

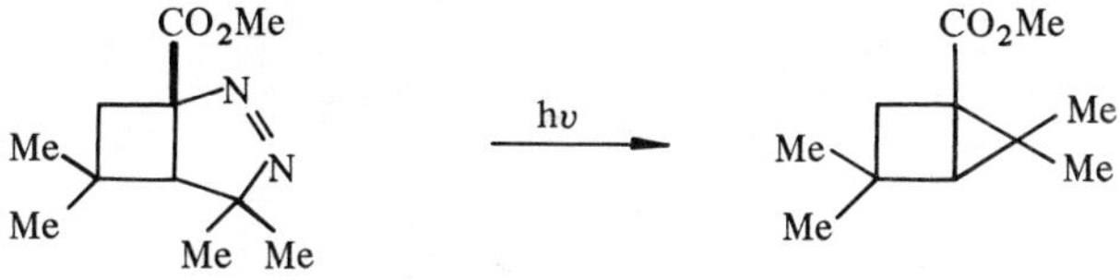

(Ref. 41)

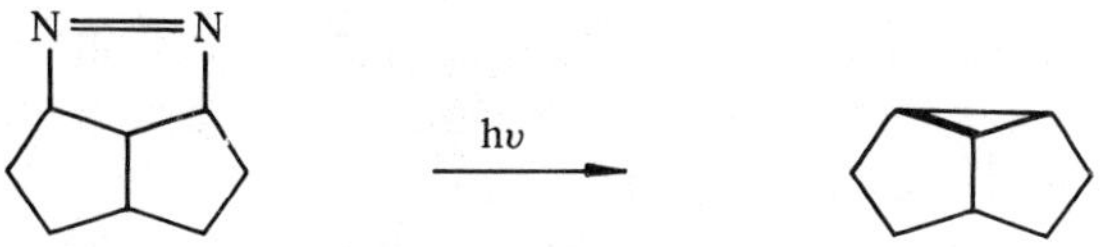

(Ref. 42)

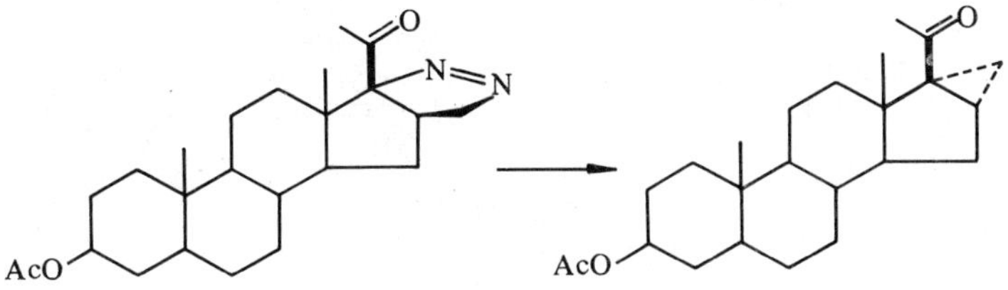

(Ref. 43)

The oldest and most common approach to pyrazolines involves the addition of diazoalkanes to olefins[32] of wide structural variety. This is a typical 1,3-dipolar addition process to a dipolarophile (the olefin) which makes up a large area of cycloaddition reactions studied by Huisgen and his students.[44] In this

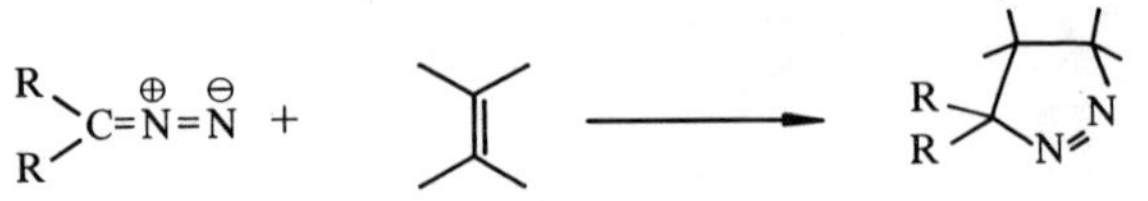

fashion many appropriately substituted pyrazolines may be prepared. The cycloaddition proceeds at its best when the olefin contains electron withdrawing substituents and the diazoalkane is equipped with alkyl or aryl groups. Thus **62** is obtained in excellent yields, whereas the formation of **63** occurs in 50% yield in a much slower process. A more versatile method for forming 1-

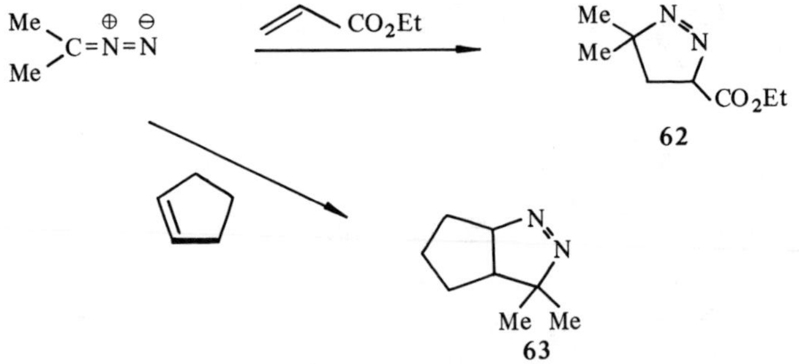

62

63

pyrazolines would be in hand if the easily obtainable 2-pyrazolines (**64**) could be utilized as precursors. The latter are synthesized from α,β-unsaturated aldehydes and hydrazine.[45] Such a method was investigated and found to be feasible.[46] Catalytic reduction (platinum) of 2-pyrazolines (**64**) was accom-

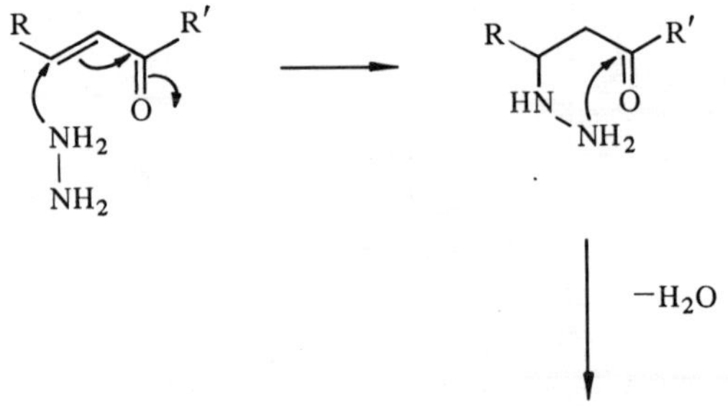

$-H_2O$

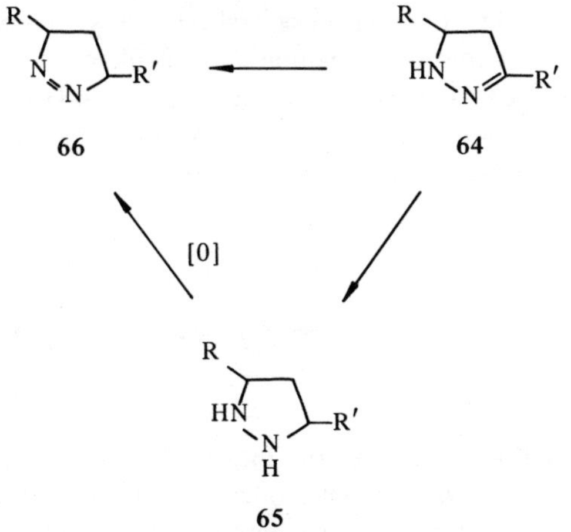

plished in 50-70% yields to **65** along with some concomitant ring cleavage. Though the pyrazolidines (**65**) were difficult to obtain pure because of the presence of acyclic products, they were smoothly oxidized to 1-pyrazolines (**66**) using mercuric oxide in pentane solvent.

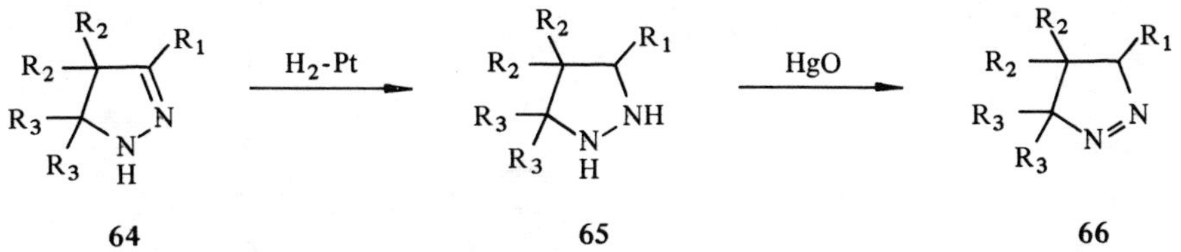

R_1	R_2	R_3	% Yield 65	% Yield 66
H	H, H	H, H	70	60
CH$_3$	H, H	H, H	75	75
H	CH$_3$, H	H, H	50	50
H	CH$_3$, CH$_3$	H, H	65	60

It, therefore, appears that sufficient techniques are available to construct various 1-pyrazolines to serve as useful precursors to cyclopropanes.

VIII.   FROM OXIRANES

The ring strain present in the oxirane system should manifest itself through a multitude of ring opening reactions. Indeed, this is the normal course of events in this small ring system. However, in several rare cases, the oxirane molecule has performed as a useful heterocycle leading to cycloalkanes.

The conversion of oxiranes (67) to cyclopropanes (68) was achieved in moderate yield using phosphonate anions.[47] The reaction required the presence of a strong electronegative substituent on the phosphonate such as CN or CO_2Et.

The reaction undoubtedly involves the intermediate formation of a pentavalent phosphorus heterocycle (69), which collapses either in concert or stepwise to form the cyclopropane. Thus we have here two useful heterocycles, the epoxide

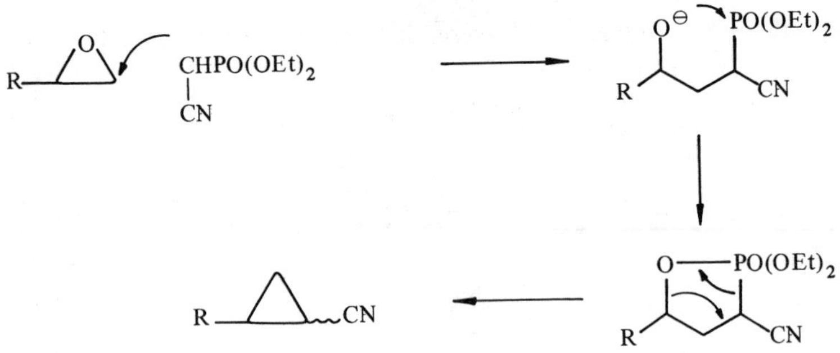

and the temporarily formed 69 demonstrating their synthetic abilities. The following represents typical formations.

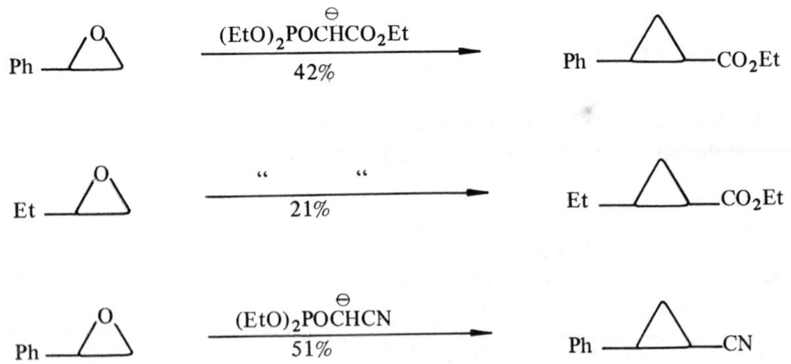

A highly useful and efficient method for preparing cyclopropanes from epoxides involves the photochemical rupture of certain oxirane derivatives.[48] Phenyl (or polyphenyl) substituted epoxides (70) serve as excellent precursors to phenyl (or diphenyl) carbenes (71) which, if generated in an olefin solvent, produce the phenyl-substituted cyclopropanes (72). The simplicity of the reaction may be appreciated by the fact that irradiation in a quartz tube of a solution of the epoxide in neat olefin gave the cyclopropane which is obtained pure by a single short path distillation.

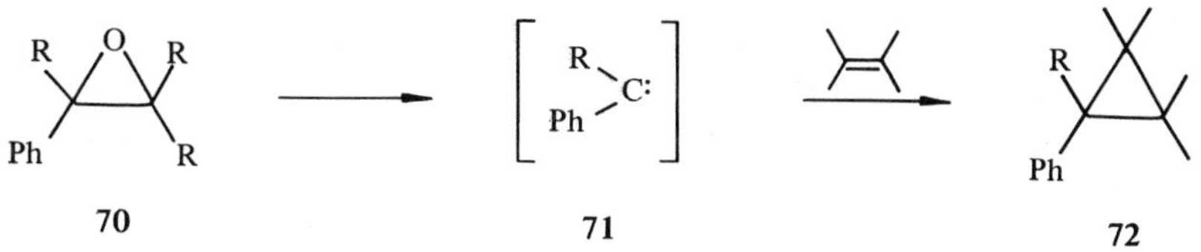

The nature of the substituents investigated, and the yields of cyclopropanes obtained, are given by the following:

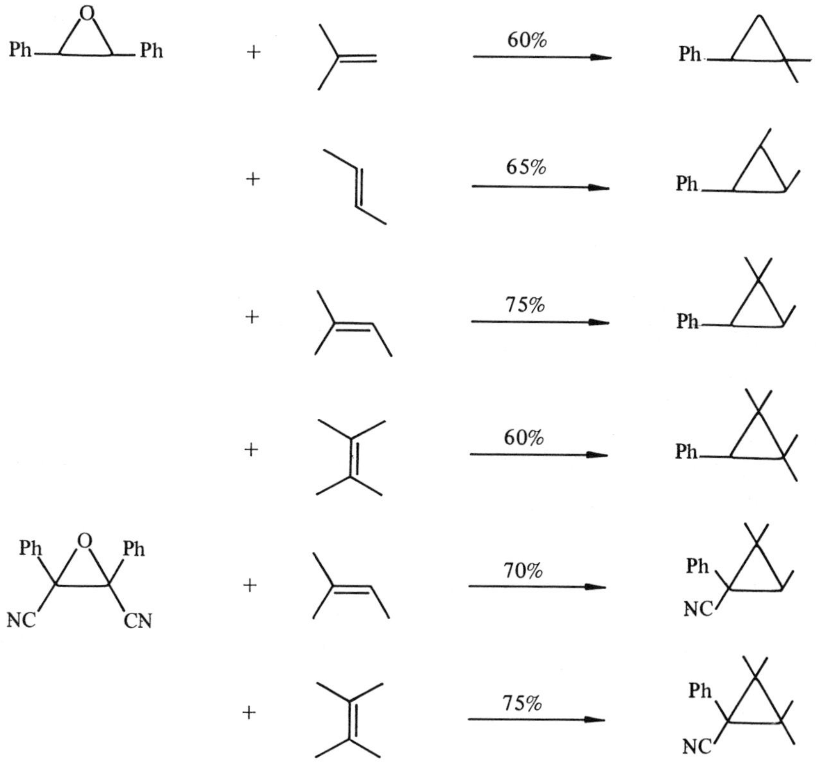

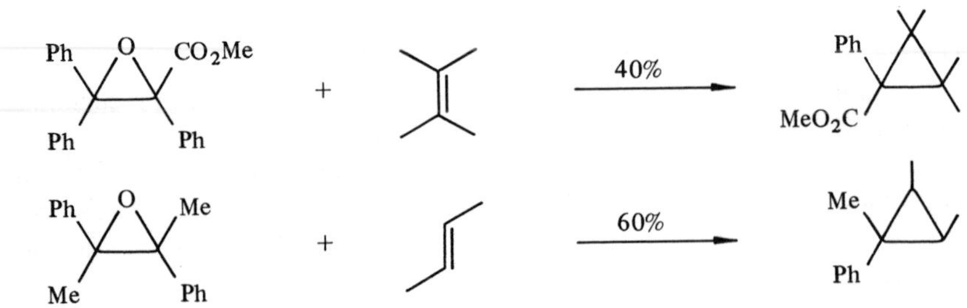

The mechanism of this reaction has been the subject of considerable study and it appears that even though the exact electronic nature of the intermediates is still an open question, the intermediary of the carbonyl ylids (73) is consistent with the observed facts.[49]

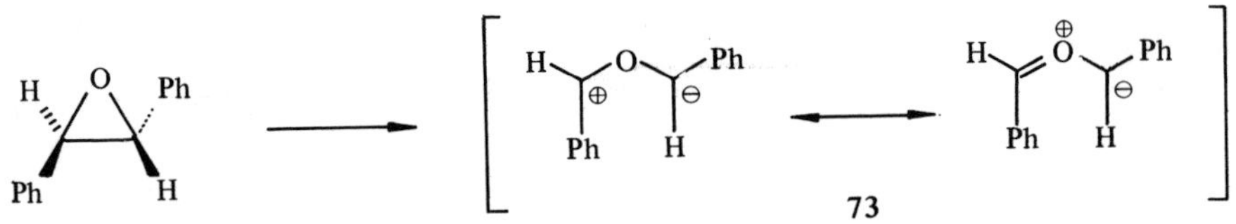

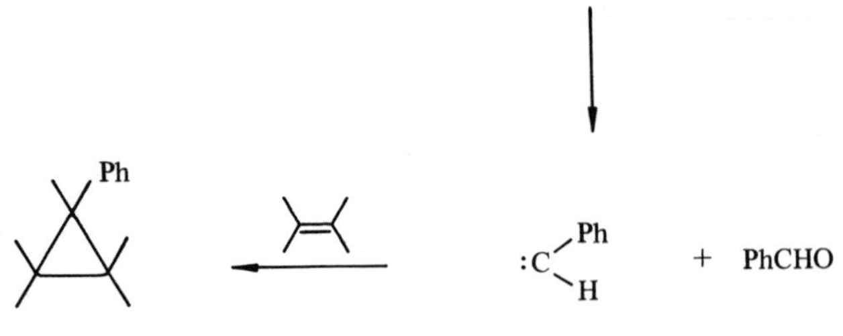

These reactions, as expected, give rise to epimeric cyclopropanes, although there is considerable stereoselectivity. The predominant products formed are those with fewest nonbonded interactions.

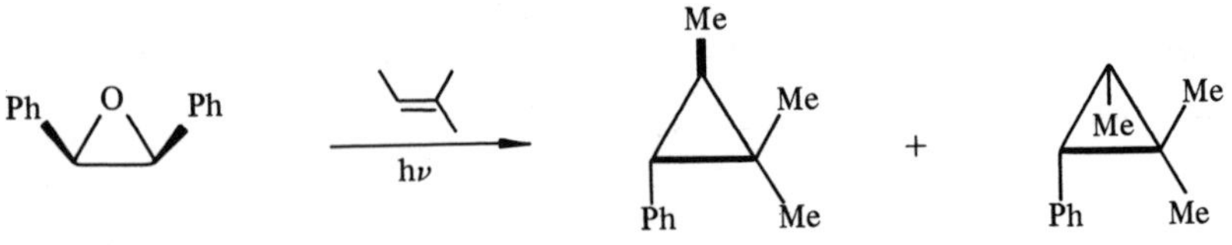

IX. FROM 2-MERCAPTOPYRIDINES

The stable anion derived from 2-(alkylthio)pyridines (74) was recently shown to react with alkyl halides affording the elaborated derivative (75).[50] This technique allowed the total synthesis of *cis*-bergamotene (77) to be achieved after reductive removal of the thiopyridine "riveting agent" (76). For other examples of heterocyclic riveting agents, the reader is referred to previously mentioned examples (pp. 22).

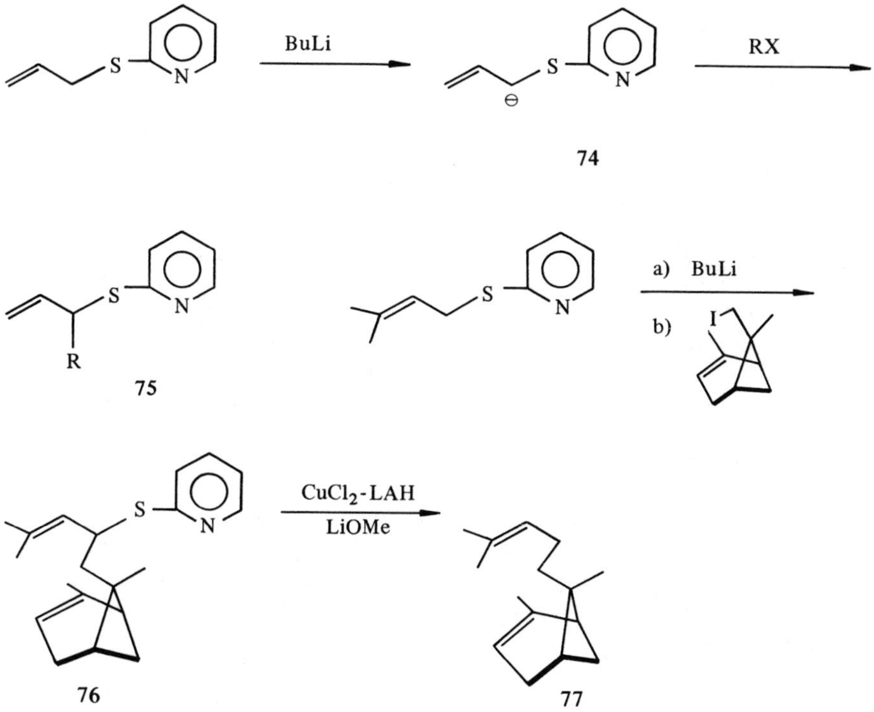

74

75

76   77

REFERENCES

1. J. March, *Advanced Organic Chemistry*, McGraw-Hill, 1968, pp. 353-355; R. T. Morrison and R. N. Boyd, *Organic Chemistry*, 3rd ed., Allyn and Bacon, 1973, pp. 90-95.
2. Coupling using hexamethylphosphoramide [H. Normant, *Bull. Soc. Chim., France*, 791 (1968)] and organocuprates [J. F. Normant, *Synthesis*, 63 (1972)] have shown some promise in solving this problem.
3. An introductory text in organic chemistry that embodies this concept has recently

appeared: N. L. Allinger, M. P. Cava, D. C. DeJongh, C. R. Johnson, N. A. LeBel, and C. L. Stevens, *Organic Chemistry,* Worth, 1971.

4. Y. L. Goldfarb and I. S. Korsakova, *Proc. Acad. Sci., U.S.S.R.* (English translation), **96,** 283 (1954).

5. N. P. Buu-Hoi, M. Sy, and N. D. Xuong, *Rec. Trav. Chim.,* **75,** 463 (1956).

6. M. M. Martin and J. G. MacConnell, *Tetrahedron,* **26,** 307 (1970).

7. H. Wynberg, G. L. Hekkert, J. P. M. Houbiers, and H. W. Bosch, *J. Am. Chem. Soc.,* **87,** 2635 (1965).

8. D. Sullivan and R. Pettit, *Tetrahedron Lett.,* 401 (1963).

9. Y. L. Goldfarb, S. Z. Taits, and L. I. Belinka, *Bull. Acad. Sci. SSR, Div. Chem. Sci., SSR,* 1287 (1957), English.

10. S. Z. Taits and Y. L. Goldfarb, *Bull. Acad. Sci., SSR, Div. Chem. Sci. SSR,* 1574 (1960), English.

11. Detailed reviews on desulfurization with Raney nickel have appeared: (a) G. R. Pettit and E. E. VanTamelen, *Organic Reactions,* Vol. 12, Wiley, New York, 1962, pp. 356-529; (b) H. Hauptmann and W. F. Wailer, *Chem. Rev.,* **62,** 347 (1962).

12. For an excellent discussion on thiophenes, see S. Gronowitz, "Chemistry of Thiophenes," in *Advances in Heterocyclic Chemistry,* A. Katritzky, ed., Vol. 1, Academic Press, New York, 1963, pp. 1-124.

13. For introduction of substituents in the 3-position of thiophene see S. Gronowitz, *Acta. Chem. Scand.,* **13,** 1045 (1959).

14. Reviewed with additional ring closure methods in D. E. Wolf and K. Folkers, *Organic Reactions,* Vol. 6, Wiley, New York, 1951, p. 410.

15. H. D. Hartung, "Thiophene and its Derivatives," in *The Chemistry of Heterocyclic Compounds,* A. Weissberger, ed., Interscience, New York, 1952.

16. I. Ognjanov, V. Herout, M. Horak, and F. Sorm, *Coll. Czech. Chem. Comm.,* **24,** 2371 (1959).

17. J. D. Roberts, W. T. Moreland, and W. Frazer, *J. Am. Chem. Soc.,* **75,** 637 (1953).

18. C. Djerassi, O. Halpern, G. R. Pettit, and G. H. Thomas, *J. Org. Chem.,* **24,** 1 (1959).

19. For a recent summary on metallation and subsequent nucleophilic reactions of 2-lithio-1,3-dithiane see D. Seebach, *Synthesis,* **1,** 17 (1969).

20. T. Hylton and V. Boekelheide, *J. Am. Chem. Soc.,* **90,** 6887 (1968).

21. D. Seebach, N. R. Jones, and E. J. Corey, *J. Org. Chem.,* **33,** 300 (1968).

22. D. S. Breslow and H. Skolnik, *The Chemistry of Heterocyclic Compounds,* Part Two, A. Weissberger, ed., Interscience, New York, 1966, p. 979.

23. D. J. Cram and M. Gordon, *J. Am. Chem. Soc.,* **77,** 1810 (1955); D. J. Cram and L. K. Gaston, *ibid.,* **82,** 6386 (1960).

24. B. M. Trost, W. L. Schinski, and I. B. Mantz, *J. Am. Chem. Soc.,* **91,** 4320 (1969), *ibid.,* **93,** 676 (1971).

25. For a review, see M. Sander, *Chem. Rev.,* **66,** 341 (1966); P. L. F. Chang and D. C. Dittmer, *J. Org. Chem.,* **34,** 2791 (1969); S. Searles, H. R. Hays, and E. F. Lutz, *ibid.,* **27,** 2828 (1962); J. D. Downer and J. E. Colchester, *J. Chem. Soc.,* 1528 (1965); R. M. Dodson and G. Klose, *Chem. Ind. (London),* 450 (1963).

26. A. Winkler and J. Gosselck, *Tetrahedron Lett.,* 2433 (1970).

27. A discussion of tetrahedrothiophenes has been given by D. E. Wolf and K. Folkers,

Organic Reactions, Vol. 6, Ch. 9, Wiley, New York, 1951, p. 443.

28. K. Hirai, H. Matsuda, and Y. Kishida, *Tetrahedron Lett.,* 4359 (1971).

29. A. Holy, *ibid.,* 585 (1972).

30. For additional methods to convert hydroxyl groups to their methyl derivatives see *Compendium of Organic Synthetic Methods,* I. T. Harrison and S. Harrison, Wiley-Interscience, New York, 1971, p. 359.

31. Protection of hydroxyl groups has been reviewed: J. W. McOmie, "Protective Groups," in *Advances in Organic Chemistry,* Vol. 3, R. A. Raphael, E. C. Taylor, and H. Wynberg, ed., Interscience, New York, 1963, p. 191.

32. H. Zollinger, *Azo and Diazo Chemistry,* Interscience, New York, 1961; R. Huisgen, *Angew. Chem. Int. Ed.,* **2,** 565 (1963); G. W. Cowell and A. Ledwith, *Quart. Rev.,* **24,** 119 (1970).

33. R. J. Crawford and A. Mishra, *J. Am. Chem. Soc.,* **88,** 3963 (1966).

34. P. B. Condit and R. G. Bergman, *Chem. Commun.,* **4,** (1971).

35. E. D. Andrews and W. E. Harvey, *J. Chem. Soc.,* 4636 (1964).

36. M. P. Schneider and R. J. Crawford, *Can. J. Chem.,* **48,** 628 (1970).

37. (a) T. V. VanAuken and K. L. Rinehart, *J. Am. Chem. Soc.,* **84,** 3736 (1962); (b) L. A. Paquette and L. M. Leichter, *ibid.,* **93,** 5128 (1971).

38. C. G. Overberger and C. Yaroslavsky, *Tetrahedron Lett.,* 4395 (1965); C. G. Overberger, N. Weinshenker, and J.-P. Anselme, *J. Am. Chem. Soc.,* **87,** 4119 (1965).

39. K. Wiberg and A. deMeijere, *Tetrahedron Lett.,* 59 (1969).

40. M. Franck-Neumann, *ibid.,* 2979 (1968).

41. T. H. Kinstle, R. L. Welch, and R. W. Exley, *J. Am. Chem. Soc.,* **89,** 3660 (1967); P. G. Gassman and K. T. Mansfield, *J. Org. Chem.,* **32,** 915 (1967).

42. M. Schwarz, A. Besold, and E. R. Nelson, *J. Org. Chem.,* **30,** 2425 (1965).

43. O. Jeger, D. Arigoni, P. G. Ferrini, and K. Kocsis, *Helv. Chim. Acta,* **43,** 2178 (1960).

44. R. Huisgen, R. Grashey, and J. Sauer, *The Chemistry of Alkenes,* S. Patai, ed., Interscience, New York, 1964, p. 739.

45. R. H. Wiley and P. Wiley, "Pyrazolones, Pyrazolidones, and Derivatives," *The Chemistry of Heterocyclic Compounds,* A. Weissberger, ed., Interscience, New York, 1964.

46. R. J. Crawford, A. Mishra, and R. J. Dummel, *J. Am. Chem. Soc.,* **88,** 3959 (1966).

47. W. S. Wadsworth, Jr., and W. D. Emmons, *ibid.,* **83,** 1733 (1961); see also W. E. McEwen, A. Blade-Font, and C. A. VanderWerf, *ibid.,* **84,** 677 (1962); L. Horner, H. Hoffmann, and V. G. Toscano, *Chem. Ber.,* **95,** 538 (1962); D. B. Denney, J. J. Vill, and M. J. Boskin, *J. Am. Chem. Soc.,* **84,** 3944 (1962); Y. Inoue, T. Sugita, and H. M. Walborsky, *Tetrahedron,* **20,** 1695 (1964).

48. H. Kristinsson and G. W. Griffin, *J. Am. Chem. Soc.,* **88,** 1579 (1966); P. C. Petrellis and G. W. Griffin, *Chem. Commun.,* 691 (1967); R. S. Becker, R. O. Bost, J. Kolc, N. R. Bertoniere, R. L. Smith, and G. W. Griffin, *J. Am. Chem. Soc.,* **92,** 1302 (1970); H. Kristinsson, *Tetrahedron Lett.,* 2343 (1966).

49. D. R. Arnold and L. A. Karnischky, *J. Am. Chem. Soc.,* **92,** 1404 (1970); G. W. Griffin, *Angew. Chem. Int. Ed.,* **10,** 540 (1971).

50. K. Narasaka and T. Mukaiyama, *Chem. Lett. (Japan),* 259 (1972).

3 ALKENES AND CYCLOALKENES

The formation of the carbon-carbon double bond has occupied the interest of organic chemists as much as any other functional group. Indeed, methods are available that encompass a wide range of techniques and these have been amply described in the literature.[1a,b] Rare among the popular methods for alkene syntheses is the use of heterocyclic compounds as starting materials or intermediates, and it is the intent of this chapter to illuminate the reader in this regard.[1c]

I. FROM THIIRANES

Although the reactions of thiiranes (episulfides) (1) with nucleophiles leading to alkenes (2) is now a well-known process,[2] the synthetic utility of such a transformation has only recently been appreciated. The fact that sulfur elimination

PhLi (ref. 2)
Ph$_3$P (ref. 3)
(EtO)$_3$P (refs. 3, 4)
(LiAlH$_4$ (ref. 5)

using phenyllithium, triphenylphosphine, or triethylphosphite is virtually stereospecific (>98%) producing *trans*-olefins from *trans*-thiiranes (3) and *cis*-olefins from *cis*-thiiranes (4) represents a further significant aspect of the process.[5a] Desulfurization reactions have also been carried out in poor yield using methyl iodide; however, the olefin produced was the result of >97% stereoselective elimination.[6]

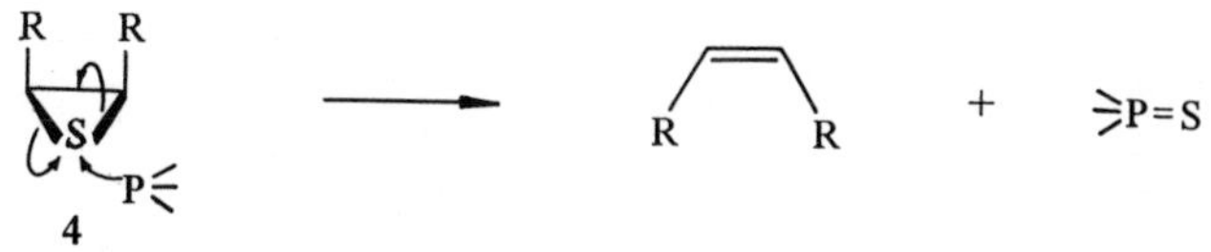

Since the conversion of thiiranes to alkenes is, in itself, not a profoundly useful process because of the manner in which thiiranes are prepared (from olefins,[7] halohydrins,[6] or epoxides[8]), a synthetic sequence that passes through a thiirane intermediate would be valuable. Such a method has recently been reported,[9] based upon the known[10] condensation of carbonyl compounds with hydrazine and hydrogen sulfide which produce excellent yields of the tetrahydrothiadiazole (5). Oxidation of the latter with lead tetraacetate to the azosulfide (6) followed by heating in the presence of triphenylphosphine gives the olefin (7) in good yield. The thiirane (8) is a confirmed intermediate as shown by pyrolytic studies[11] on the azosulfides, 6. By simply adding triphenylphosphine to the reaction prior to pyrolysis, it was possible to sequentially desulfurize the thiirane intermediate and proceed directly to the olefin. In this fashion high yields of bicyclohexylidine (9) and the previously unreported bicyclobutylidine (10)[12] were obtained.

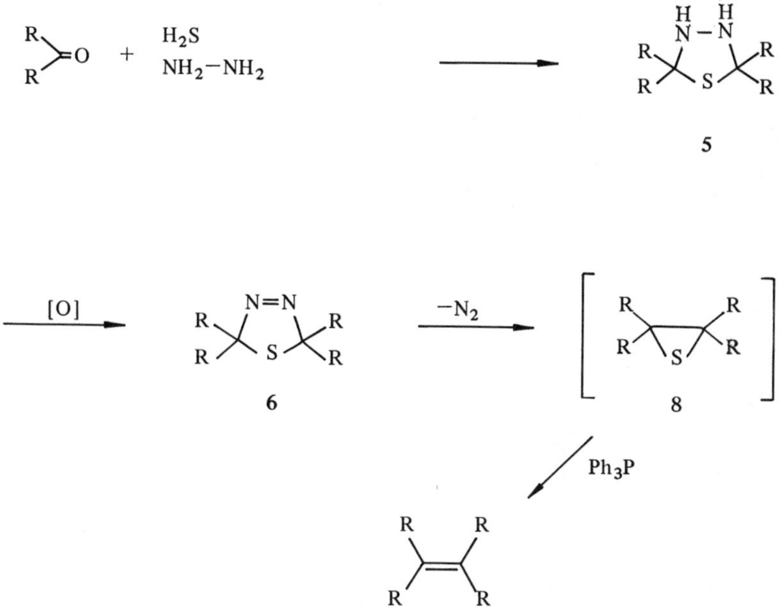

Another olefin synthesis using the thiirane intermediate in a "twofold extrusion process"[13] stemmed from the thermally labile oxathiolan-5-ones (11). In this instance carbon dioxide was previously shown[14] to eliminate producing thiiranes. Once again, the presence of a tervalent phosphorous nucleophile [P(NEt$_2$)$_3$] during the thermal decomposition of 11 resulted in direct formation of hindered tetrasubstituted olefins (12). This method, via the heterocycle (11), is a general reaction that leads to highly hindered olefins from readily available materials. The oxathiolan-5-ones are readily prepared from thiobenzilic acid and carbonyl compounds by acid catalyzed azeotropic removal of water.

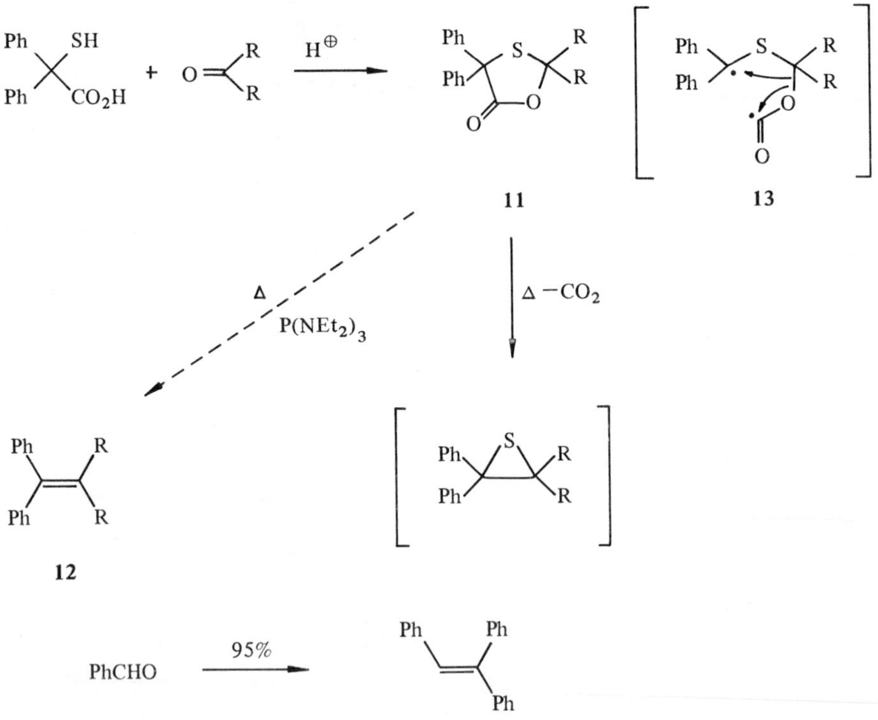

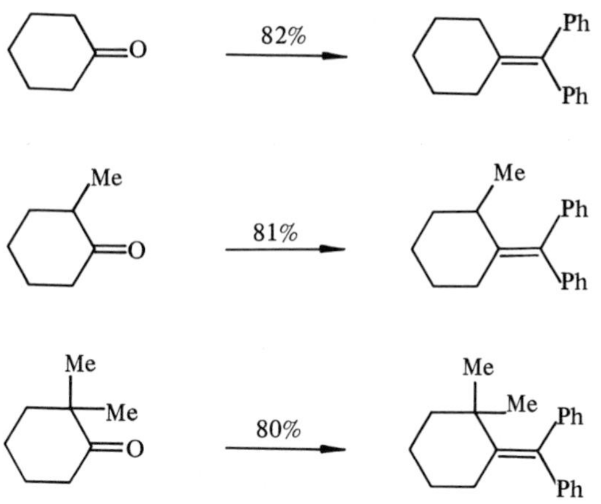

The reaction above, involving the oxathiolan-5-ones, presently appears to be limited to phenyl substitution in (11), since a stable radical intermediate (13) is a necessary requisite to loss of carbon dioxide. In the absence of conjugative substituents the reaction takes a different course.

The transformation of a divinyl sulfide to a conjugated diene[15] via a thiirane intermediate is another example of the sulfur contraction process mentioned earlier (p. 9). Irradiation of *cis* or *trans-β,β'* diphenyldivinyl sulfide gave the hitherto unknown cyclobutene episulfide (14) in 34% yield. Heating with

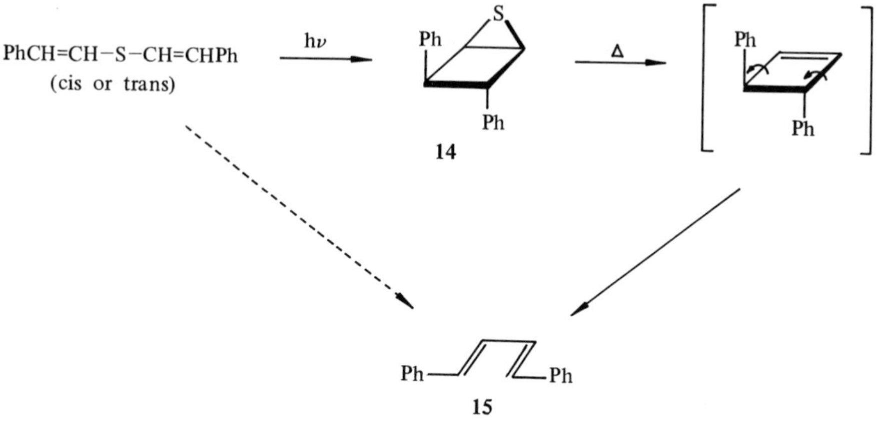

trivalent phosphorous derivatives transformed the thiirane to *trans-trans*-1,4-diphenyl butadiene (15), which undoubtedly arose from the conrotatory ring opening of the initially formed cyclobutene.

II. FROM THIIRANE 1,1-DIOXIDES

The thiirane 1,1-dioxide (episulfone) system has been known for over half a century to disintegrate readily to an alkene and sulfur dioxide (**16 → 17**).[16] The synthetic value of this extrusion process was not fully appreciated until 1940 when Ramberg and Backlund[17] discovered that thiirane dioxides, formed

in situ from α-halosulfones, gave rise to good yields of olefins (Scheme 1). In recent years there has been considerable activity[18] directed toward the synthesis of complex olefins using this general technique. The unique features of

Scheme 1

the Ramberg-Backlund reaction have led to the preparation of a variety of interesting structures. Treatment of the α-halotricyclic sulfone (**18**) with potassium *tert*-butoxide afforded the [4.4.2] propella-3,8,11-triene (**19**) in good yield.[19] A synthesis of $\Delta^{1,5}$-bicyclo[3.3.0]octenes (**21**) was reported[20] which should have considerable utility. The classic Ramberg-Backlund reaction was modified in this instance to involve a transannular displacement of halide ion in **20**. Thus a route is available from 1,5-cyclooctadiene to bicyclic olefins by passing through the thiirane dioxide and using a series of standard manipulations. It has been shown[18,21,22] that the α-halosulfones exchange their α-protons faster than displacement of halide ion occurs (to give the thiirane dioxide). Thus a method for preparing deuterated alkenes is also in hand (Scheme 2).

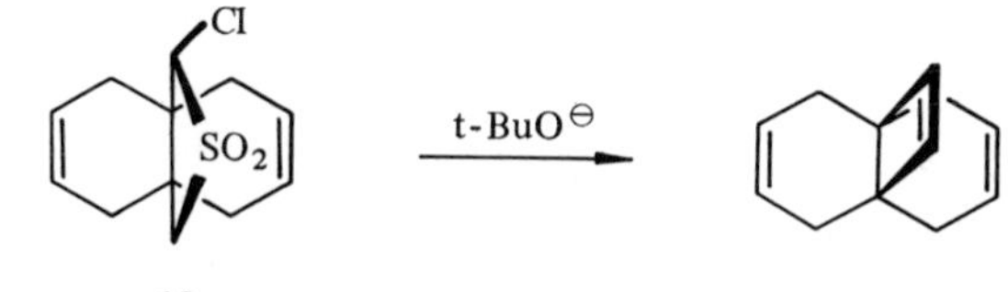

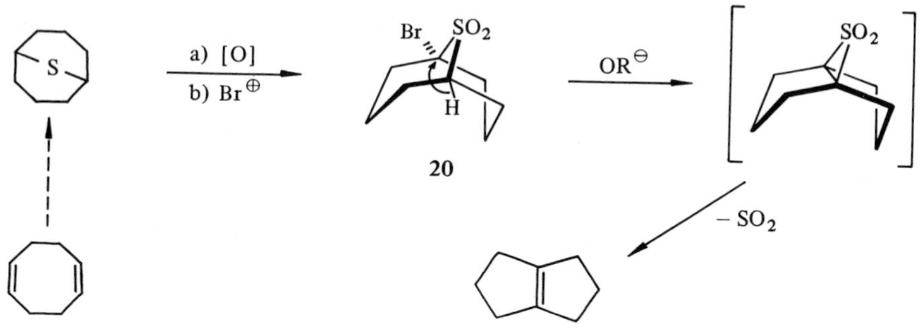

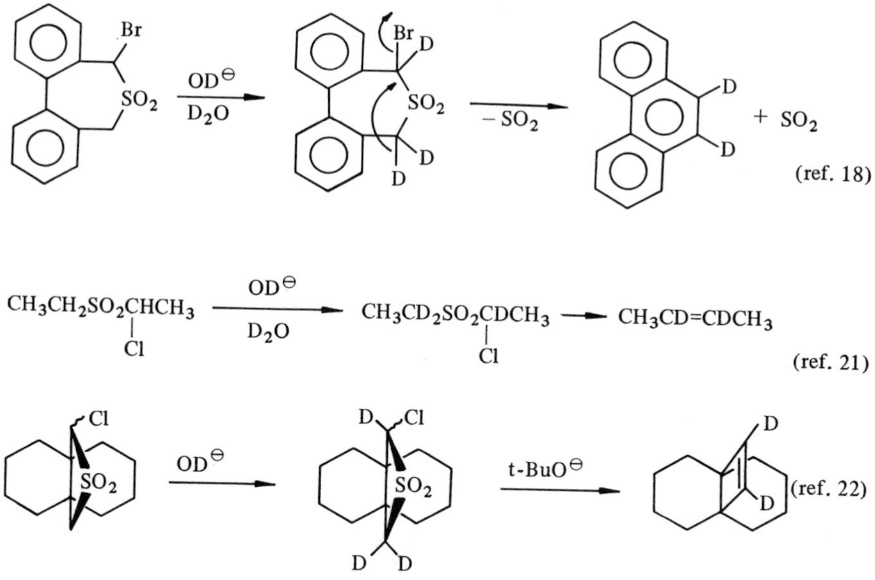

(ref. 18)

(ref. 21)

(ref. 22)

Scheme 2

III.   FROM AZIRIDINES

Since deamination of aziridines to olefins has been reported using difluoro-amine,[23] N-nitrosocarbazole,[24] and other nitrosating agents,[25] a potentially useful synthesis of alkenes is suggested provided that a variety of aziridines could be made accessible. The fact that the deamination proceeds with an exceedingly high degree of stereoselectivity is also of no minor consequence.

The reaction of cis-2,3-dimethyl aziridine with nitrosyl chloride leads to 99% cis-2-butene while the trans-alkene is obtained from the corresponding trans-aziridine.[25] Recently a method has been developed[26] that allows the formation of a variety of 2,3-cis-disubstituted aziridines (22) from lithium aluminum hydride recudtion of ketoximes. The reaction is considered to pass through the azirine intermediate (23) which is consistently reduced from the least hindered side of the molecule. Treatment of these aziridines with a nitrosating agent (e.g., n-butyl nitrite) leads to the thermodynamically less stable cis-olefin (24).[27] Transformation of ketoximes to aziridines appears to be sensitive to the configuration of the hydroxyl group. That is, the α-carbon syn to the hydroxyl in the oxime eventually becomes part of the aziridine ring. The synthesis[27] of cis-stilbene (24 R=R'=Ph) was accomplished in 58% yield using this method. Many other cis-olefins could conceivably be prepared by simple modification of the starting ketoxime. The lithium aluminum hydride

reduction of 2-isoxazolines (25) derived readily from dipolar addition of nitrile oxides and olefins has also been shown to produce aziridines[28] by a ring opening-ring closing mechanism. The reaction is most efficient when the

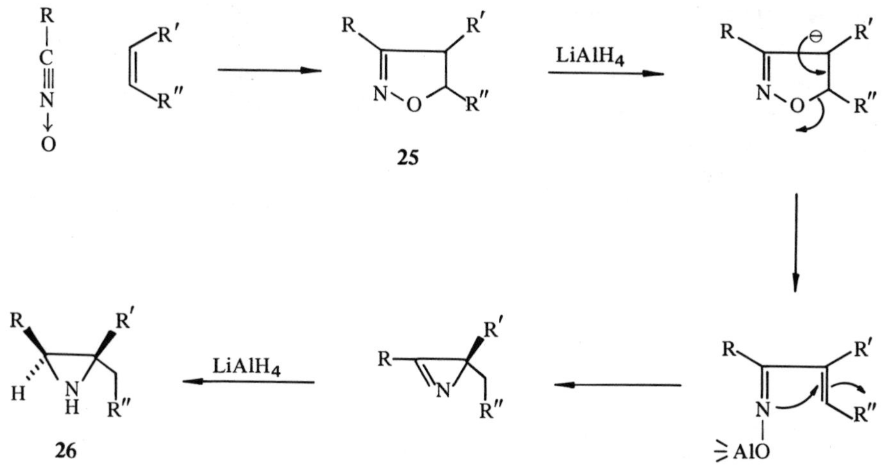
LiAlH4
25
LiAlH4
26

initially formed carbanion is stabilized by an aryl group (R′=Ph) and proceeds poorly when R′ is an alkyl group. A series of trisubstituted aziridines **26** were prepared in variable yields, but more importantly, this sequence represents a

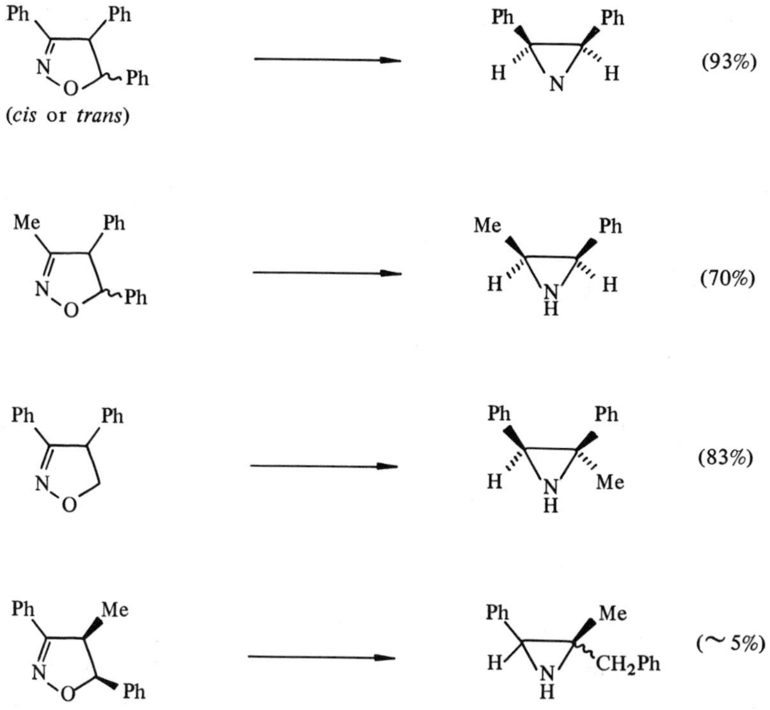
(cis or trans)
(93%)
(70%)
(83%)
(~5%)

feasible method for transforming disubstituted olefins into their trisubstituted derivatives by temporarily constructing two successive heterocyclic systems (Scheme 3). Examination of the mechanism for this sequence readily reveals that the stereochemistry of the starting disubstituted olefin is of no concern, since the aziridine **26** is derived from the azirine which is itself devoid of geometric isomers. Furthermore, since the deamination is essentially stereo-specific, the resulting trisubstituted olefin will be geometrically pure.

$$R'CH=CHR'' \xrightarrow{R-C\equiv N\rightarrow O} \boxed{2\text{-Isoxazoline}} \xrightarrow{LiAlH_4}$$

$$\boxed{\text{Aziridine}} \xrightarrow{\text{deamination}} \underset{H}{\overset{R}{\diagdown}}C=C\underset{CH_2R''}{\overset{R}{\diagup}}$$

Scheme 3

The successful synthesis[29] of azirines (**27**) from olefins via decomposition of vinyl azides (**28**) now renders this once elusive class of compounds as useful precursors to substituted aziridines. The addition of various Grignard reagents

28

27

to 1,2-diphenylazirine (**29**) has been reported[27] to give products derived from attack at the less hindered side and, after deamination with butyl nitrite, gave good yields of *cis*-stilbene derivatives (**30**).

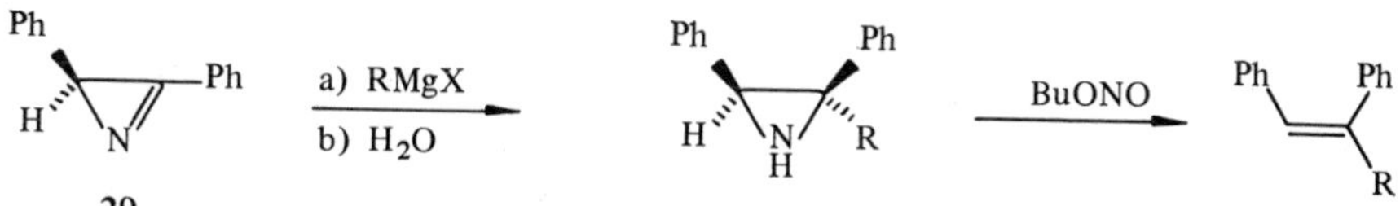

$$R = CH_3 \ (80\%) \qquad R = CH_3 \ (87\%)$$
$$= C_2H_5 \ (100\%) \qquad = C_2H_5 \ (83\%)$$
$$= C_6H_5 \ (100\%) \qquad = C_6H_5 \ (73\%)$$

A somewhat related study involving the stereospecific deamination of N-aminoaziridines (31) was shown to produce geometrically pure olefins in high yield.[29a]

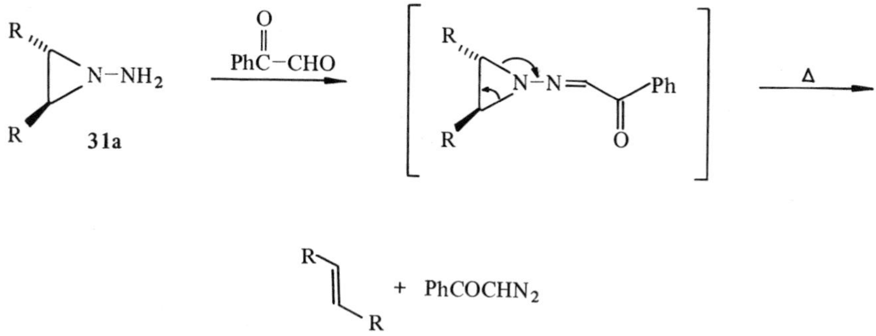

As studies continue on versatile aziridine syntheses, the subsequent deamination process could well provide some of the most innovative olefin syntheses yet known.

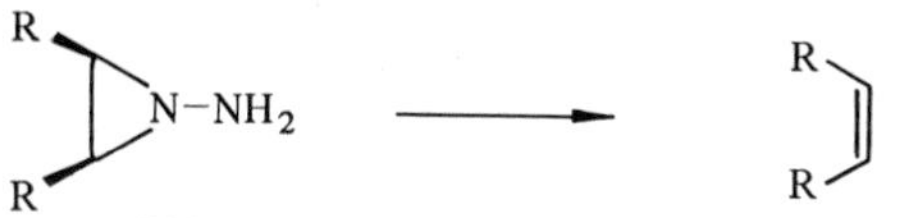

IV. FROM THIONCARBONATES AND TRITHIONCARBONATES

The numerous methods[30] available for preparing 1,2-glycols with *cis* or *trans* configuration provide an unusual opportunity for the synthesis of the corresponding *cis* or *trans* alkenes if the vicinal hydroxyl groups could be stereo-specifically removed. The temporary conversion of 1,2-diols to their thion-carbonate (**32a, 32b**) derivatives allows this very useful elimination to take place. Thus a tervalent phosphorous derivative was employed to effect the

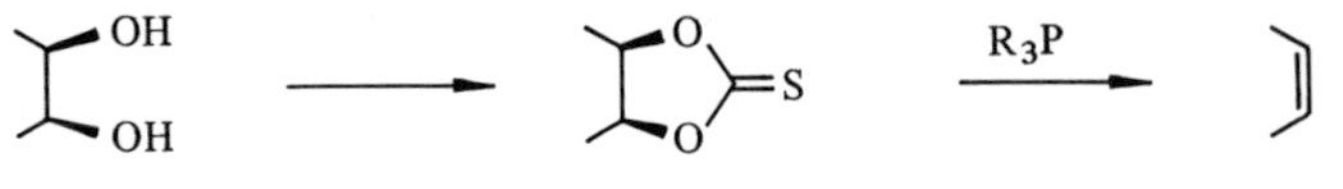

desulfurization and elimination of carbon dioxide leading to the corresponding alkenes.[31] The thioncarbonates were prepared in excellent yield by the reaction of diols with thioncarbonyl diimidazole (**33**)[32] by simple heating in an inert solvent or treatment of the glycol with *n*-butyllithium followed by addition of carbon disulfide and methyl iodide. Once formed, the thioncarbonate hetero-cycle serves to convert a variety of diols to configurationally pure alkenes as exemplified by the following:

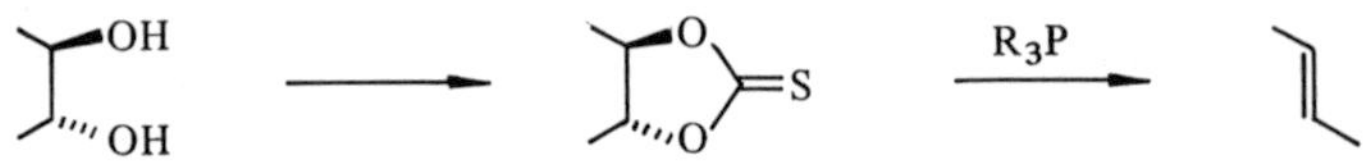

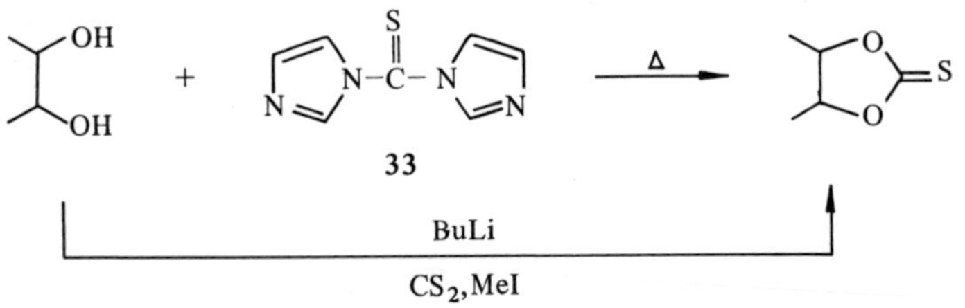

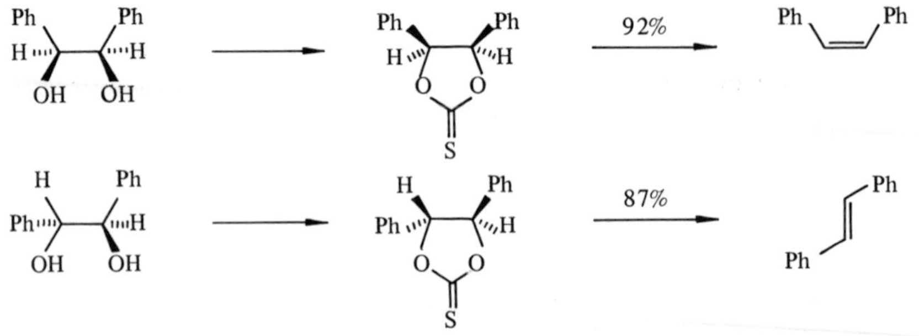

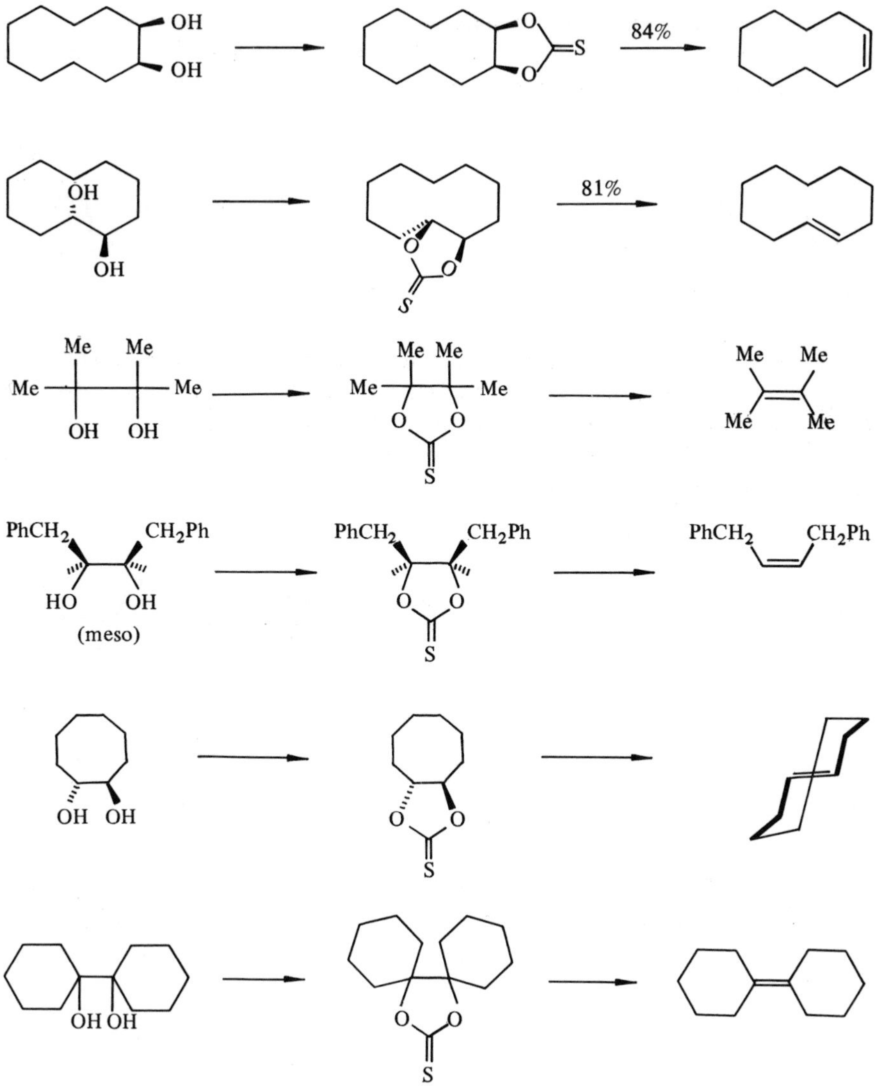

A recent application of the thioncarbonate route to alkenes was reported[33] to lead to cyclobutenes (34).

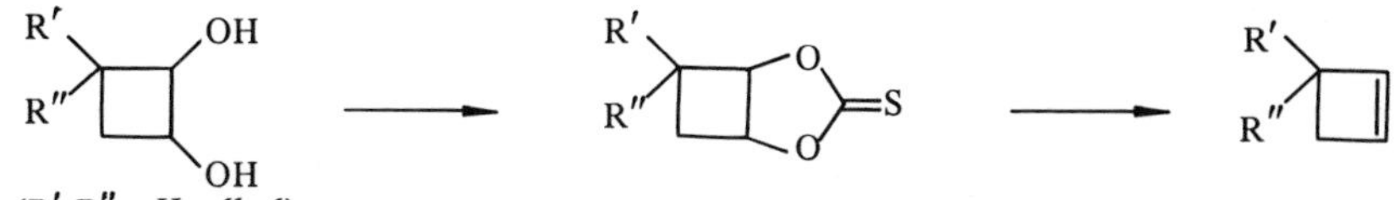

34

Elimination of 1,2-trithioncarbonates (35) using trialkylphosphites is another potentially useful route to olefins and this was demonstrated[31] by the conversion of *cis*-cyclooctene to its *trans*-isomer.

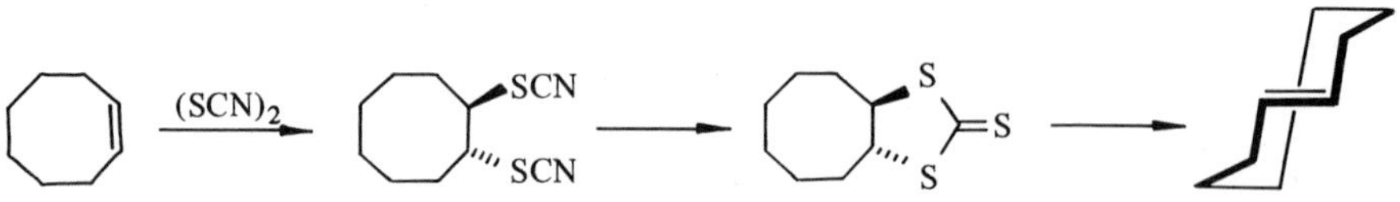

35

The trithioncarbonates may be prepared from olefins using thiocyanogen or from epoxides using xanthates. The latter method has been shown to give *trans*-trithioncarbonates (36) from *cis*-epoxides and vice versa.[34] In this fashion *cis*-epoxides may be converted to *trans*-alkenes or, more generally, *cis*-alkenes are converted to their *trans* isomer.

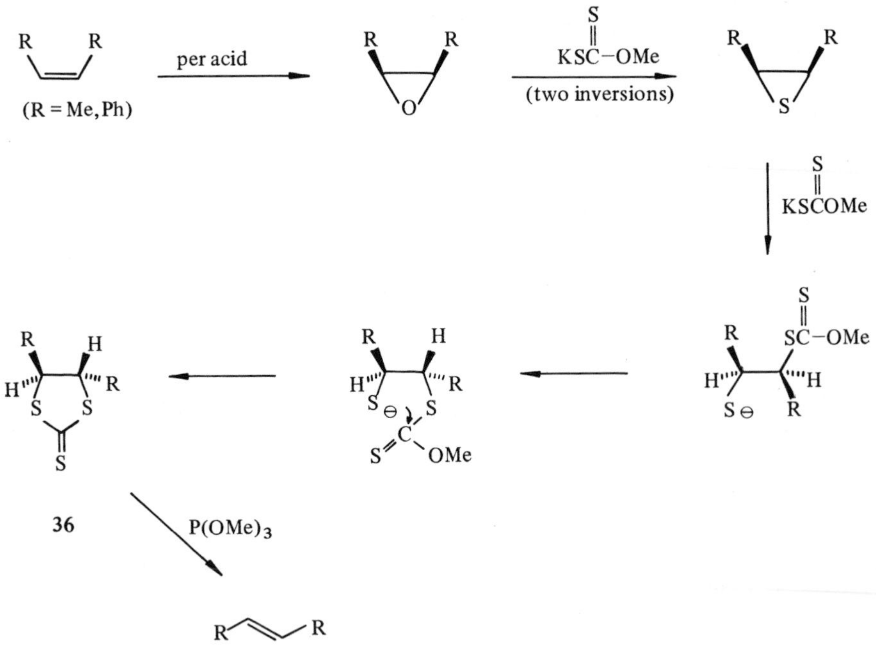

An interconversion of *trans*-alkenes to their *cis*-derivatives, which also starts from epoxides and shows considerable utility, is the ring opening with lithium diphenyl phosphide.[35] The sequence proceeds through the betaine (37), formed *in situ* by addition of methyl iodide, which collapses in the expected manner to the *cis*-olefin. Using this method, the preparation of *cis*-2-octene from *trans*-2-octene was reported to proceed in 75% yield with greater than 99% stereoselectivity. The conversion of *cis*-cyclooctene to *trans*-cyclooctene was also

achieved in 95% yield (>99.5% selective). The value of using the heterocycle (epoxide) lies in the fact that the ring opening is assured to give the *trans*-disposed oxygen and phosphorous functions. Since the betaine will collapse to the alkene only when these two functions are eclipsed, the isomeric alkene usually observed in Wittig coupling is avoided. Epoxides have also been reported to be deoxygenated to the alkenes using magnesium amalgam-magnesium bromide,[36] lithium diamidophosphites,[37] zinc-copper couple,[38] and chromium(II) ethylenediamine complexes.[39]

Somewhat related to the thioncarbonate method for conversion of glycols to alkenes is the amino dioxolane (38) decomposition which produces *trans*-alkenes (39) from *trans*-glycols.[40] An example in the carbohydrate field was offered to demonstrate the utility of this intermediate in complex molecules as well as in simpler systems.

38   39 (99% trans)

V. FROM 3-PYRAZOLIDINONES

The ready formation of 3-pyrazolidinones (40) from α,β-unsaturated acids and hydrazine provides a useful vehicle[41] for transforming unsaturated acids to their corresponding alkenes (42). By warming the pyrazolidinone in dichloromethane in the presence of mercuric oxide, the azocarbonyl compound (41) is produced which rapidly loses carbon monoxide and nitrogen affording the *trans*-alkene (42).

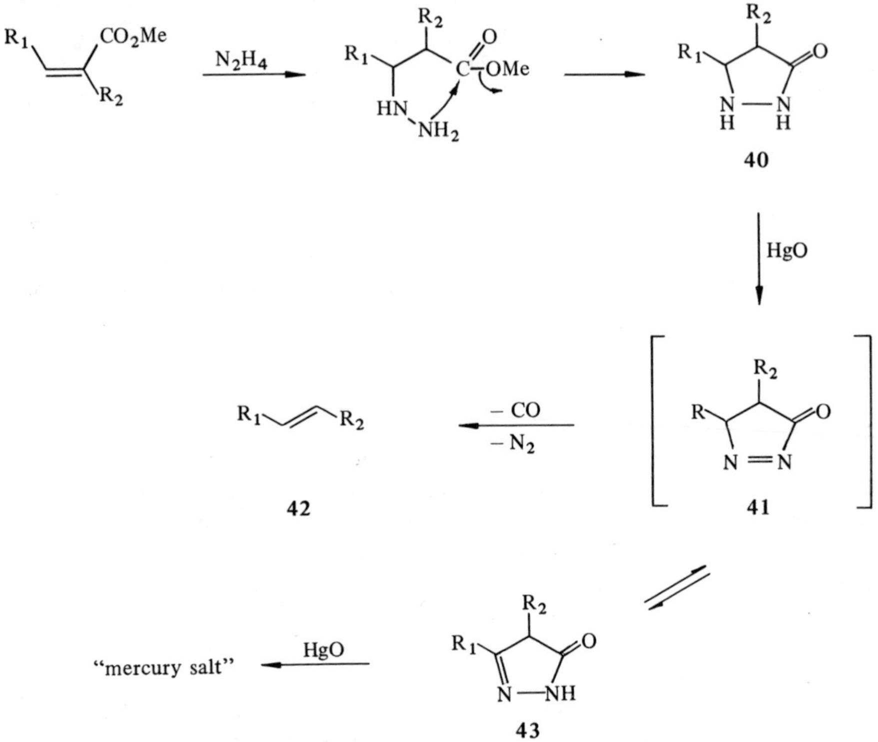

R_1	R_2	% Alkene (42)
Ph	H	42
PH	Ph	73
p-Cl–Ph	p-MeO–Ph	48
p-Cl–Ph	Ph	79

Of interest in this process is the fact that alkene formation competes well with isomerization to the more stable pyrazolinone (43). The latter was presumed to be stable to the mercuric oxide treatment producing only a high melting mercury salt. This method, although demonstrated with relatively few unsaturated acids, is considerably milder than the classic approaches to oxidative decarboxylation.

VI. FROM *N*-NITROSOOXAZOLIDONES

The base catalyzed decomposition of *N*-nitrosooxazolidones **(44)** gives rise to either vinyl cations **(45)** or vinyl carbenes **(46)** depending on the reaction conditions.[42] This useful property of the readily available heterocycle opens

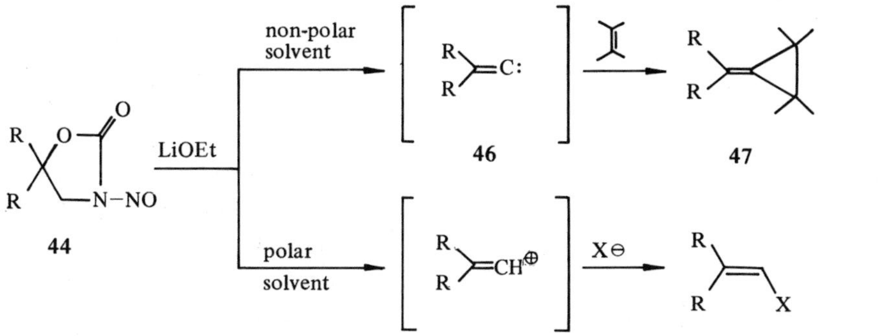

broad avenues to methylene cyclopropanes **(47)** and vinyl halides **(48)**. The carbene **(46)** also reacts with alcohols to produce vinyl ethers via HO insertion. If **(44)** is treated with lithium alkoxides dissolved in their corresponding alcohols,[43] ring rupture proceeds through the intermediates *A-E* affording good yields of vinyl ethers **(49)**. Alternatively, if a suspension of lithium ethoxide in

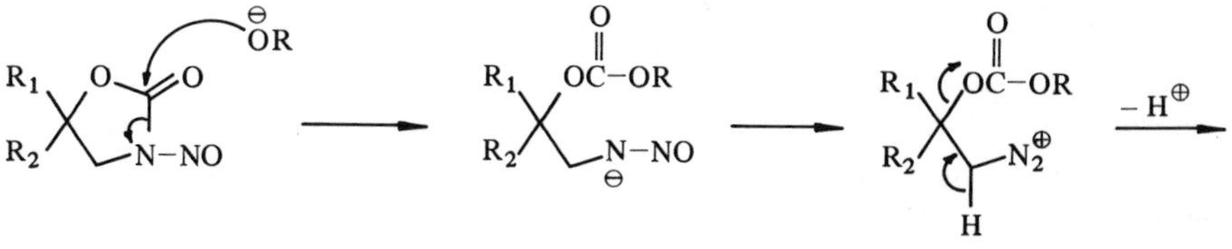

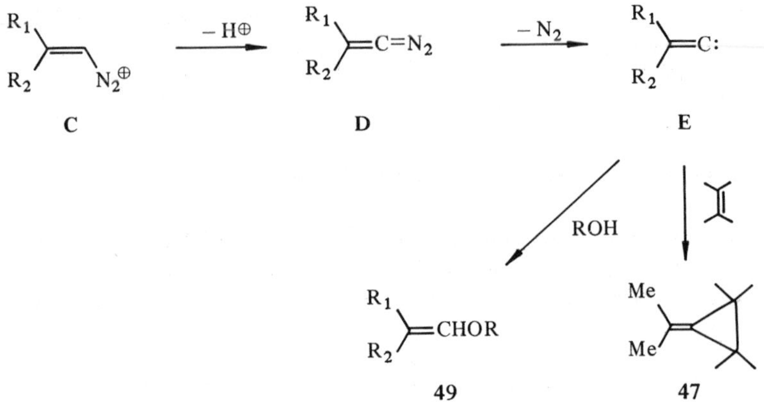

R_1	R_2	R	%49
(cyclohexyl)		Me	85
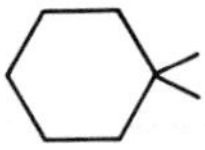 (cyclohexyl)		i-Pr	82
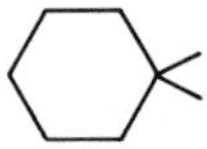 (cyclohexyl)		t-Bu	72
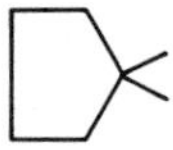 (cyclopentyl)		n-Bu	64
Ph	H	PhC≡CCH$_2$	65
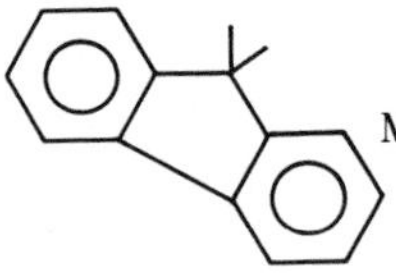 (fluorenyl)		Me	85

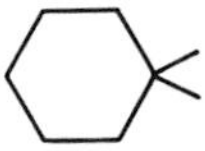

Alkene	%47
tetramethylethylene	21
cyclohexene	56
cyclooctene	18
cycloheptene	43
styrene	32
t-butoxyethylene	31
1-octene	31
4-methyl-2-pentene	34
tetrachloroethylene	0
tetramethylallene	14

the alkene (as solvent) is utilized to decompose the nitrosooxazolidinone, the resulting carbene is trapped as the dialkylmethylene cyclopropane (47). The carbene has also been shown to react smoothly with trialkylsilanes producing the vinyl trialkyl silane (50) via Si-H insertion. The fact that the vinyl species derived from (44) inserted into the silane rather than abstracting a hydride ion

$$\left[\begin{array}{c} R \\ \diagdown \\ R \diagup \end{array} =C: \right] \quad + \quad HSiEt_3 \quad \longrightarrow \quad \begin{array}{c} R \\ \diagdown \\ R \diagup \end{array} \diagdown SiEt_3$$

50

lent support that a carbene was indeed generated.[42] Of additional synthetic value was the high degree of stereoselectivity observed when the *t*-butyl carbene (**51**) was allowed to form in the presence of triethyl silane. The products consisted of a 10:1 mixture of vinyl silanes in which the less stable isomer (**52**) was the predominant product.[44]

$$\begin{array}{c} Me \\ \diagdown \\ t\text{-}Bu \diagup \end{array} =C: \quad + \quad HSiEt_3 \quad \longrightarrow \quad \begin{array}{c} Me \\ \diagdown \\ t\text{-}Bu \diagup \end{array} \diagdown SiEt_3$$

51   **52**

The synthetic utility of the *N*-nitrosooxazolidinones has also been extended to the *in situ* preparation of vinyl cations (**45**) and their conversion to a series of vinyl halides.[42] Thus basic decomposition of the heterocycle with lithium 2-methoxyethanolate in 2-methoxyethanol saturated with lithium halide produced the corresponding vinyl halides (**48**) in good yield. If the decomposition is performed in a polar solvent, the formation of the vinyl cation (**45**) is favored over the vinyl carbene (**46**). When aqueous sodium hydroxide is

$$\begin{array}{c} R \\ \diagdown \\ R \diagup \end{array}=C=N_2 \quad \xleftarrow{-H^\oplus} \quad \begin{array}{c} R \\ \diagdown \\ R \diagup \end{array}\diagdown N_2^{\oplus} \quad \xrightarrow{-N_2} \quad \begin{array}{c} R \\ \diagdown \\ R \diagup \end{array}=CH^\oplus$$

45

$$\begin{array}{c} R \\ \diagdown \\ R \diagup \end{array}=C:$$

46

$$\begin{array}{c} R \\ \diagdown \\ R \diagup \end{array}\diagdown OH$$

$$\begin{array}{c} R \\ \diagdown \\ R \diagup \end{array}\diagdown X$$

48

$$\begin{array}{c} R \\ \diagdown \\ R \diagup \end{array}-CHO$$

53

employed as the medium for nitrosooxazolidinone decomposition, aldehydes (53) are produced but are accompanied by other products.[42]

The latent functionality of N-nitrosooxazolidines as precursors to vinyl carbene and vinyl cations is indicative of the value of heterocycles in synthesis. A variety of alkenyl systems may be prepared by constructing the appropriate heterocycle from any number of β-hydroxy esters and hydrazine.[45] The acylhydrazide (54) on nitrosation and rearrangement to the carbamic acid (55) cyclizes to the oxazolidinone (56). Nitrosation of the latter leads to the desired nitroso derivatives (44).

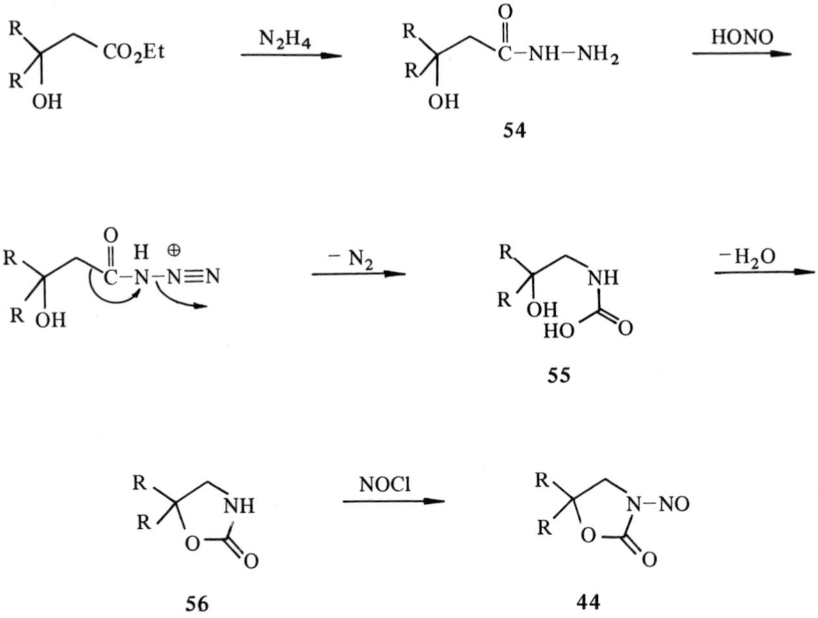

VII.   FROM 2-OXAZOLINES

Alpha-metallation of isocyanides (57) with butyllithium followed by introduction of a carbonyl component at −70° affords the alkenes (58) after hydrolysis.[46] The reaction passes through the temporarily formed 2-oxazoline (59) which eliminates lithium isocyanate upon warming to room temperature. Trisubstituted alkenes were prepared via this route in variable yields. Proof that the lithio oxazoline was indeed the heterocyclic vehicle for this carbonyl olefination reaction was shown by treatment of the oxazoline (60) with

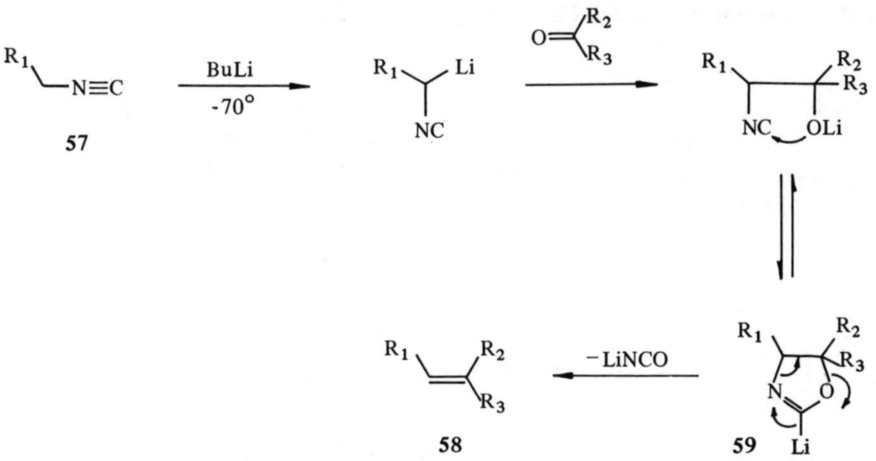

R_1	R_2	R_3	% Alkene
H	Ph	CH_3	65
H	PhCH=CH−	H	28
H	Ph	Ph	40
CH_2=CH−	Ph	H	41
CH_2=CH−	Ph	Ph	38
Ph	Ph	CH_3	44
Ph	Ph	Ph	74
Ph	$(CH_2)_5$		10

n-butyllithium at 0° followed by addition of deuterium oxide.[47] The lithio salt was trapped by deuterium oxide producing 50% of deuterated oxazoline (61) and ∿50% of isobutylene. Of interest is the fact that the lithio oxazoline is in equilibrium with the isonitrile (62) making two modes of reaction possible.[48]

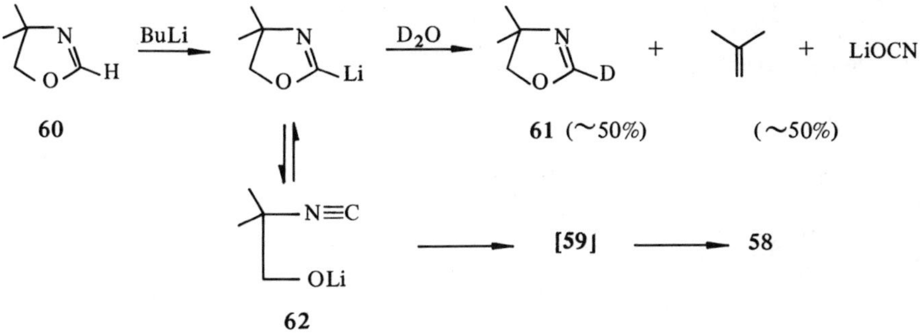

VIII.   FROM 1,3,2-BENZODIOXABOROLE

The ready hydroboration[49] of alkynes by 1,3,2-benzodioxaborole (63) to the alkenyl benzodioxaborole (64) provides a new stereospecific and regioselective conversion of alkynes to vinylmercurial salts (65). This facile synthesis of mercurial salts should be quite valuable in syntheses of various other alkenyl metallic compounds.

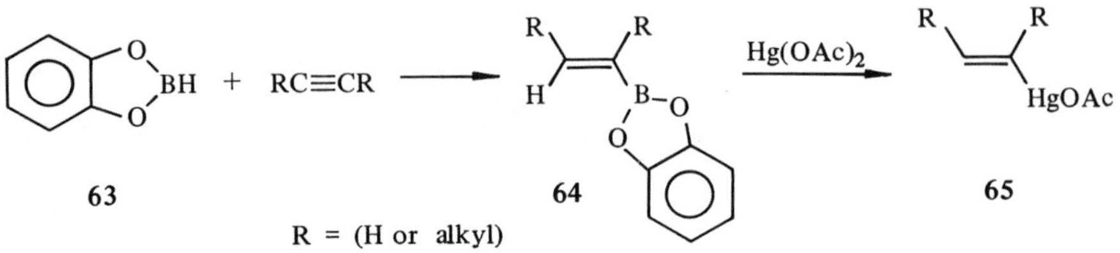

63   64   65

R = (H or alkyl)

IX.   FROM 1,3-DITHIANES

A useful heterolytic fragmentation was described[50] that transformed the dithiane system (66) to the alkene (67), thus providing a route to alicyclic unsaturated acids from cyclohexane derivatives. The fact that the dithiane moiety can be readily introduced into a molecule containing an aldehyde function now renders the formyl proton acidic. Treatment with a suitable strong base generates the dithiane carbanion which rearranges in the manner depicted.

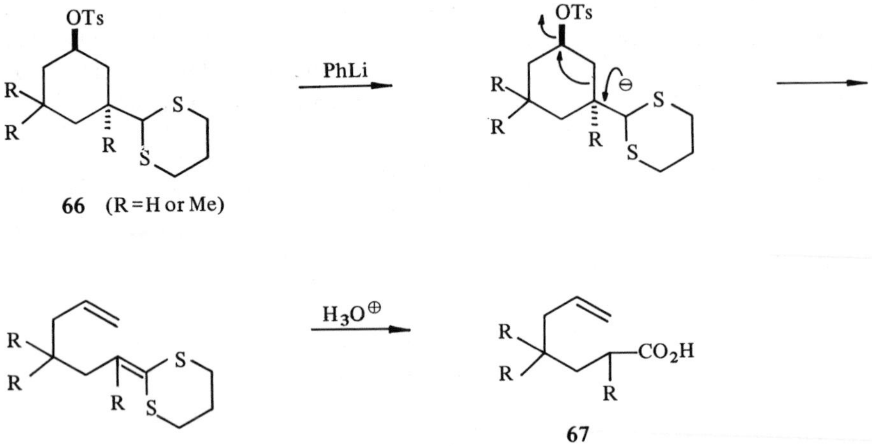

66   (R=H or Me)

67

X. FROM DIHYDROPYRANES

Cyclohexene derivatives (68) were found to be accessible from the thermal reorganization of 3,4-dihydro-2H-pyranylethylenes (69).[51] The latter are available from the dimerization of a,β-unsaturated carbonyl compounds which lead to acyl dihydropyranes (70) followed by Wittig coupling. The cyclohexenes formed in this process are geometrically pure since the rearrangement is stereospecific (69a → 68a). The presence of *cis*-substituents (69, $R_2=CH_3$) hinders the rearrangement presumably by crowding in the transition state (69a, $R_2=CH_3$). Rearrangement of the *trans*-pyranyl ester (71) gave, at 230°, an 88% yield of the cyclohexenyl keto ester (72) with *cis*-configuration, whereas treatment of the *cis*-pyranyl ester (73) produced, at 285°, only 26% of 72 along with 70% of 74. Thus a loss of stereoselectivity accompanies the thermolysis

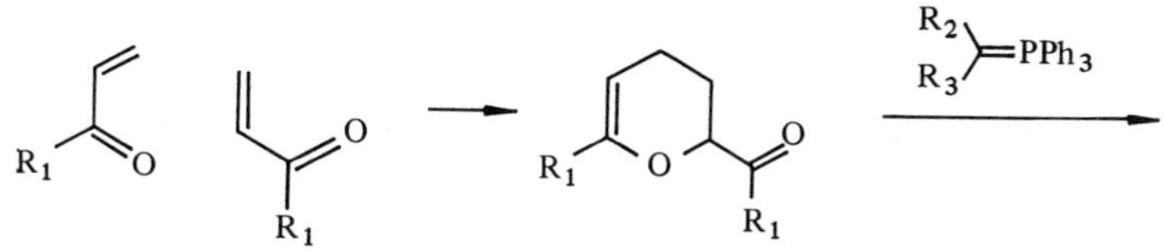

70

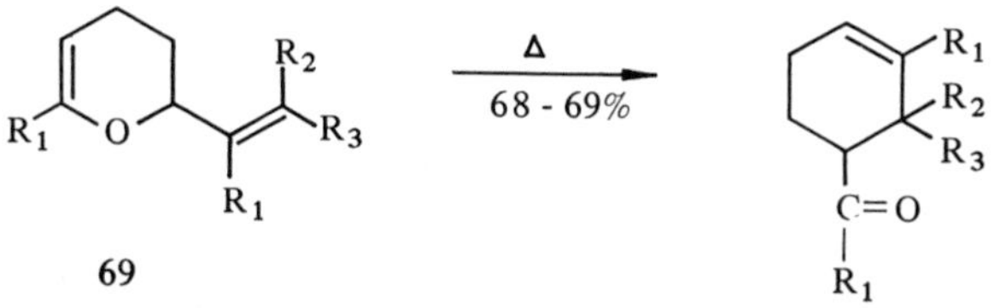

69   68

69a   68a

$R_1 = R_2 = R_3 = H$
$R_1 = R_3 = CH_3, R_2 = H$

of *cis*-pyranylethylenes. The synthetic utility of this process was demonstrated by the multigram synthesis of **75** from the tetraene **76**. This method comple-ments the Diels-Alder method for preparing Δ^3-cyclohexene carboxaldehydes

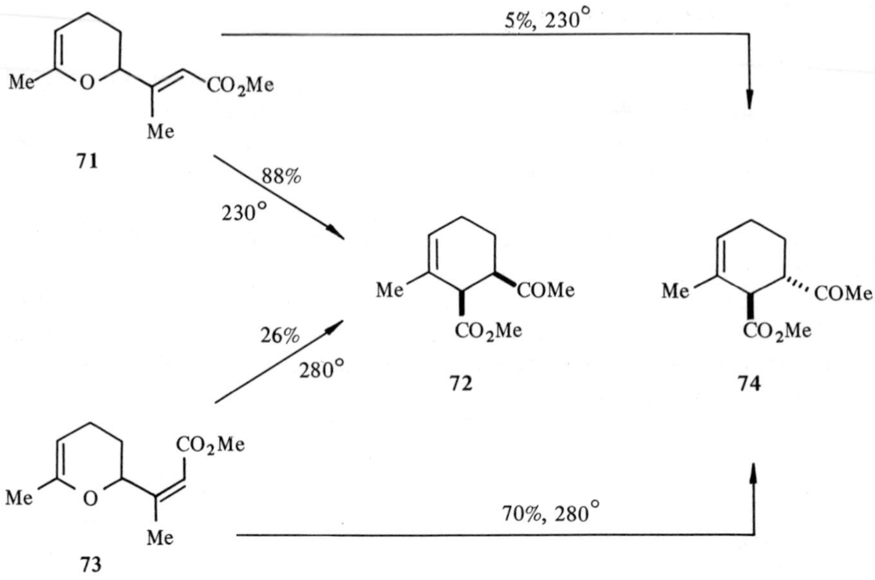

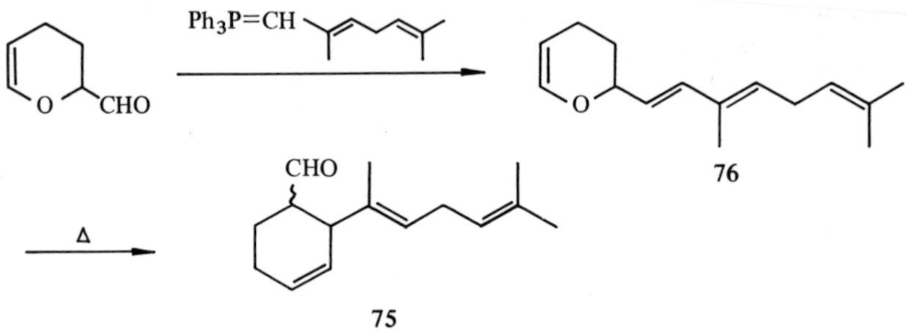

by allowing unsymmetrical substituents to be incorporated without the usual danger of product mixtures.

XI. FROM 2-MERCAPTOPYRIDINES

An interesting olefin synthesis was recently reported by Mukaiyama[52] which required the use of alkylated 2-mercapto pyridines **77**. These derivatives are readily obtained by alkylation of 2-mercaptopyridine and alkyl halides under alkaline conditions. Treatment of **77** with phenyllithium in the presence of cuprous iodide led to the cuprate **78** which undergoes rapid elimination to the

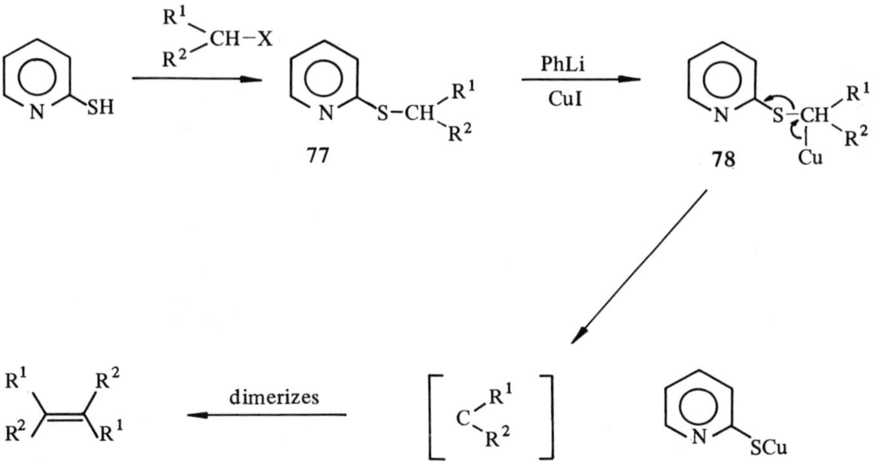

R_1	R_2	%
Ph	H	86
Ph	Ph	33
PhCH=CH–	H	10

carbenoid species **79** which ultimately dimerizes to the olefin **80**. The study has only been reported in preliminary form offering the three examples given.

XII. FROM 4-CHLOROPYRIDINES

Alcohols may be dehydrated[53] to olefins under essentially neutral conditions by employing 4-chloropyridine **81** to prepare an intermediate ether **82**. The latter, on conversion to its *N*-methyl quaternary salt **83**, readily fragments to the olefin **84** on heating. Olefins have been prepared in this fashion in excellent yields by fragmentation at temperatures ranging from 25° to 185°. The procedure is simply to heat the methiodide salt **83** until it melts and to collect the volatile olefin in a cold receiver. A representative list of olefins, obtained via this method, is presented below. It is evident that the dehydration produces the most stable olefin and the *trans*-isomer is heavily favored. A lack of stereospecificity is noted by the fact that *cis* and *trans*-methylcycloalkanols give

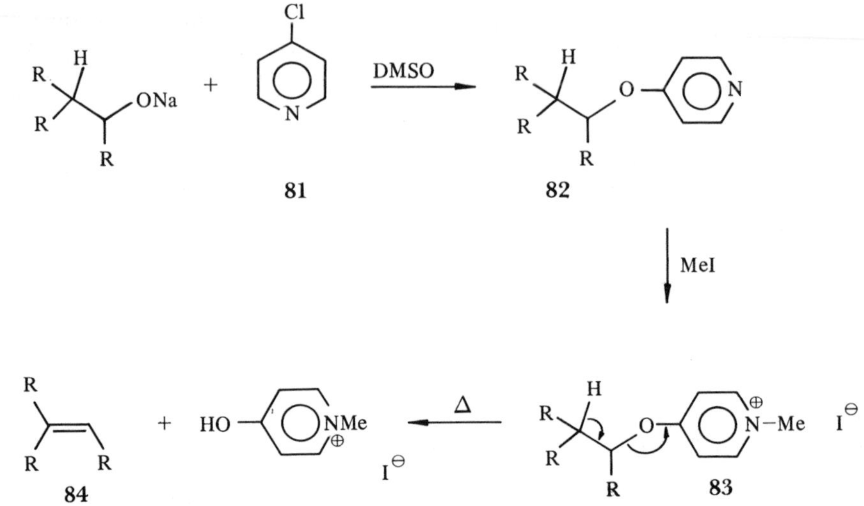

Alcohol	t°	Alkene	% yield
Cyclopentanol	165	Cyclopentene	100
Cyclohexanol	165	Cyclohexene	90
Cycloheptanol	165	Cycloheptene	87
Cyclooctanol	145-155	Cyclooctene (cis)	76
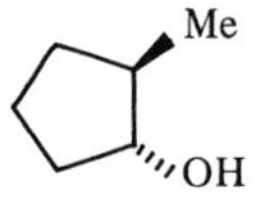	145	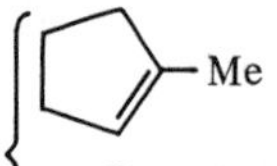	98.3
		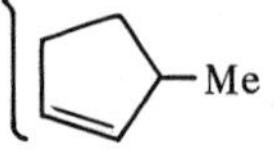	0.7
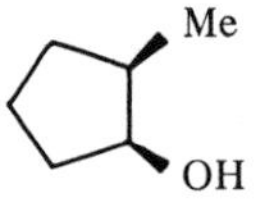	145	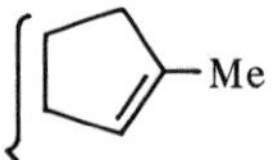	85.2
		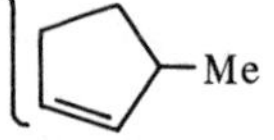	13.8
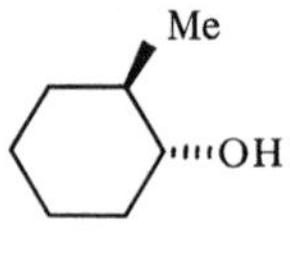	160-165	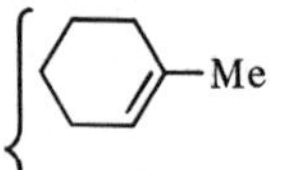	86.2
		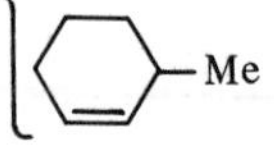	8.7

Alcohol	$t°$	Alkene	% yield
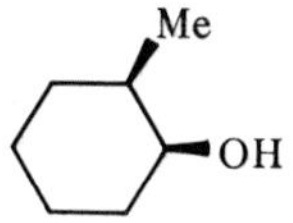	160-165	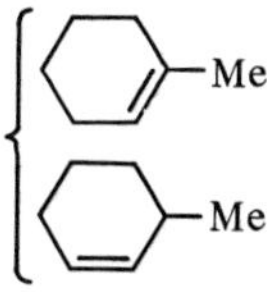 —Me	95.1
		—Me	4.9
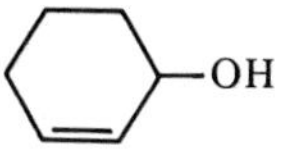 —OH	25		100
2-Octanol	165	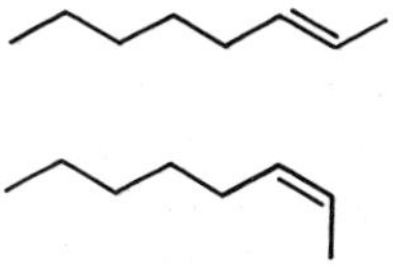	68.3
			23.9
			7.8
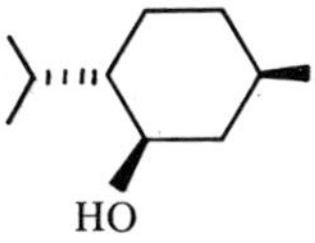	170	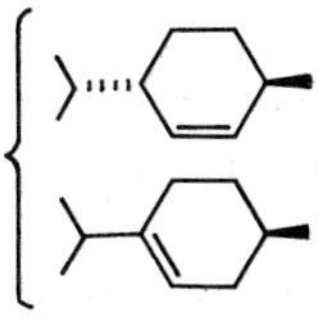	11.1
			88.9
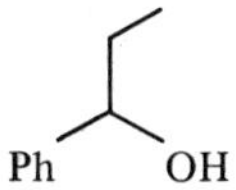	150	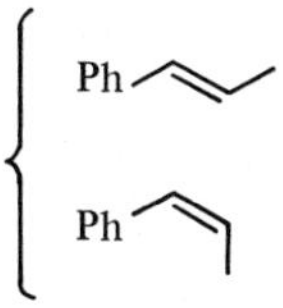 Ph	98
		Ph	1.7
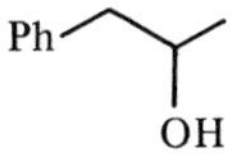	150	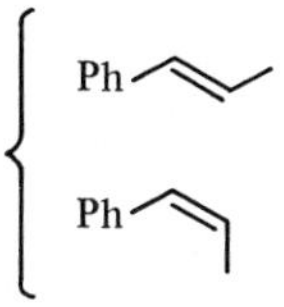 Ph	91.7
		Ph	6.7
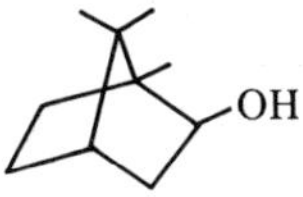 —OH	150	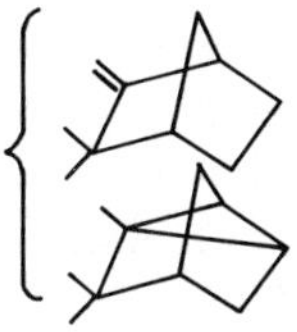	99.1
			0.9

comparable mixtures of olefins with the more highly substituted derivatives predominating. It was concluded that the olefins are formed under kinetic control in view of the limited time that the olefin is in contact with the acid formed. The facile nucleophilic displacement of the 4-chlorosubstituent by sodium alkoxides in DMSO renders this technique widely applicable to alcohols of diverse structure. Unfortunately, dehydration of tertiary alcohols via this sequence is precluded because of the bulky nature of their alkoxide derivatives.

REFERENCES

1a. S. Patai, ed., *The Chemistry of Alkenes,* Wiley-Interscience, 1964.

1b. I. T. Harrison and S. Harrison, *Compendium of Organic Synthetic Methods,* Wiley-Interscience, 1971.

1c. An excellent monograph that describes the extrusion of simple molecules (N_2, SO_2, CO_2, etc.) from a variety of heterocyclic compounds has touched on this subject and many examples of synthetically useful processes have been discussed: B. P. Stark and A. J. Duke, *Extrusion Reactions,* Pergamon Press, 1967.

2. L. Goodman and E. J. Reist, *The Chemistry of Organosulfur Compounds,* Vol. 2, Pergamon Press, New York, 1966, p. 105; F. G. Bordwell, H. M. Andersen, and B. M. Pitt, *J. Am. Chem. Soc.,* **76,** 1082 (1954).

3. R. E. Davis, *J. Org. Chem.,* **23,** 1767 (1958); R. D. Schuetz and R. L. Jacobs, *ibid.,* **26,** 3467 (1960).

4. N. P. Neureiter and F. G. Bordwell, *J. Am. Chem. Soc.,* **81,** 578 (1959).

5. N. Latif, N. Mishriky, and I. Zeid, *J. Prakt. Chem.,* **312,** 421 (1970).

5a. B. M. Trost and S. D. Ziman, *J. Org. Chem.,* **38,** 932 (1973).

6. G. K. Helmkamp and D. J. Pettitt, *J. Org. Chem.,* **25,** 1754 (1960).

7. D. J. Pettitt and G. K. Helmkamp, *ibid.,* **28,** 2932 (1963).

8. F. G. Bordwell and H. M. Anderson, *J. Am. Chem. Soc.,* **75,** 4959 (1953). Formation of olefins from epoxides have also been reported: E. Vedejs and P. L. Fuchs, *J. Am. Chem. Soc.,* **93,** 4070 (1971); D. L. J. Clive and C. V. Denyer, *Chem. Commun.,* 253 (1973).

9. D. H. R. Barton, E. H. Smith, and B. J. Willis, *Chem. Commun.,* 1226 (1970).

10. K. Rühlmann, *J. Prakt. Chem.,* **8,** 285 (1959).

11. R. M. Kellog and S. Wassenaar, *Tetrahedron Lett.,* 1987 (1970).

12. J. W. Everett and P. J. Garratt, *Chem. Commun.,* 642 (1972); see also A. P. Schaap and G. R. Faler, *J. Org. Chem.,* **38,** 3061 (1973).

13. D. H. R. Barton and B. J. Willis, *Chem. Commun.,* 1225 (1970).

14. C. T. Pedersen, *Acta. Chim. Scand.,* **22,** 247 (1968).

15. E. Block and E. J. Corey, *J. Org. Chem.,* **34,** 896 (1969).

16. H. Staudinger and F. Pfenninger, *Berichte,* **49,** 1941 (1916).

17. L. Ramberg and B. Backlund, *Arkiv. Kemi, Mineral. Geol.,* **13A,** No. 27 (1940); Chem. Abstr., **34,** 4725 (1940).

18. For a review of the mechanistic details concerning this reaction see L. A. Paquette, *Acct. Chem. Res.,* **1,** 209 (1968).

19. L. A. Paquette, J. C. Philips, and R. E. Wingard, *J. Am. Chem. Soc.,* **93,** 4516 (1971).

20. E. J. Corey and E. Block, *J. Org. Chem.,* **34,** 1233 (1969).

21. N. P. Neureiter, *J. Am. Chem. Soc.,* **88,** 558 (1966).

22. L. A. Paquette and R. E. Wingard, *ibid.*, **94**, 4398 (1972).

23. C. L. Bumgardner, K. J. Martin, and J. P. Freeman, *ibid.*, **85**, 98 (1963); J. P. Freeman and W. H. Graham, *ibid.*, **89**, 1761 (1967).

24. C. L. Bumgardner, K. S. McCallum, and J. P. Freeman, *ibid.*, **83**, 4417 (1961).

25. R. D. Clark and G. K. Helmkamp, *J. Org. Chem.*, **29**, 1316 (1964).

26. For a review of this reaction see K. Kotera and K. Kitahonoki, *Org. Proced. Prep.*, **1**, 305 (1969).

27. R. M. Carlson and S. Y. Lee, *Tetrahedron Lett.*, 4001 (1969).

28. K. Kotera, Y. Takano, A. Matsuura, K. Kitahonoki, *Tetrahedron*, **26**, 539 (1970).

29. A. Hassner and F. W. Fowler, *J. Am. Chem. Soc.*, **90**, 2869 (1968).

29a. R. K. Muller, D. Felix, J. Schreiber, and A. Eschenmoser, *Helv. Chim. Acta.*, **53**, 1479 (1970).

30. H. O. House, *Modern Synthetic Reactions*, Benjamin, Menlo Park, Calif., **1972**, pp. 168, 275, 300.

31. E. J. Corey and R. A. E. Winter, *J. Am. Chem. Soc.*, **85**, 2677 (1963); E. J. Corey, F. A. Carey, and R. A. E. Winter, *ibid.*, **87**, 934 (1965).

32. H. A. Staab and G. Walther, *Annals*, **657**, 98 (1962).

33. W. Hartmann, H. M. Fischder, and H. G. Heine, *Tetrahedron Lett.*, 853 (1972).

34. C. G. Overberger and A. Drucker, *J. Org. Chem.*, **29**, 360 (1964).

35. E. Vedejs and P. L. Fuchs, *J. Am. Chem. Soc.*, **93**, 4070 (1971).

36. F. Bertini, P. Grasselli, G. Zubiani, and G. Cainelli, *Chem. Commun.*, 144 (1970).

37. E. J. Corey and D. E. Cane, *J. Org. Chem.*, **34**, 3053 (1969).

38. S. M. Kupchan and M. Maruyama, *ibid.*, **36**, 1187 (1971).

39. J. K. Kochi, D. M. Singleton, and L. J. Andrews, *Tetrahedron*, **24**, 3503 (1968).

40. F. W. Eastwood, K. J. Harrington, J. S. Josan, and J. L. Pura, *Tetrahedron Lett.*, 5223 (1970).

41. R. H. Kent and J.-P. Anselme, *Can. J. Chem.*, **46**, 2322 (1968); see also B. T. Gillis and R. Weinkam, *J. Org. Chem.*, **32**, 3321 (1967).

42. M. S. Newman and C. D. Beard, *J. Am. Chem. Soc.*, **92**, 4309 (1970). [For a recent modification c.f. *J. Org. Chem.*, **38**, 547 (1973).]

43. M. S. Newman and A. O. M. Okorodudu, *J. Org. Chem.*, **34**, 1211 (1969).

44. M. S. Newman and T. B. Patrick, *J. Am. Chem. Soc.*, **92**, 4312 (1970).

45. M. S. Newman and A. Kutner, *ibid.*, **73**, 4199 (1951).

46. U. Schollkopf and F. Gerhart, *Angew. Chem.*, **80**, 842 (1968).

47. A. I. Meyers and E. W. Collington, *J. Am. Chem. Soc.*, **92**, 6676 (1970).

48. F. Gerhart and U. Schollkopf, *Tetrahedron Lett.*, 6231 (1968).

49. R. C. Larock, S. K. Gupta, and H. C. Brown, *J. Am. Chem. Soc.*, **94**, 4372 (1972).

50. J. A. Marshall and J. L. Belletire, *Tetrahedron Lett.*, 871 (1971).

51. G. Buchi and J. E. Powell, *J. Am. Chem. Soc.*, **92**, 3126 (1970).

52. T. Mukaiyama, K. Narasaka, and M. Furasato, *Bull. Soc. Chem. Japan*, **45**, 652 (1972).

53. G. H. Schmid and A. W. Wolkoff, *Can. J. Chem.*, **50**, 1181 (1972).

4 DIENES, POLYENES, ALLENES, AND ACETYLENES

A number of routes to polyunsaturated molecules are available through hetero-cyclic precursors. These methods result from some unique property of the heterocyclic molecule which enables the chemist to prepare, in many instances by stereospecific destruction, a variety of conjugated and nonconjugated dienes, allenic systems, and acetylenes.

I. FROM SULFOLENES

The reaction of sulfur dioxide with conjugated dienes has long been known[1] to produce the cycloaddition products, sulfolenes (1). That the reaction may be reversed by heating to regenerate the diene suggests that the sulfolene molecule, if it could be elaborated, would provide a useful vehicle for diene synthesis. In fact, sulfolene itself (R=H) may be used as a source of butadiene in the Diels-Alder reaction[2] and this was later demonstrated[3] by its reaction with benzyne to form 1,4-dihydronaphthalene (2).

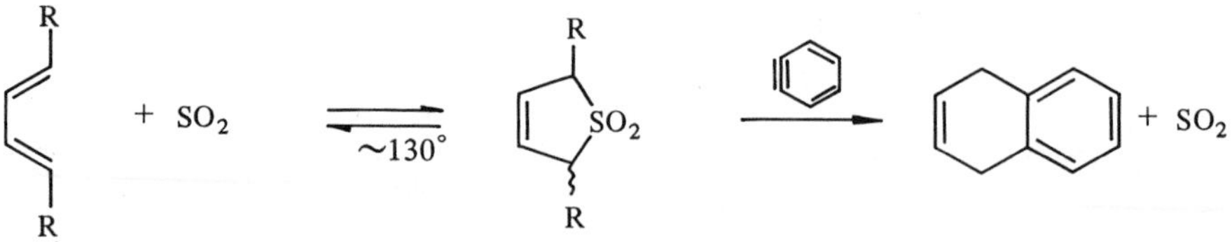

The preparation of 2-bromomethyl-1,3-butadiene (4) was achieved[4] using the sulfolene derivative (3) and bromination of the methyl group by N-bromosuccinimide. Thermolytic extrusion of sulfur dioxide returned the modi-

fied diene in good yield. The complex mixture of bromo derivatives expected
from direct bromination of isoprene was thus avoided. Additional examples

concerned with elaboration of sulfolenes were provided by Mock[5] who described their conversion to the cyclopropyl derivatives **(5)** by addition of
diazomethane. After heating, the homo conjugated diene **(6)** was cleanly
produced. The utility of this scheme is obvious when one considers that a

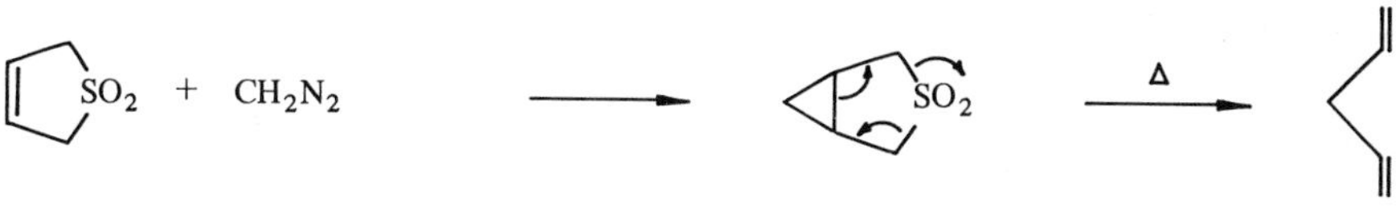

methylene group is inserted, via the heterocyclic intermediate, between the
vinyl groups of a conjugated diene. Further studies have shown[5,6] that this
transformation is not limited to methylene insertion. Both oxygen and nitrogen
may likewise be incorporated into the 1,3-diene system (**7** → **8** and **9** → **10**). If

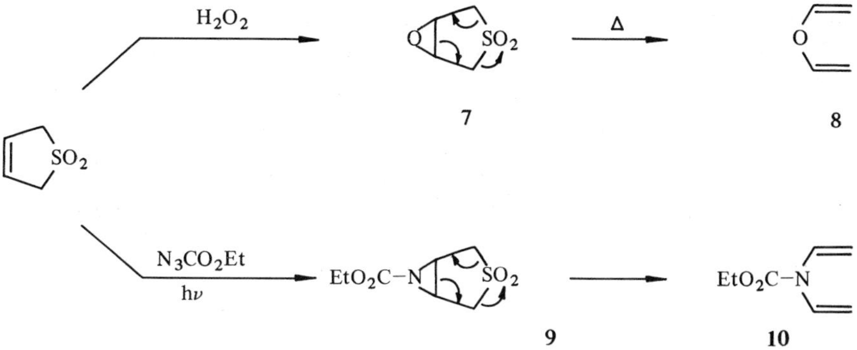

the sulfolene contains alkyl groups, derived from substituted butadienes, it is possible to prepare geometrically pure dienes (11, 12, 13) or their hetero derivatives. This is the result of stereospecific extrusion of sulfur dioxide in a disrotatory process.

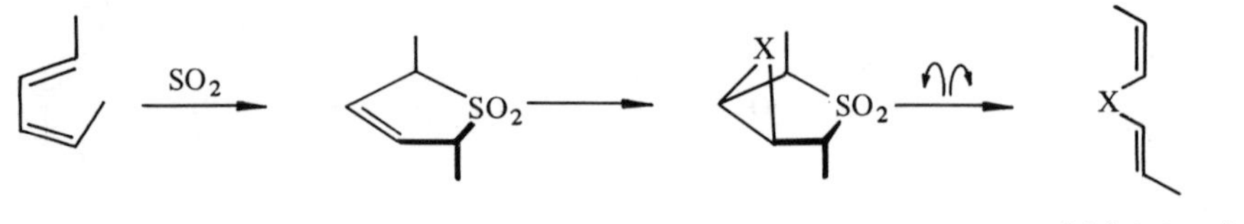

11 (*cis-trans*)

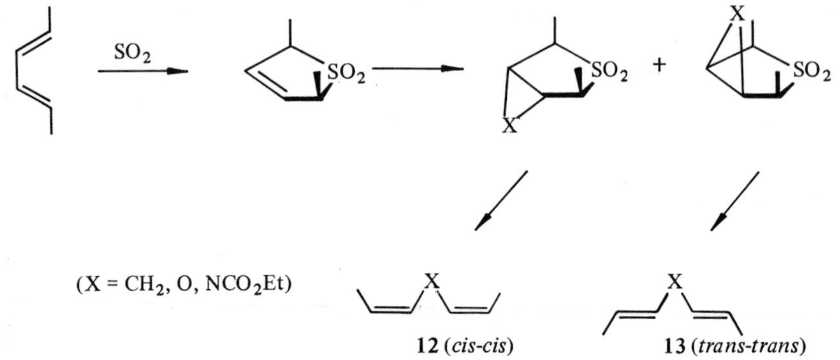

(X = CH$_2$, O, NCO$_2$Et)

12 (*cis-cis*)   13 (*trans-trans*)

II.   FROM 2,5-DIHYDROTHIOPHENES

Irradiation[7] of 2,5-dihydrothiophenes (14), formed by thermal addition of thiocarbonyl ylids (15) to acetylenic esters, results in quantitative isomerization to the unstable vinyl episulfides (16), which are efficiently transformed into

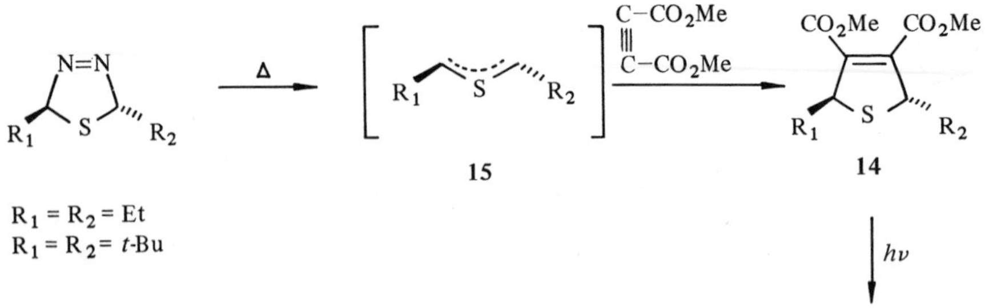

15

14

R$_1$ = R$_2$ = Et
R$_1$ = R$_2$ = *t*-Bu

$h\nu$

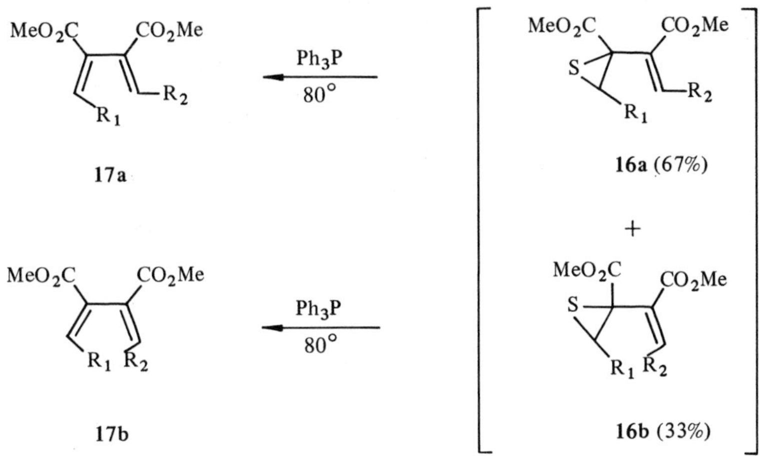

MeO$_2$C, CO$_2$Me

Ph$_3$P
—————
80°

17a

MeO$_2$C, CO$_2$Me

Ph$_3$P
—————
80°

17b

$$\left[\begin{array}{c} \text{MeO}_2\text{C, CO}_2\text{Me} \\ \text{S} \\ R_1 \quad R_2 \\ \\ \textbf{16a} (67\%) \\ \\ + \\ \\ \text{MeO}_2\text{C, CO}_2\text{Me} \\ \text{S} \\ R_1 \; R_2 \\ \\ \textbf{16b} (33\%) \end{array}\right]$$

a 1:1 mixture of the isomeric dienes (17).

III. FROM CYCLIC PHOSPHONIUM SALTS (PHOSPHOLENIUM, PHOSPHORANIUM, AND DIPHOSPHORANIUM SALTS)

The availability of heterocyclic phosphonium salts 18,[8] 19,[9] and 20[10] has stimulated interest in their synthetic utility. Treatment of 18 with potassium tert-butoxide in the presence of aromatic aldehydes led to a series of

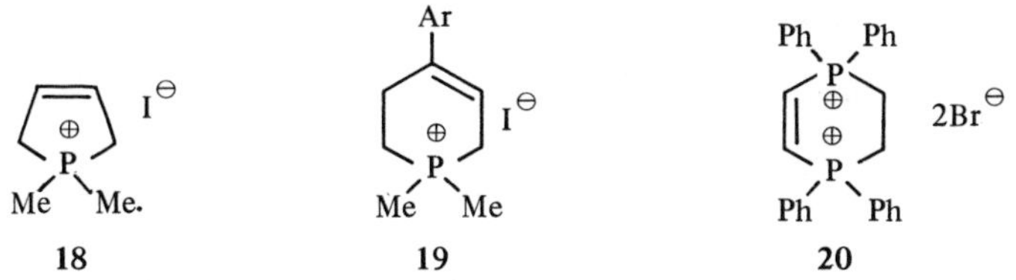

18 19 20

1,6-diarylhexadienes (24) in 10-26% yield.[11] The starting heterocycle (18) possesses the necessary structural features to engage in two successive Wittig condensations (21 → 22; 23 → 24). The reaction, although briefly explored,

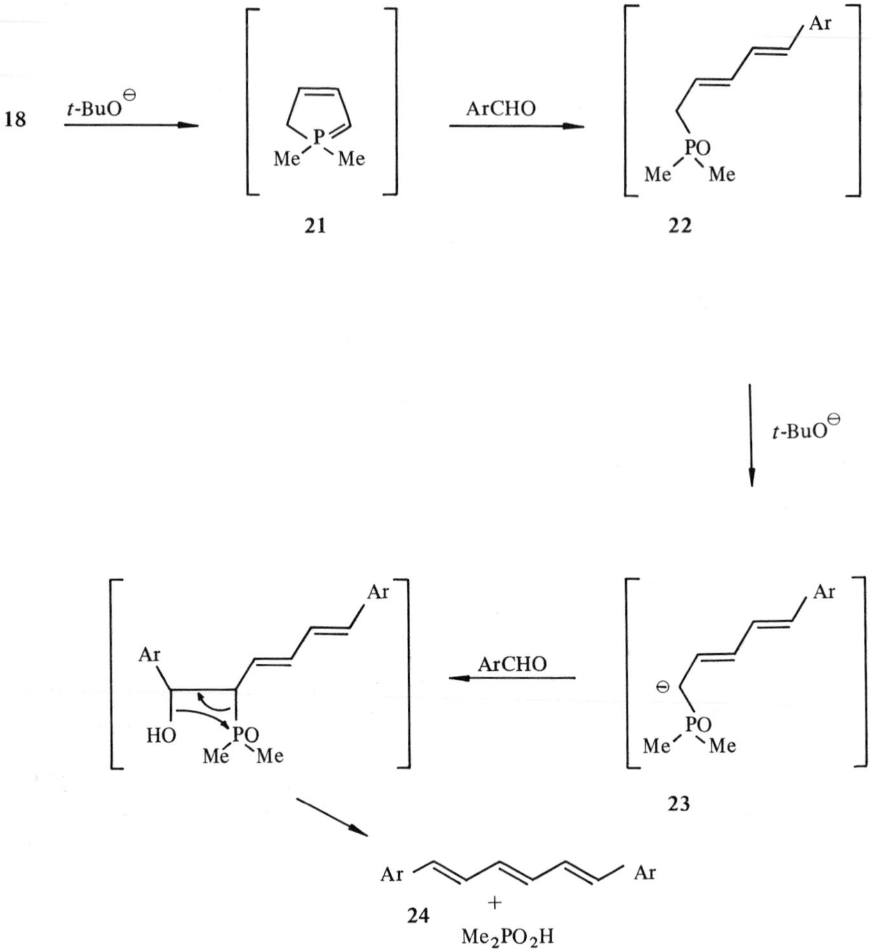

ArCHO = benzaldehyde
p-tolualdehyde
p-methoxy benzaldehyde

shows promise in that the phosphorus heterocycle may be a useful precursor to this class of polyolefins.

In a related study,[12] the phosphorinium salt (19) was treated with butyllithium and then benzaldehyde to afford the diene (25). The latter products were obtained in overall yields of 13-50% depending upon the substituents present on the aldehyde or the starting phosphorinium salt (19). A synthesis

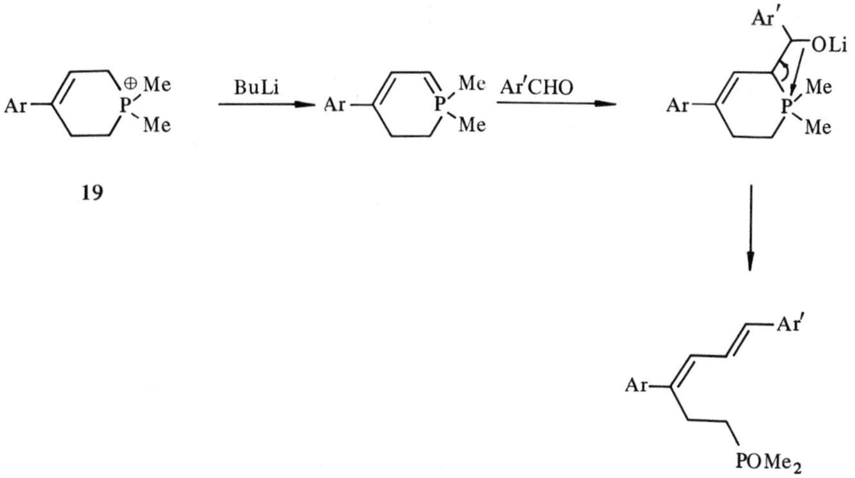

19

25

of 1,3-butadienes derived from the proposed 1,2-bisylide of 1,1,4,4-tetraphenyl-1,4-diphosphoniacyclohexene-2-dibromide (**20**) was reported by Vedejs.[13] Thus

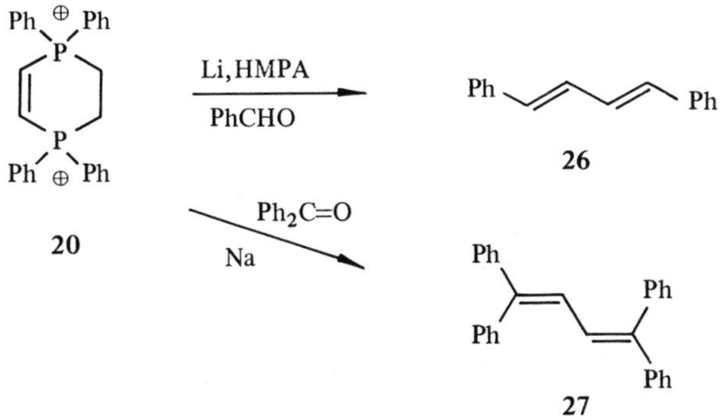

20

26

27

addition of lithium metal in hexamethylphosphoramide to **20** and benzalde-hyde produced *trans*-1,4-diphenyl-1,3-butadiene (**26**) while addition of sodium to **20** in the presence of benzophenone afforded the tetraphenylbutadiene (**27**). The reaction failed when cyclohexanone was utilized as the carbonyl com-ponent. Although the details of the mechanism are still vague, the process appears to involve an initial electron transfer to the carbonyl group followed by nucleophilic addition of the radical ion to give **28**. Repetition of this process

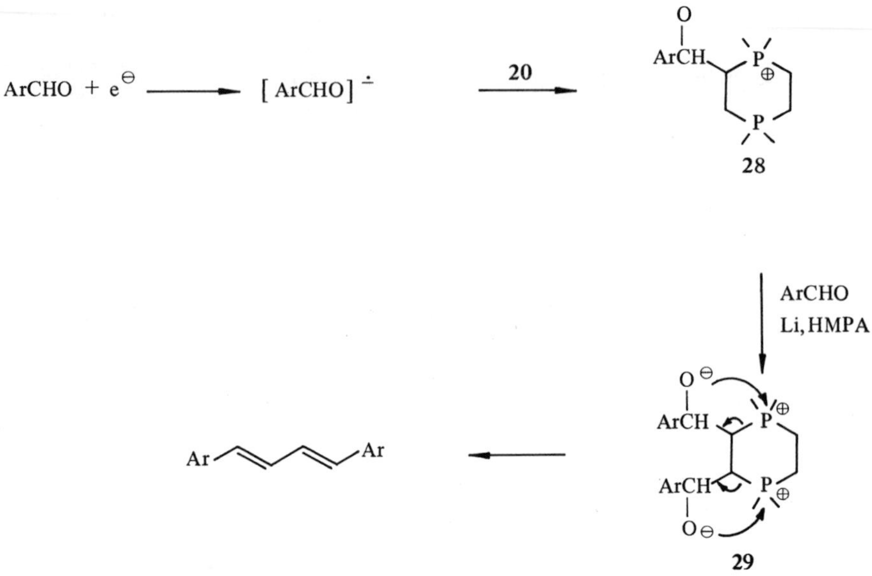

$$ArCHO + e^{\ominus} \longrightarrow [ArCHO]^{\pm} \xrightarrow{\;20\;}$$

28

ArCHO
Li, HMPA

29

Ar⌒⌒Ar

leads to **29** and ultimately to the diene by collapse of the betaine. This method possesses considerable potential and could provide a useful technique for diene synthesis.

IV.   FROM 3-PYRROLINES

Dienes are generated in high yield from 3-pyrrolines (**30**) by treatment with nitrohydroxylamine. The availability of 3-pyrrolines from pyrroles makes this reaction synthetically useful particularly since the process is accompanied by complete stereospecificity[14]. By reduction of 2,5-dimethyl pyrrole, the resulting mixture (*cis:trans*, 1:3.5) was separated by fractional crystallization of the

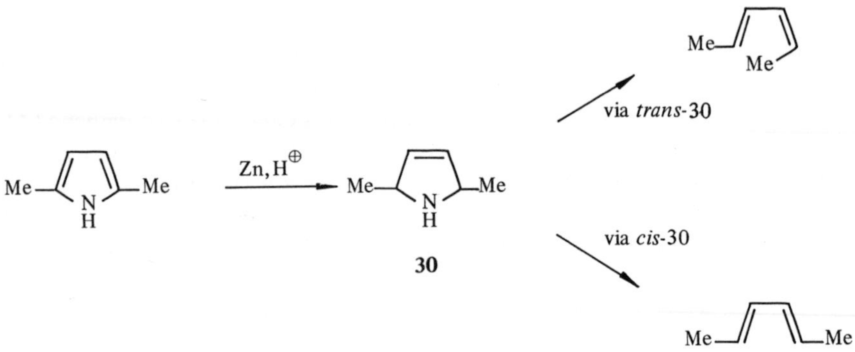

Me⌒Me
via *trans*-30

$$Me\text{—}\underset{\underset{H}{N}}{\diamond}\text{—}Me \xrightarrow{\; Zn, H^{\oplus}\;} Me\text{—}\underset{\underset{H}{N}}{\diamond}\text{—}Me$$

30

via *cis*-30

Me⌒⌒Me

p-toluene sulfonamides. The geometrically pure 3-pyrrolines were recovered by reductive cleavage with sodium in liquid ammonia. Deamination to the dienes was postulated to proceed through the diazene intermediate (31) with subsequent loss of nitrogen in a disrotatory process. The susceptibility of

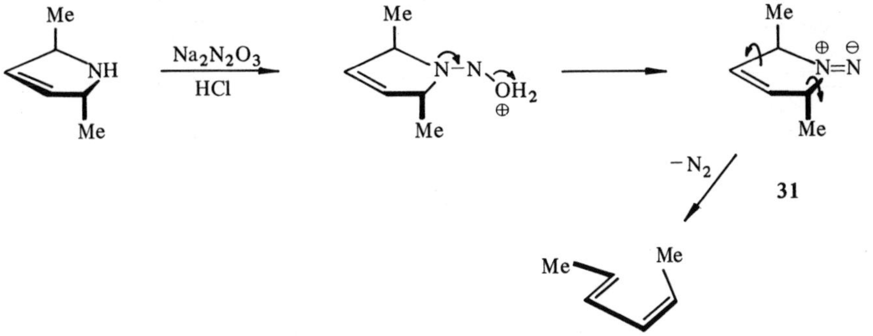

pyrroles toward electrophilic substitution should make many 2- and 2,5-disubstituted derivatives available as starting materials for this diene synthesis (Scheme 1).

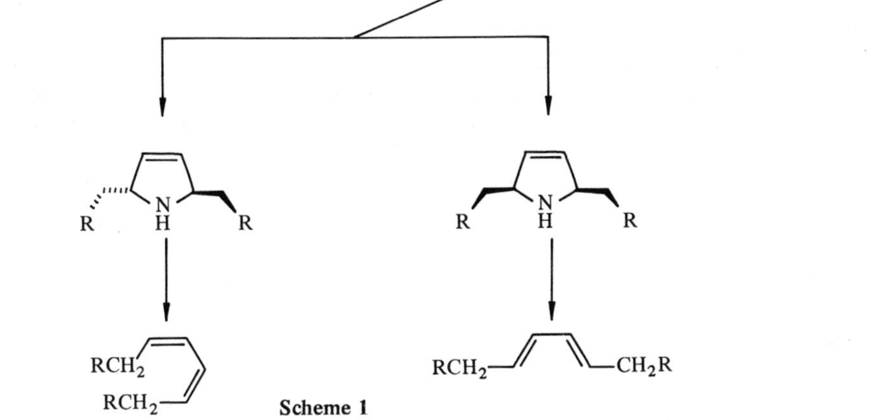

Scheme 1

V. FROM TETRAHYDROFURANS AND TETRAHYDROPYRANS

An efficient synthesis of acetylenic alcohols **34** and **35** is possible by the treatment of 2-chloromethyltetrahydrofuran **(32)** and 2-chloromethyltetrahydropyran **(33)** respectively with sodium amide in liquid ammonia.[15] The yields are reported to be in the range of 80-85%, and the reaction provides acetylene derivatives otherwise difficult to obtain. Internal triple bonds **(37)** may also be

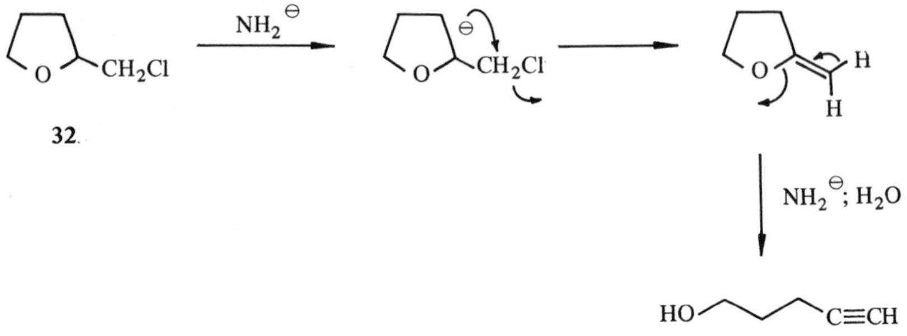

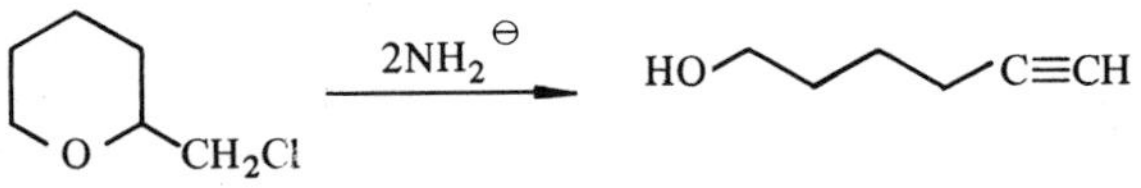

produced by the utilization of the 2-alkyl-3-chlorotetrahydrofurans **(36)**, as starting materials although this reaction gave poor yields. A clever approach to acetylenic alcohols was reported[16] from 2,3-dibromotetrahydropyran **(38)** (or its dehydrohalogenated derivative) using butylsodium. Halogen-metal interchange undoubtedly leads to the sodio derivative **(39)** which undergoes elimination to the acetylenic alcohol. An extension of this approach was also success-

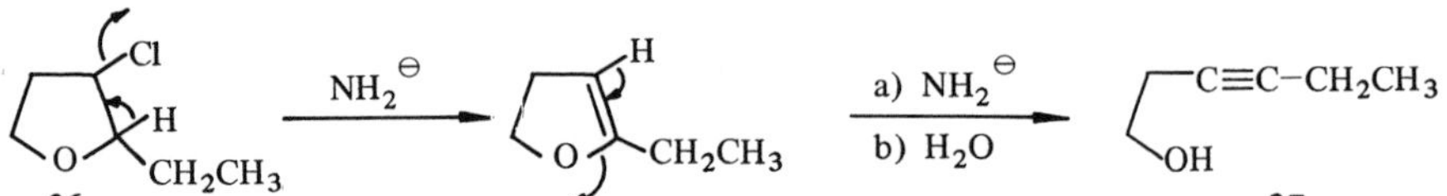

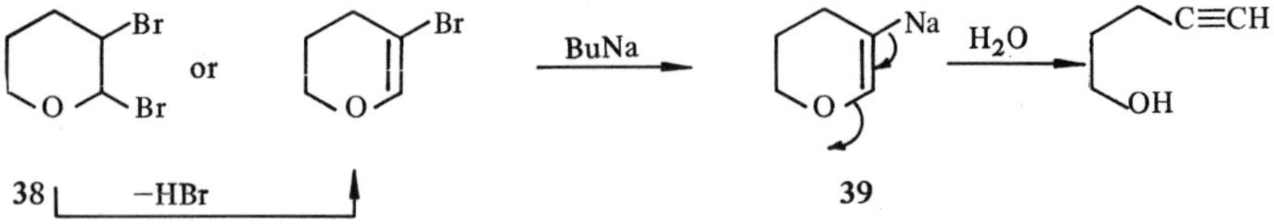

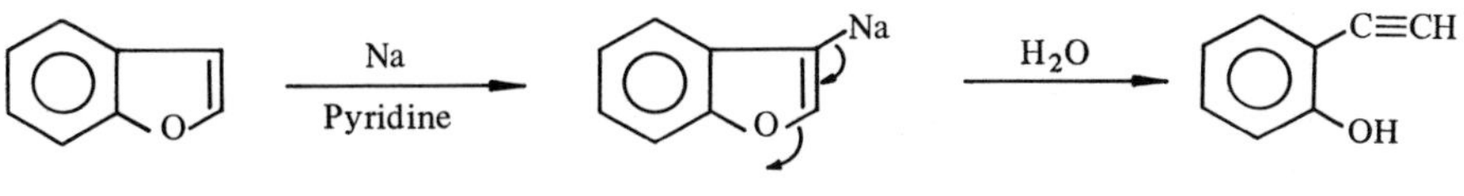

40

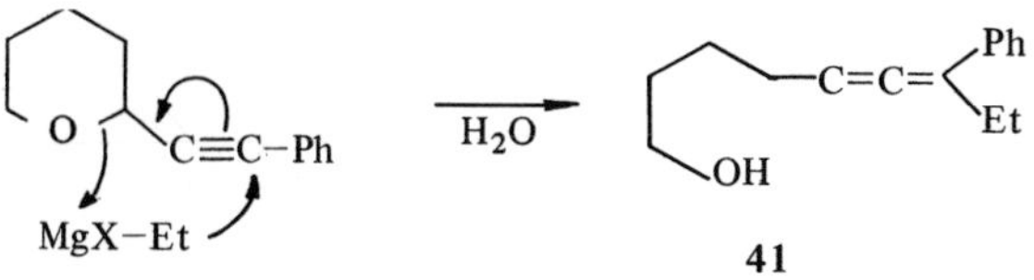

41

fully employed to transform benzofuran to o-ethynylphenol (40)[17a] and the acetylenic ether into the allenic alcohol (41).[17b]

VI. FROM PYRIDAZINE *N*-OXIDES

The use of pyridazine *N*-oxides (42) as precursors to ene-ynes (43) is an excellent example of heterocyclic utility in synthesis.[18,19] The addition 1.5-2.0 equivalents of an organometallic, in THF, to the pyridazine *N*-oxide is believed to take place at the electrophilic 6-position which proceeds to cleave the heterocyclic ring to the open chain azo-diene (44). The presence of excess organometallic causes the removal of the vinyl proton resulting in elimination of molecular nitrogen and the formation of the ene-yne. The products were shown to possess *trans* configuration and this was presumed to be the result of the conformation of the intermediate (44). When the reaction is carried out in ether or benzene solvent, considerable quantities of dienes (45) were obtained (15-30%). This was attributed to the poorer basic properties of the organo-

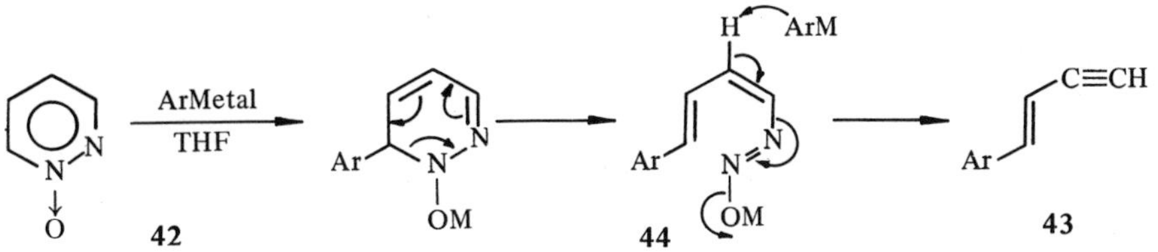

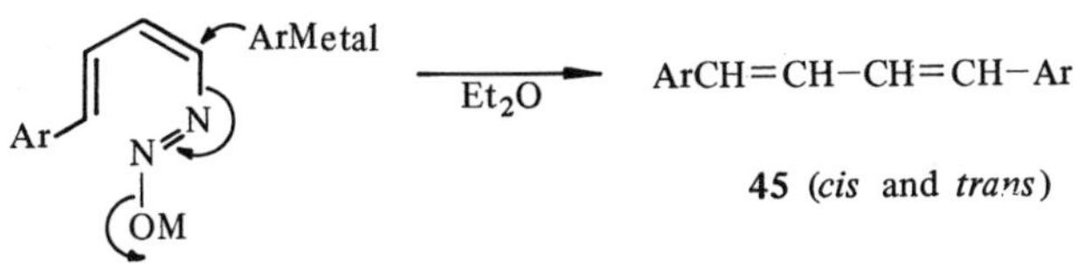

$$ArCH=CH-CH=CH-Ar$$

45 (*cis* and *trans*)

44

metallic in solvents less polar than THF resulting in substitution on **44** rather than proton abstraction. Although the scheme outlined indicates that the organometallic adds to **44** displacing nitrogen, the possibility of a vinyl cation **(46)** as an intermediate must also be considered. These species have already been discussed in the preceding chapter where their existence and synthetic

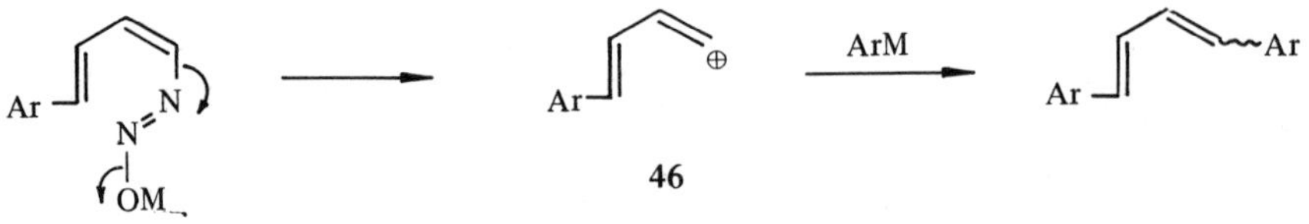

utility were amply demonstrated. Furthermore, since addition of aryl Grignard reagents to 3-methylpyridazine N-oxide **(47)** and the 4-methyl derivative **(48)** produced the same ene-yne **(49)**, a mechanism based on vinyl cations is rather valid. The rearrangement[20] of the vinyl diazonium salt **(50)** to the cyclopropyl

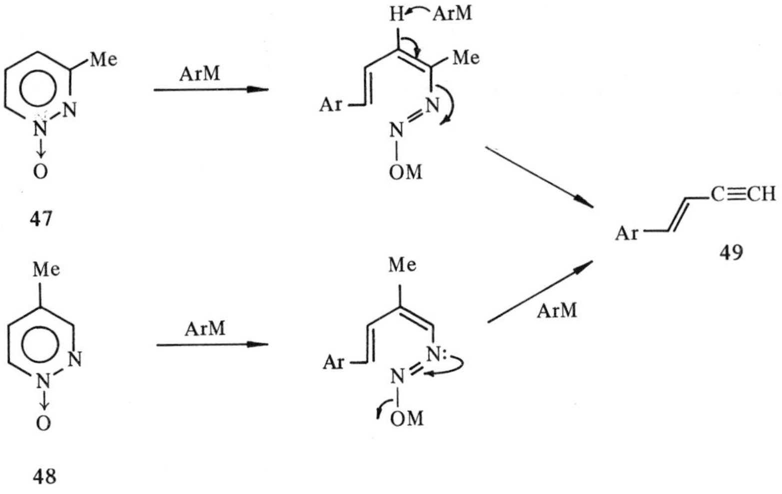

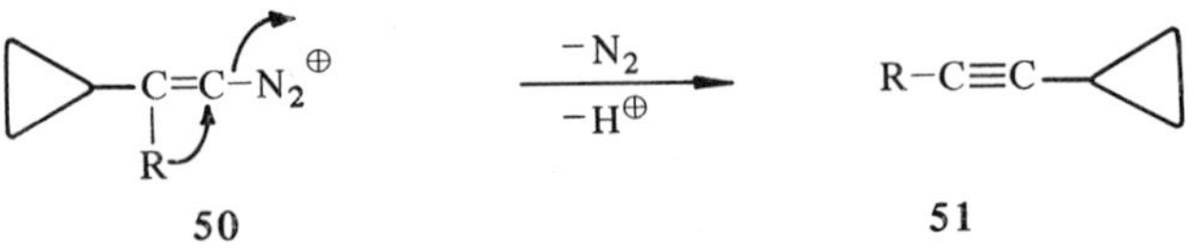

acetylene **(51)** provides a suitable analogy for this reaction (c.f. next section). It should be noted that the cyclopropyl acetylenes were formed both in aqueous methanolic potassium hydroxide and in lithium ethoxide in cyclohexene solvent. It is, therefore, the belief of this author that organometallic additions to pyridazine *N*-oxides proceeds through vinyl cations (and even perhaps vinyl carbenes). The use of vinyl cations to form acetylenes is mentioned in the next section. Despite the question concerning the mechanistic aspects, this method of forming ene-ynes possesses considerable synthetic potential as can be seen from the following:

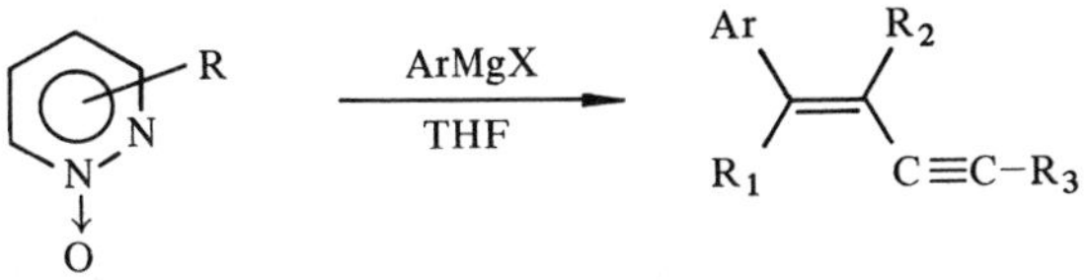

R	Ar	R_1	R_2	R_3
H	Ph	H	H	H
3-Me	Ph	H	H	Me
4-Me	Ph	H	H	Me
5-Me	Ph	H	Me	H
6-Me	Ph	Me	H	H
H	*p*-Tolyl	H	H	H
H	*o*-Tolyl	H	H	H
H	*p*-Anisyl	H	H	H

VII.   FROM   5,5-DISUBSTITUTED   *N*-NITROSOOXAZOLIDONES

The *in situ* generation of vinyl carbenes and vinyl cations from *N*-nitroso-oxazolidones **(52)** was already discussed in Chapter 2. This extraordinary hetero-cyclic system has also demonstrated its versatility toward the synthesis of allenic[21] **(53)** and acetylenic[20] **(54)** derivatives.

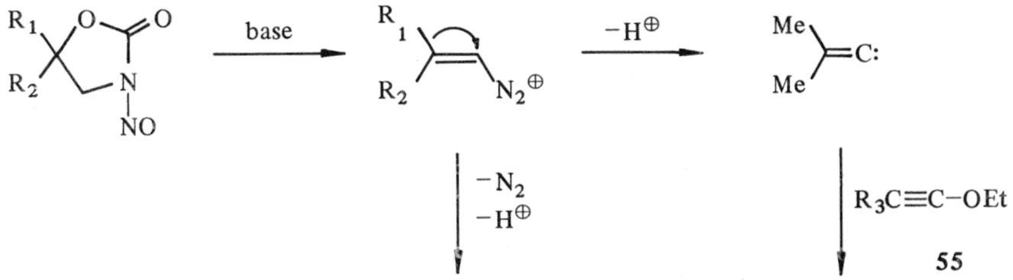

$R_1 = R_2 = \triangle$
$R_1 = Ph, R_2 = \triangle$
$R_1 = Me, R_2 = \triangle$

$$R_2-C\equiv C-R_1$$

54

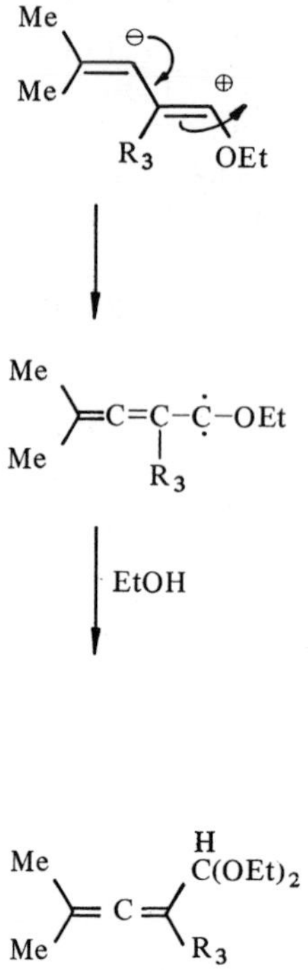

53

$$(R_3 = H, Me, Et)$$

Thus the loss of nitrogen from the vinyl diazonium salt is accompanied by migration of the substituent *trans* to the departing nitrogen molecule leading to the acetylenes.[20] This sequence constitutes an efficient method for preparing arylcyclopropylacetylenes. On the other hand, when the nitrosooxazolidone is treated with base in the presence of an ethoxyacetylene (55), the allenic acetal (53) is formed by the route shown. Although the yields of allenic acetals are only 30-40%, the ease with which they are obtained represent a significant new approach to these systems.

VIII.   FROM 1,2,3,6-TETRAHYDROPYRIDAZINES

The cycloaddition of azocarboxylates to dienes provides an easy entry into the tetrahydropyridazine system (56). Hydrolysis and oxidation of the latter leads

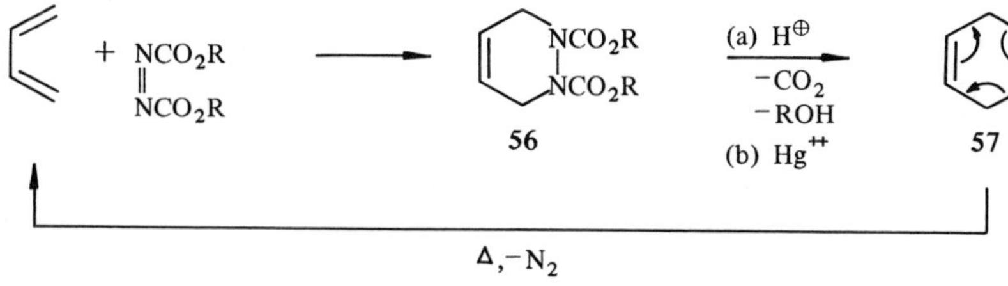

$$\Delta, -N_2$$

to the dihydropyridazine (57), a substance that smoothly decomposes to the starting diene. On the surface, this sequence does not seem to be of any synthetic value unless one considers the various modifications that are possible for (56) prior to reentry into the diene. The fact that the extrusion process (57 → diene) occurs in a stereospecific manner is also of no minor consequence.[22] For example, *trans-trans*-2,4-hexadiene (58) is obtained in >99% yield by disrotatory decomposition of the *cis*-3,6-dimethylpyridazine (59) while the *cis-trans*-hexadiene (60) is similarly produced from the *trans*-dimethylpyridazine (61). By elaborating the intermediate tetrahydropyridazine (62) with diazo-

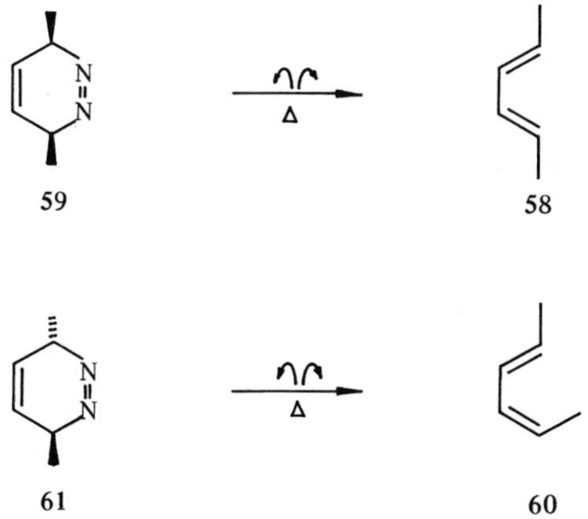

methane, the cyclopropano derivative (63) may be obtained, albeit as a mixture of two isomers (A, B). Hydrolysis, decarboxylation, and oxidation provide pure *cis-cis*-heptadiene (64) from 63A and *trans-trans*-heptadiene (65) from 63B.

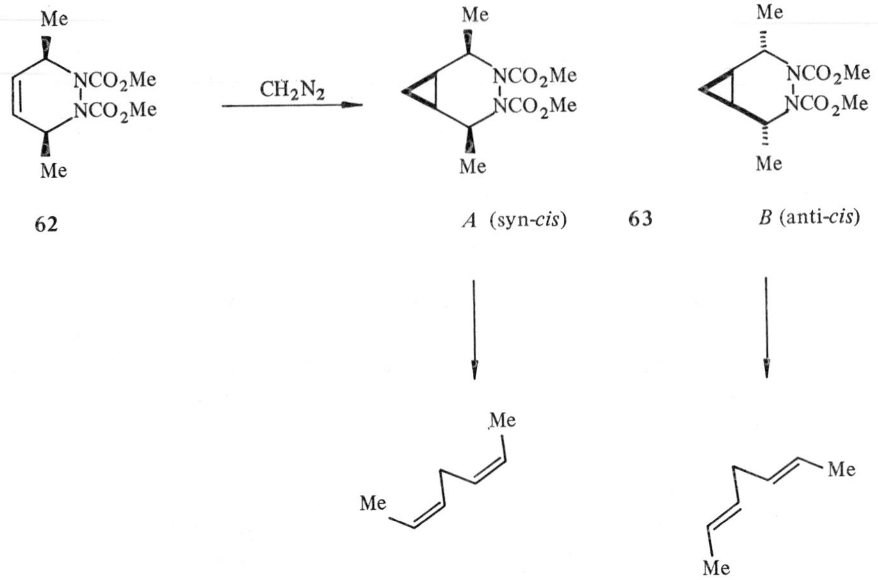

62   **CH₂N₂**   *A* (syn-*cis*)   **63**   *B* (anti-*cis*)

64   **65**

Modification for synthetic purposes on the Diels-Alder adducts **56** and **62** has only rarely been considered and this would seem to be a fruitful area for further investigation. For example, the transformation of the bicyclopyridazine ester **(66)**, formed from cyclopentadiene and di-*tert*-butylazocarboxylate, into its aziridino derivative **(67)** served as a novel route to dihydropyridine[23a]

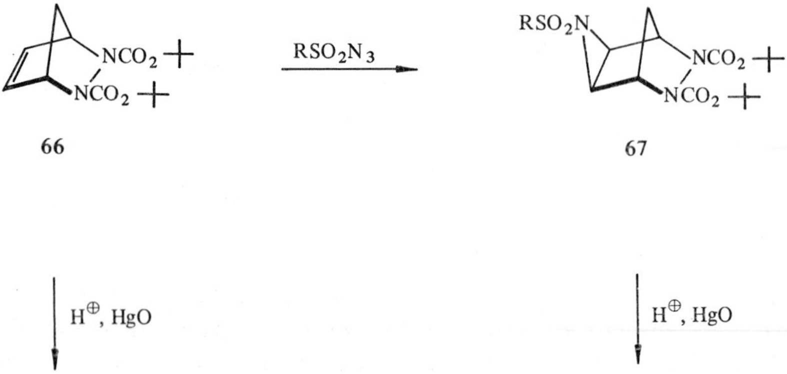

66   RSO₂N₃   **67**

H⊕, HgO   H⊕, HgO

$$\left[\text{structure} \right] + 2CO_2 + 2 \text{(alkene)}$$

$$\left[\text{RSO}_2\text{N structure} \right] + 2CO_2 + 2 \text{(alkene)}$$

$$\downarrow -N_2 \qquad \qquad \downarrow -N_2$$

(cyclopentadiene structure) - - - - - - - - - - - - - → (pyridine structure with N–SO$_2$R)

68   R = Me, Ph

derivatives (68) after hydrolysis, decarboxylation, and oxidation in a single step. Had this modification of 66 not been performed, oxidative decarboxylation would have only returned the starting diene. Presumably, if 66 were to be treated with diazoalkanes (as was the case with 62), the resulting cyclopropano system would serve as a source of 1-alkyl-1,4-cyclohexadienes (69).[23b] This would be of synthetic value solely on the fact that Birch reductions of alkylbenzenes usually give the isomeric alkyl cyclohexadienes (70).[24]

(structure) $\xrightarrow{\text{RCHN}_2}$ (structure) $\longrightarrow$ (structure)

66   (R = t-Bu, Et, etc.)   69

(structure) $\xrightarrow{\text{Birch}}$ (structure)

70

Another illustration that demonstrates the importance of modification of temporarily constructed heterocycles comes from the synthesis of semibullvalene (71).[25,26] By reaction of cyclooctatetraene with azocarboxylates, the basketane derivative (72) is formed. The latter is, in effect, a tetracyclic-tetrahydro pyridazine derivative and was modified with silver fluoroborate to 73. Hydrolysis, decarboxylation, and air oxidation afforded the unstable azo derivative (74) which expelled nitrogen producing semibullvalene (71).

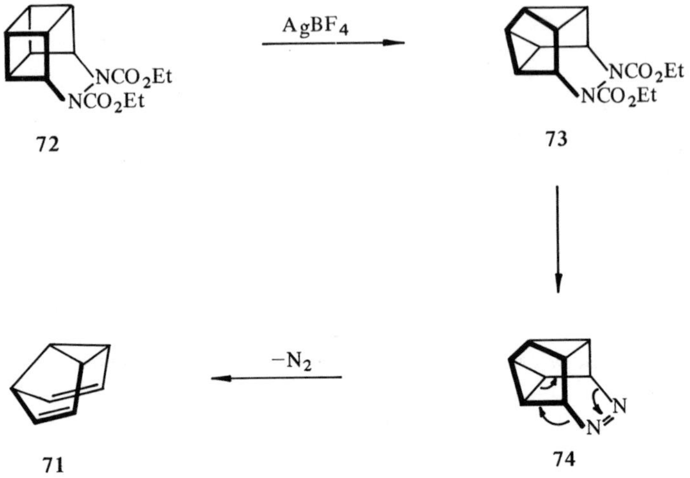

IX.   FROM 4-KETOPYRAZOLINES

A unique synthesis of allenes (75) was briefly reported[27] to arise by the extrusion of 2 moles of nitrogen from the hydrazone of the 4-ketopyrazoline (76). The ketopyrazoline, obtained by oxidation of the diaminoketone,[28] was converted to its hydrazone and then subjected to nickel peroxide in ether. The intermediate diazo compounds (77) vigorously evolved nitrogen producing tetramethylallene in 87-91% yield.

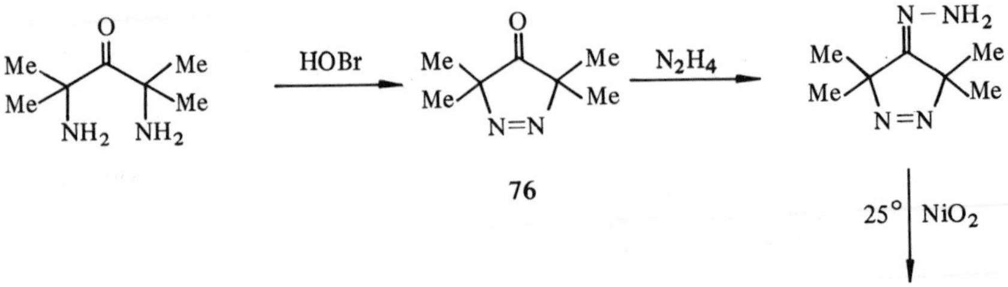

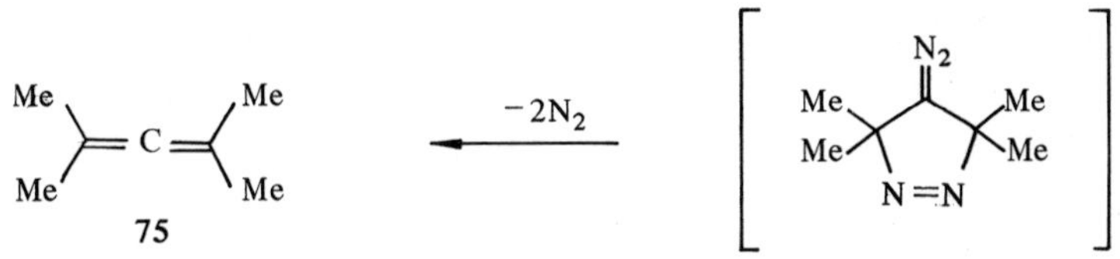

75    $-2N_2$ 77

X. FROM 5-KETOPYRAZOLINES (5-PYRAZOLONES)

A novel preparation of allenic[29] **(79)** and acetylenic[30] **(81)** esters was reported to arise from the thallium(III) nitrate (TTN) oxidation of 5-ketopyrazolines **79** and **80**, respectively. The latter are readily obtained from reaction of hydrazine with various β-ketoesters. This remarkable transformation is presented in detail in the chapter covering carboxylic acid syntheses (Chapter 10, Section VI.A).

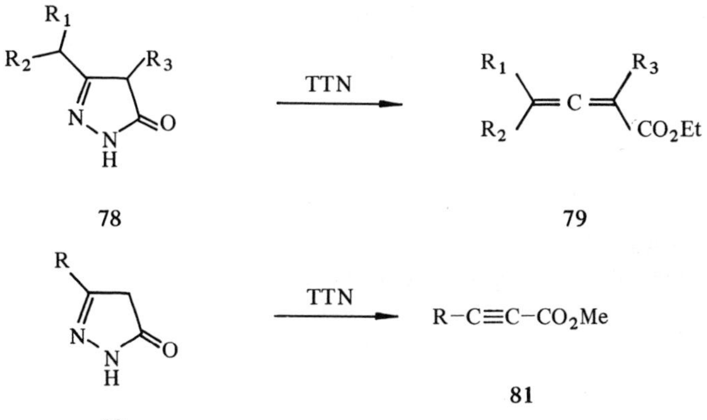

78   TTN   79

80   TTN $R-C{\equiv}C-CO_2Me$ 81

XI. FROM 1,2,3-SELENADIAZOLES

The thermal instability of 1,2,3-selenadiazoles **(82)** opens a versatile route to acetylenes **(83)**,[31] particularly in view of the fact that this heterocycle is easily available from selenium dioxide oxidation of simple semicarbazones **(84)**.[32]

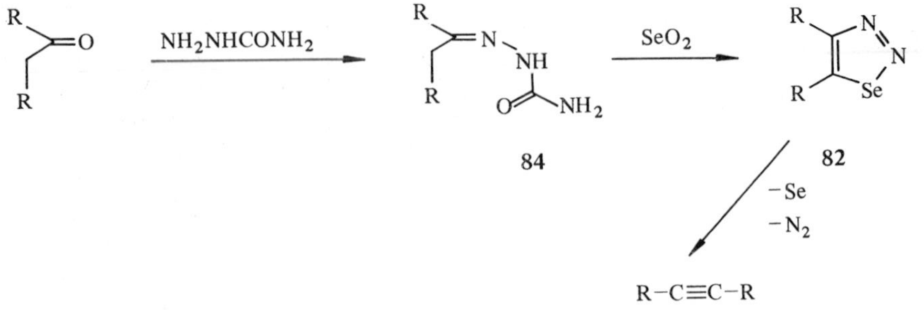

84   **82**

$-Se$
$-N_2$

$R-C{\equiv}C-R$

83

The yields of acetylenes containing a wide variety of substituents were 85-95% based on the selenadiazole and 40-65% based on the starting ketones. The use of diketones, via their *bis*-semicarbazones was also shown to lead to diynes **(85)**. This method also proved applicable to the synthesis of cycloalkynes **(86)** where

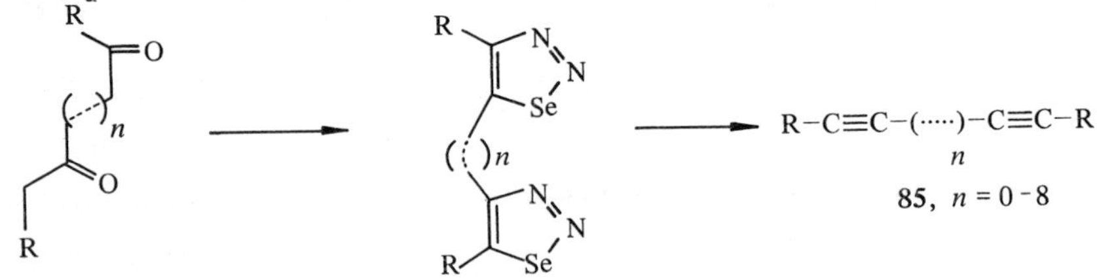

$R-C{\equiv}C-({\cdots})-C{\equiv}C-R$
n
85, $n = 0-8$

the ring size was greater than eight carbons. When the semicarbazone possessed two different α-hydrogens, the resulting product was a mixture of isomeric

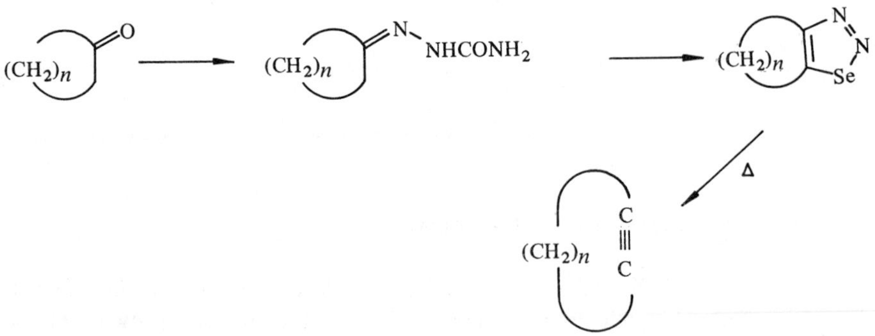

Δ

86, $n = 6-10$

selenadiazoles. The dominant product appears to be in accord with the expected radical (or carbanion) stabilities of the α-carbon atoms. The utility of this

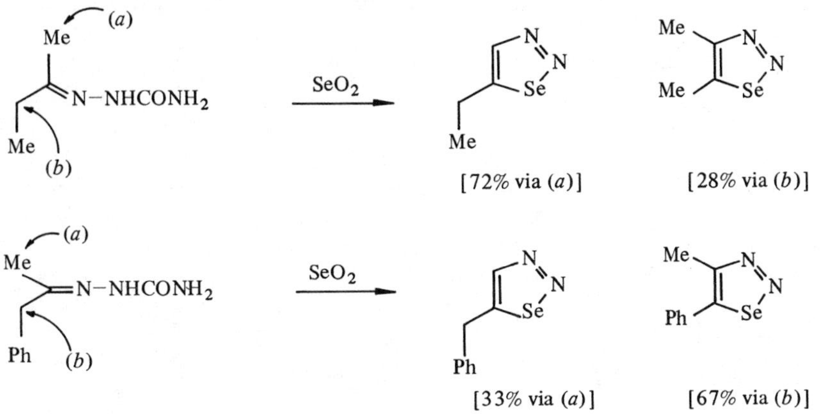

[72% via (a)]   [28% via (b)]

[33% via (a)]   [67% via (b)]

rather rare heterocycle can be appreciated by the following acetylene syntheses.[33]

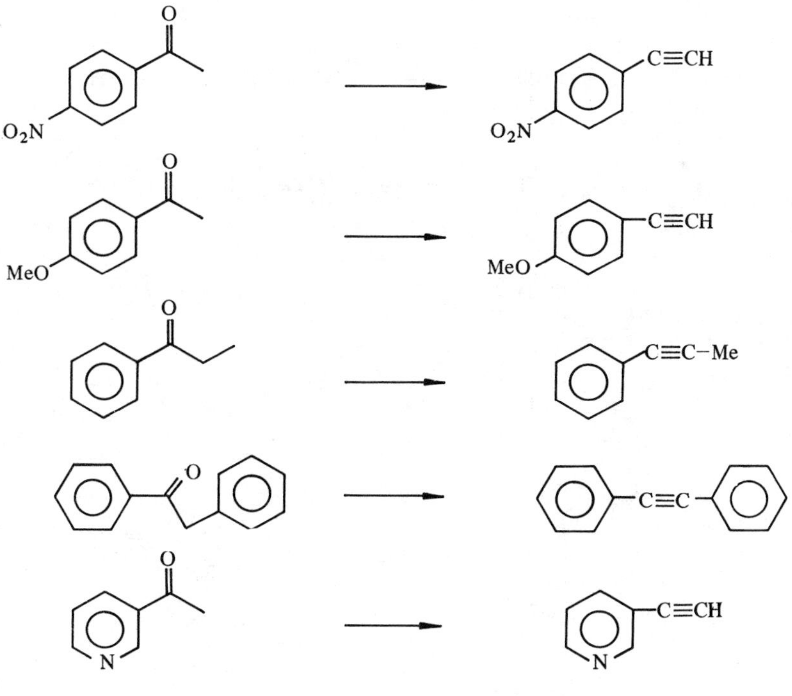

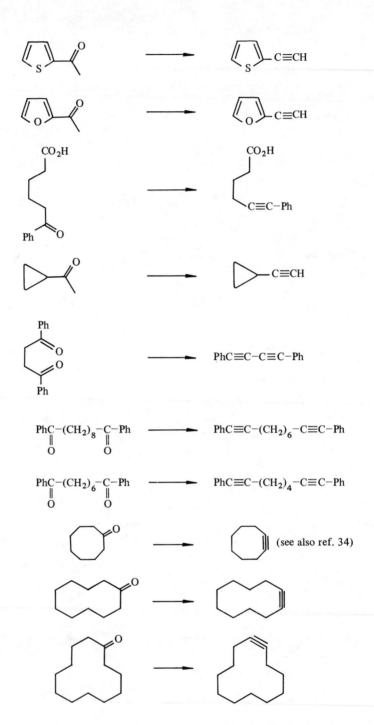

XII. FROM 2-THIOALKYLTHIAZOLINES

The versatile 2-thioalkylthiazoline **88**, prepared from commercially available 2-mercaptothiazoline **87** and propargyl halides, provides a unique approach to internal alkynes **91** as well as 1,5-diynes **93**. As shown in an earlier chapter (p. 26), the a-thio anion of these heterocycles is readily generated using butyllithium (THF, $-60°$) and may be alkylated with alkyl halides. This technique has now been extended to include the a-thio anion **89** which after alkylation to **90** leads to acetylenes **91** after Raney nickel desulfurization.[35] An interesting feature of this process is the fact that although the thioalkynyl thiazoline **88** contains a terminal acetylene linkage (R=H) and, therefore,

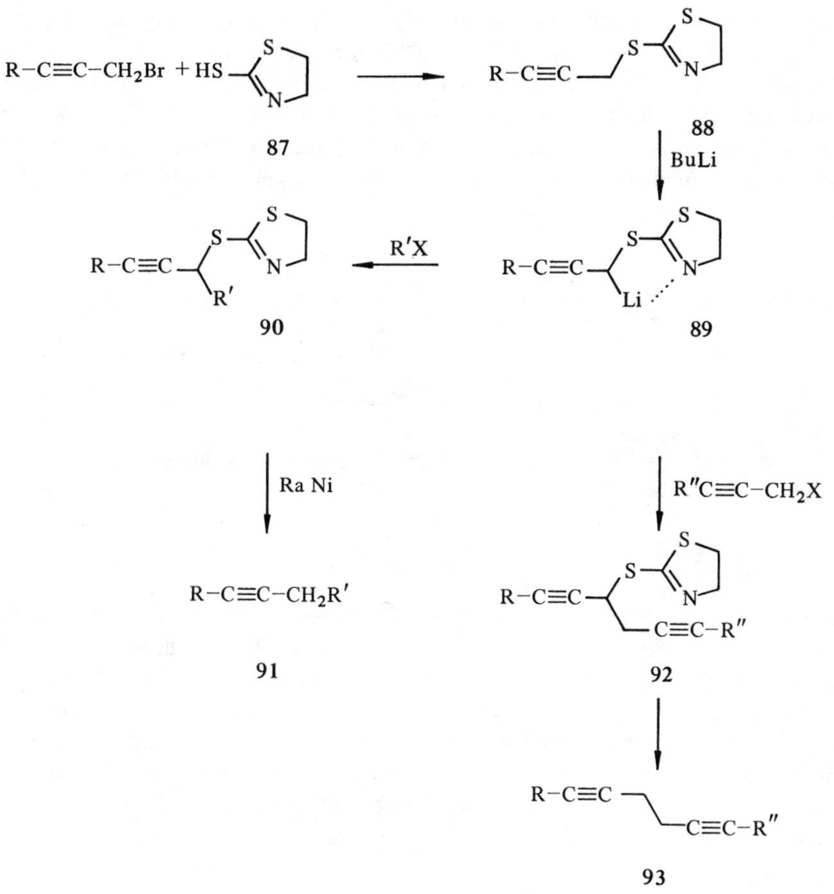

R	R'		R	R''
H	PhCH$_2$		H	Ph
H	Me		H	H
Ph	PhCH$_2$		Ph	Ph
Ph	Me		Ph	H
Me	PhCH$_2$		Me	Ph
Me	Me		Me	Me
H	Allyl		Me	H

requires two equivalents of butyllithium, alkylation proceeds only at the carbon adjacent to sulfur. This specificity has been attributed to the more nucleophilic nature of the α-carbon since it is part of a five-membered ring chelate (89). If the α-thio carbanion is treated with propargyl halides, the *bis*-alkynyl thiazoline (92) is formed and is ultimately transformed by Raney nickel to the 1,5-diynes, 93. Thus the role of 87 as a "riveting" agent to form functionalized molecules is clear. The only drawback to this method seems to lie in the Raney nickel step which leads to mixtures of acetylenes and allenes. Thus 94, upon heating in ethanol (Raney Ni), gave 95 (87%) and 96 (14%). The use of zinc dust in acetic acid was also effective in bringing about desulfurization and in this case 97 produced an 8:1 mixture of 98 and 99.

PhC≡C–CH$_2$S — [thiazoline] $\xrightarrow{\text{Ra Ni}}$ PhC≡C–CH$_3$ + PhCH=C=CH$_2$

94   95   96

PhC≡C–[thiazoline with C≡CH] $\xrightarrow{\text{Zn, H}^{\oplus}}$ PhC≡C–CH=C=CH$_2$ + PhC≡C–C≡CH

97   98   99

Another novel result was obtained in this series[36] when the α-thio alkynyl thiazoline anion 89 was treated with benzaldehyde. The expected adduct 100 was not stable and appeared to undergo further reaction to the spiro intermediate 101. This behavior was presumed to be responsible for the production of the alkynyl thiirane 102. Distillation of the latter caused thermal desulfurization to the ene-yne 103. This interesting process, although yielding 20% of 102, is worthy of further study and no doubt will become a valuable synthetic method if and when the efficiency of the transformation is enhanced.

89   PhCHO   100

102
(cis-trans)

101

MeC≡C–HC=CHPh

103
(cis-trans)

REFERENCES

1. H. Staudinger and F. Pfenninger, *Berichte*, **49**, 1446 (1916).
2. H. J. Backer and T. A. H. Blaas, *Rec. Trav. Chim.*, **61**, 785 (1942).
3. L. F. Hatch and D. Peter, *Chem. Commun.*, 1499 (1968).
4. R. C. Krug and T. F. Yen, *J. Org. Chem.*, **21**, 1082, 1441 (1956).
5. W. L. Mock, *J. Am. Chem. Soc.*, **92**, 6918 (1970).
6. A. I. Meyers and T. Takaya, *Tetrahedron Lett.*, 2609 (1971).
7. R. M. Kellogg, *J. Am. Chem. Soc.*, **93**, 2344 (1971).
8. L. D. Quin and J. A. Peters, *Tetrahedron Lett.*, 3689 (1964).
9. R. P. Welcher, G. A. Johnson, V. P. Wystrach, *J. Am. Chem. Soc.*, **82**, 4437 (1960); H. E. Shook and L. D. Quin, *ibid.*, **89**, 1841 (1967).
10. A. M. Aguiar and H. Aguiar, *ibid.*, **88**, 4090 (1966).
11. D. Lednicer, *J. Org. Chem.*, **36**, 3473 (1971).
12. D. Lednicer, *ibid.*, **35**, 2307 (1970).
13. E. Vedejs and J. P. Bershas, *ibid.*, **37**, 2640 (1972).
14. D. M. Lemal and S. D. McGregor, *J. Am. Chem. Soc.*, **88**, 1335 (1966).
15. G. Eglington, E. R. H. Jones, and M. C. Whiting, *J. Chem. Soc.*, 2873 (1952).
16. R. Paul and S. Tchelitcheff, *Compt. Rend.*, **230**, 1473 (1950); **232**, 2230 (1951).
17a. Y. Odaira, *Bull. Chem. Soc. (Japan)*, **29**, 470 (1956).
17b. D. J. Nelson, and W. J. Miller, *Chem. Commun.*, 444 (1973).
18. H. Igeta, T. Tsuchiya and T. Nakai, *Tetrahedron Lett.*, 2667 (1969); C. Kaneko, T. Tsuchiya, and H. Igeta, *ibid.*, 2347 (1973).

19. G. Okusa, M. Kumagai, and T. Itai, *Chem. Commun.*, 710 (1969).

20. M. S. Newman and S. J. Gromelski, *J. Org. Chem.*, **37**, 3221 (1972); M. S. Newman and L. F. Lee, *ibid.*, **37**, 4468 (1972); H. P. Hogan and J. Seehafer, *ibid.*, **37**, 4466 (1972); T. B. Patrick, J. M. Disher, and W. J. Probst, *ibid.*, **37**, 4467 (1972); M. S. Newman and V. Lee, *ibid.*, **38**, 2435 (1973).

21. M. S. Newman and C. D. Beard, *ibid.*, **35**, 2412 (1970).

22. J. A. Berson and S. S. Olin, *J. Am. Chem. Soc.*, **91**, 778 (1969).

23a. A. I. Meyers, T. Takaya, and D. Stout, *Chem. Commun.*, 1260 (1972); D. C. Horwell and J. A. Deyrup, *ibid.*, 486 (1972).

23b. E. L. Allred, J. C. Hinshaw, and A. L. Johnson, *J. Am. Chem. Soc.*, **91**, 3382 (1969).

24. H. O. House, *Modern Synthetic Reactions*, 2nd ed., Benjamin, 1972, p. 199.

25. L. A. Paquette, *J. Am. Chem. Soc.*, **92**, 5765 (1970).

26. R. Askani, *Tetrahedron Lett.*, 3349 (1970).

27. R. Kalish and W. H. Pirkle, *J. Am. Chem. Soc.*, **89**, 2781 (1967).

28. W. L. Mock, Ph.D. Thesis, Harvard University, Cambridge, Mass., 1964.

29. E. C. Taylor, R. L. Robey, and A. McKillop, *J. Org. Chem.*, **37**, 2798 (1972).

30. E. C. Taylor, R. L. Robey, and A. McKillop, *Angew. Chem. Int. Ed.*, **11**, 48 (1972).

31. I. Lalezari, A. Shafiee, and M. Yalpani, *Angew. Chem. Int. Ed.*, **9**, 464 (1970).

32. I. Lalezari, A. Shafiee, and M. Yalpani, *Tetrahedron Lett.*, 5105 (1969).

33. Personal communication. The author thanks Professor Lalezari for permission to use this information prior to publication.

34. H. Meier and I. Menzel, *Chem. Commun.*, 1059 (1971).

35. K. Hirai and Y. Kishida, *Tetrahedron Lett.*, 2117 (1972).

36. K. Hirai, H. Matsuda, and Y. Kishida, *Chem. Pharm. Bull. (Tokyo)*, **20**, 2067 (1972).

5 AROMATIC COMPOUNDS

A number of benzenoid nuclei may be constructed from a heterocyclic precursor that undergoes either a rearrangement or an extrusion process. The advantages of such a synthesis lies mainly in the fact that any substituents originally present in the heterocycle are incorporated into the aromatic product in a regiospecific manner. This avoids the perennial problem associated with direct introduction of substituents which usually leads to isomeric mixtures. This mode of benzenoid synthesis is presented in the first part of this chapter. Heterocyclic systems may also serve as useful reagents for direct introduction or removal of substituents on aromatic nuclei; this technique is discussed in the second part of the chapter.

I. CONSTRUCTION OF AROMATIC NUCLEUS

A. From Pyrylium Salts

Substituted pyrylium salts (**1**) have long been recognized as useful precursors to polysubstituted aromatic molecules (**2**) by a successive ring opening-ring closure process. Treatment of pyrylium salts with various nucleophilic reagents allows the production of a host of functionalized derivatives and makes this route particularly attractive.[1]

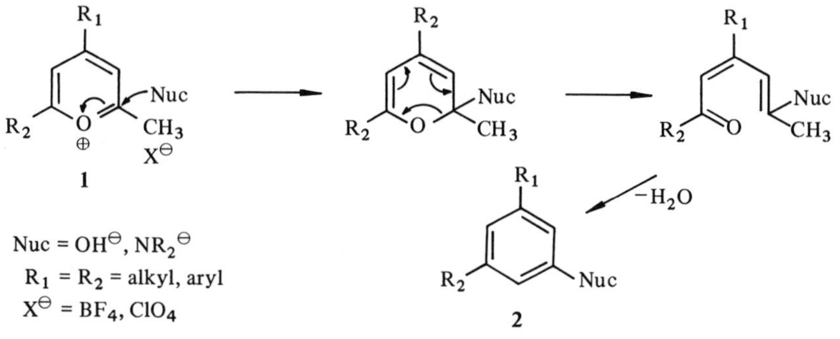

R1
R2
O
⊕
Nuc
CH3
X⊖
1
R2
Nuc
CH3
R1
Nuc
CH3
−H2O
R1
R2
Nuc
2
Nuc = OH⊖, NR2⊖
R1 = R2 = alkyl, aryl
X⊖ = BF4, ClO4

If carbanions are utilized as the nucleophile, aromatic nitro compounds (3), esters (4), ketones (5), nitriles (6), and even hydrocarbons (7 and 8) are produced. When a Grignard reagent is employed, the aromatic hydrocarbon formed

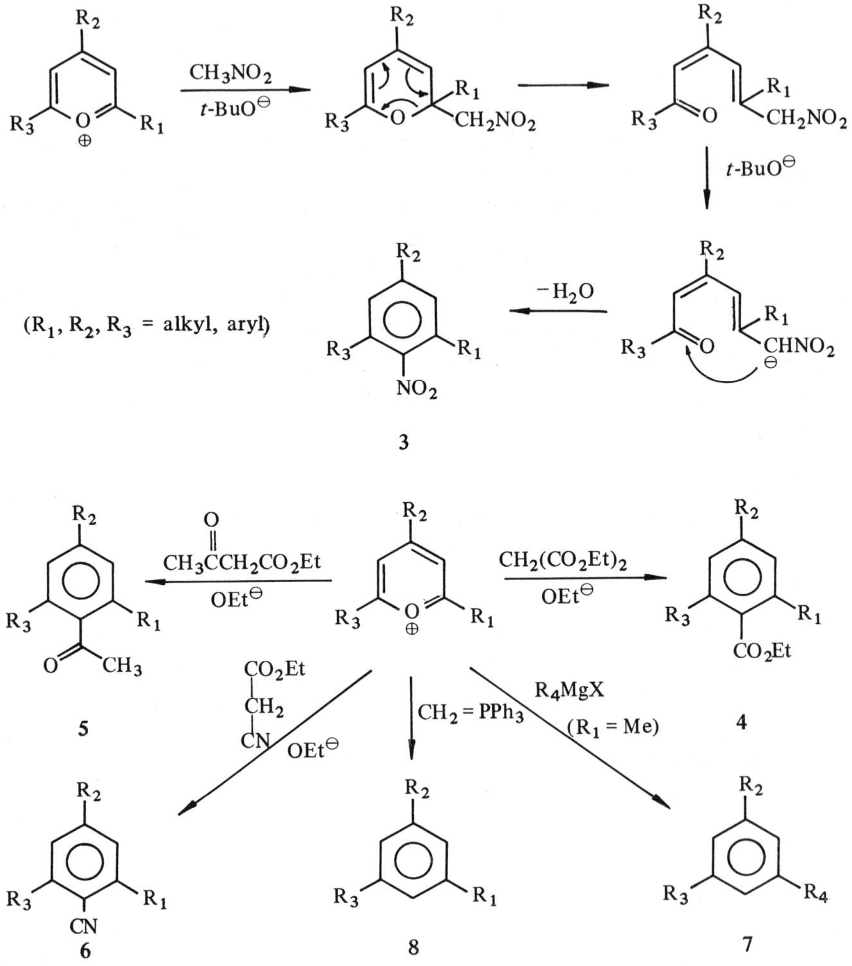

(7) has incorporated the 2-substituent of the pyrylium salt ($R_1 = CH_3$) into the aromatic ring. Therefore, one of the two substituents flanking the oxygen in the pyrylium salt must possess a proton (e.g., methyl or benzyl). However, if the Grignard reagent contains an a-proton, this too serves to allow aromatic ring formation (9 → 10).

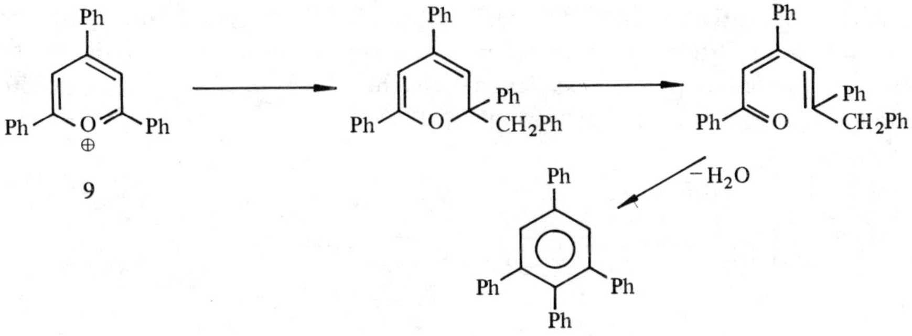

9

10

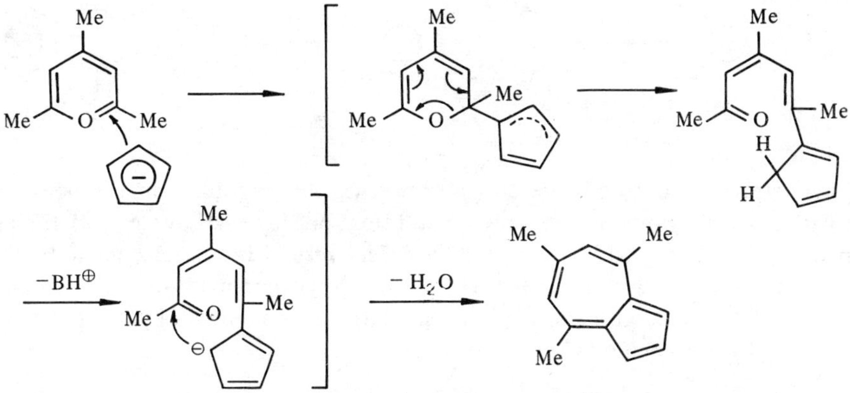

11

Sodium cyclopentadienide has been successfully utilized with pyrylium salts and led to azulenes **(11)** in 60% yield.[2] This interesting transformation takes place in a single step by merely mixing the reactants at room temperature.

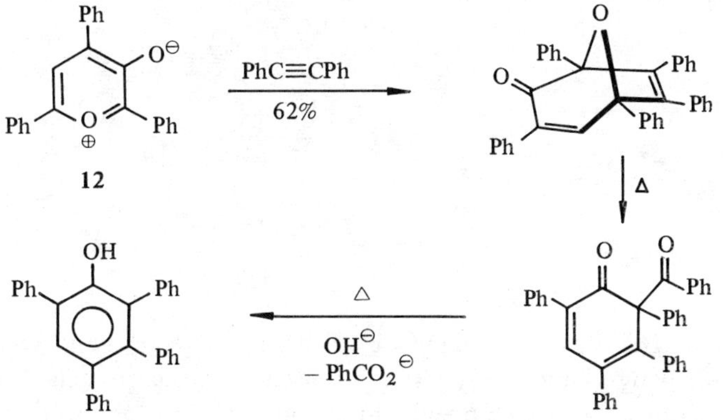

12

Potts[2a] recently described the cycloaddition of acetylenic dipolarophiles to the pyrylium betaine **12** leading to a 1:1 cycloadduct. Thermolysis of the latter furnished the cyclohexadienone system which smoothly rearranged to 2,3,4,6-tetraphenyl phenol on warming in alkali.

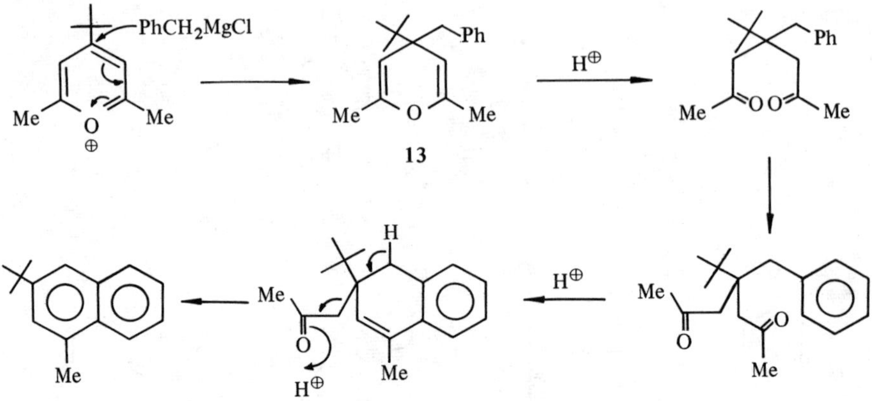

Polynuclear hydrocarbons are also prepared via the pyrylium salt precursor by isolation of the initially formed 1,4-addition product, 4H-pyran (**13**). Treatment of the latter with 70% perchloric acid results in rearrangement to the naphthalene derivatives in yields of 80-90%. Polycyclic pyrylium salts are also useful in the preparation of naphthalene derivatives as shown by the synthesis of nitro (**14**) and amino (**15**) naphthalenes.

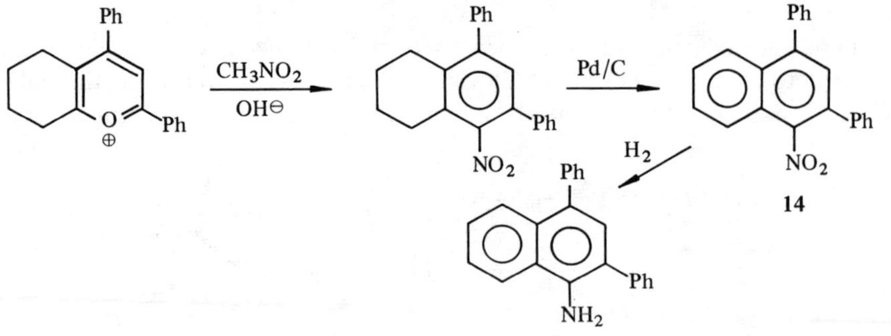

Polyphenyls (**16**) may be obtained by initiating the synthetic scheme with p-disubstituted pyrylium salts (**17**) and constructing the aromatic rings in the manner described above. The phenanthrene nucleus (**18**) has also been included

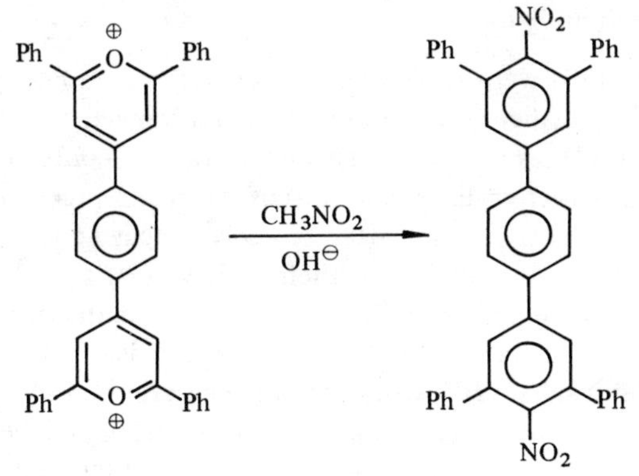

17   16

in the repertoire of aromatic systems available from pyrylium salts. This was achieved by treating the 4H-pyran **(19)**, formed by 1,4-addition of the naphthyl-

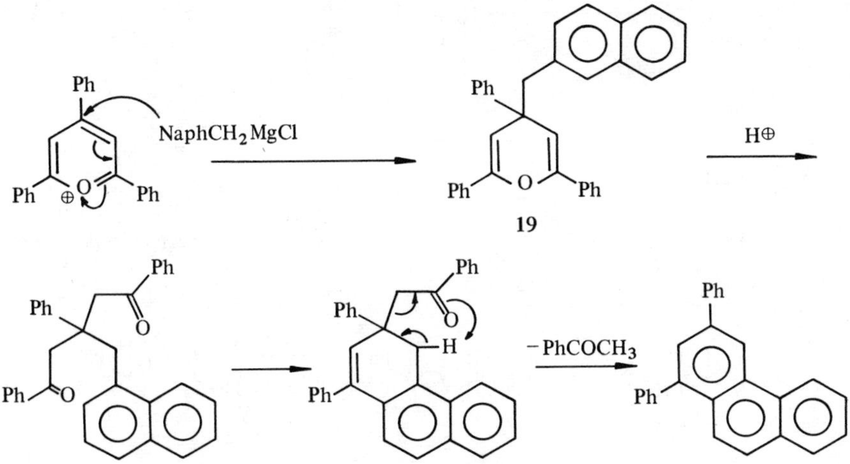

19

18

methyl Grignard reagent, with strong acid and allowing the rearrangement to occur as mentioned earlier.

There are many additional variations involving pyrylium salts as a ready source of aromatic nuclei and the reader is referred to the excellent and detailed review on the scope, limitations, and mechanism of this process.[1] Methods for preparing pyrylium salts with appropriate substituents are also discussed.

B.   From Thiapyrylium Salts

Thiapyrylium salts (20) were reported to react with organometallics producing the highly colored "thiabenzenes" (21) which rearranged to the 2H- and 4H-thiapyrans (22 and 23) respectively.[3] This behavior is in contrast to the reaction of organometallics with pyrylium salts discussed in the preceding section. The chemistry of thiapyrylium salts has been scarcely touched and only recently has there been any effort in evaluating their behavior with other nucleophilic agents. Two reports appeared describing the reaction of thiapyrylium salts with carbanions derived from active methylene compounds.[4,5] Reaction of 2,4,6-triaryl thiapyrylium salts (20) with malonitrile, ethyl cyanoacetate, diethyl malonate, and nitromethane gave the substituted triphenyl benzenes (24) in 50-75% yield. The reactions were carried out in tert-butanol-potassium tert-butoxide[5] or in ethanol using diisopropylethyl amine.[4] The latter base appeared to give slightly higher yields. The mechanism of this process appears to proceed

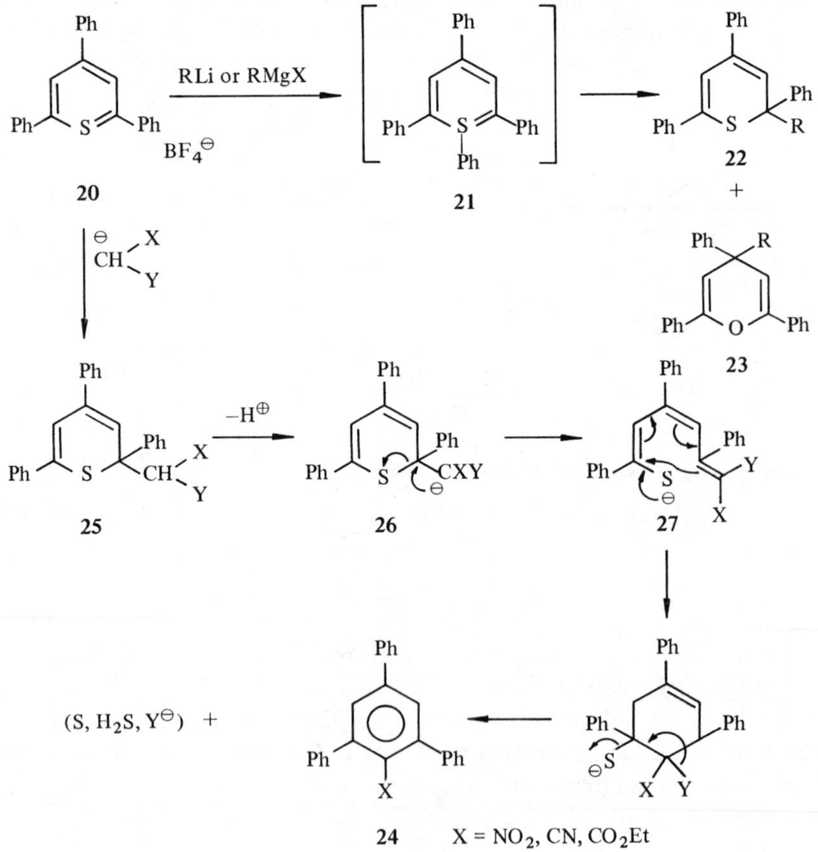

24   X = NO$_2$, CN, CO$_2$Et

in an analogous manner to that of the pyrylium salts (25 → 26 → 27). The loss of elemental sulfur, hydrogen sulfide, or other anionic species (SCN$^{\ominus}$, NO$_2$$^{\ominus}$) accompanies the final step. Support for the intermediacy of the open chain hexatriene (27) (X=Y=CN) was offered by comparison of the ultraviolet spectrum of the reaction mixture with authentic 1,1′-dicyano-2,4,6-triphenyl-hexatriene.[5]

C. From Pyridinium Salts

A new route to tropones and tropolones has been described by Katritzky[6] which originates from the *N*-methyl pyridinium salt (28). Although tropones and tropolones have been reported to possess little aromatic character,[7] their structural features are related to homocyclic conjugated polyenes and are therefore included in this section. Heating anhydro-3-hydroxy-1-methylpyridinium hydroxide (28) and acrylonitrile (X=CN) in tetrahydrofuran in the presence of hydroquinone afforded the cycloaddition product (29) in 75% yield. After conversion of the latter to the methiode salt (30), it was treated with base and

rearrangement proceeded to the dihydrotropone that air oxidized to a readily separable mixture of the dimethylaminotropone (31) and the tropolone (32). The total yield of cycloheptatrienones was 73% and is considered to be superior to the previous methods for preparing this ring system.

D. From Quinolizinium Salts

The efficient preparation of various benzoquinolizinium salts (33) from pyridine-2-carboxaldehydes and benzyl halides[8] has provided heterocyclic precursors for an elegant synthesis of 2-naphthols (37).[9] Regiospecific cycloaddition of ketene acetals to the benzoquinolizinium perchlorates at room temperature leads to the adduct (34) in 70-96% yields. The next step in the naphthol synthesis is accomplished by reduction of the pyridinium moiety in (34) using either catalytic means (PtO$_2$) or sodium borohydride. The resulting tetrahydropyridine (35) is readily transformed into the transient enol 36 by treatment with acid. Retroaddition of dihydropyridine furnishes the substituted naphthol 37. This technique has been successfully applied to a variety of poly-cyclic quinolizinium salts (38a-d) and ketene acetals producing the corresponding polycyclic phenols (39a-e).

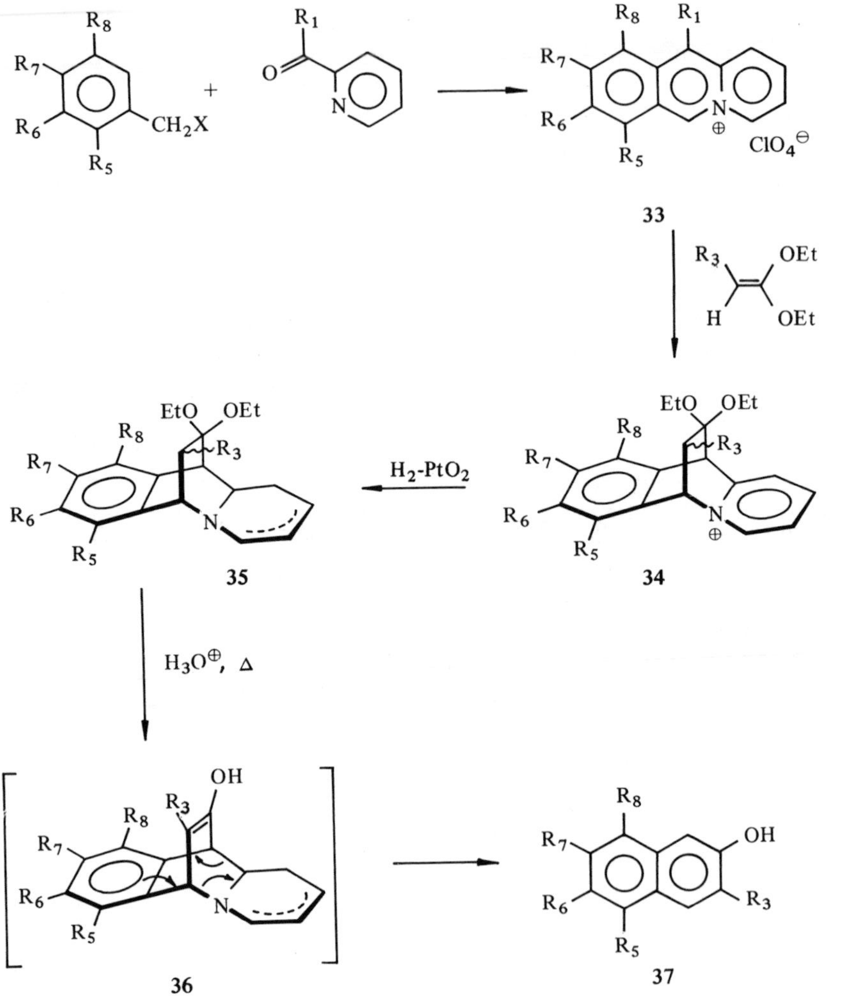

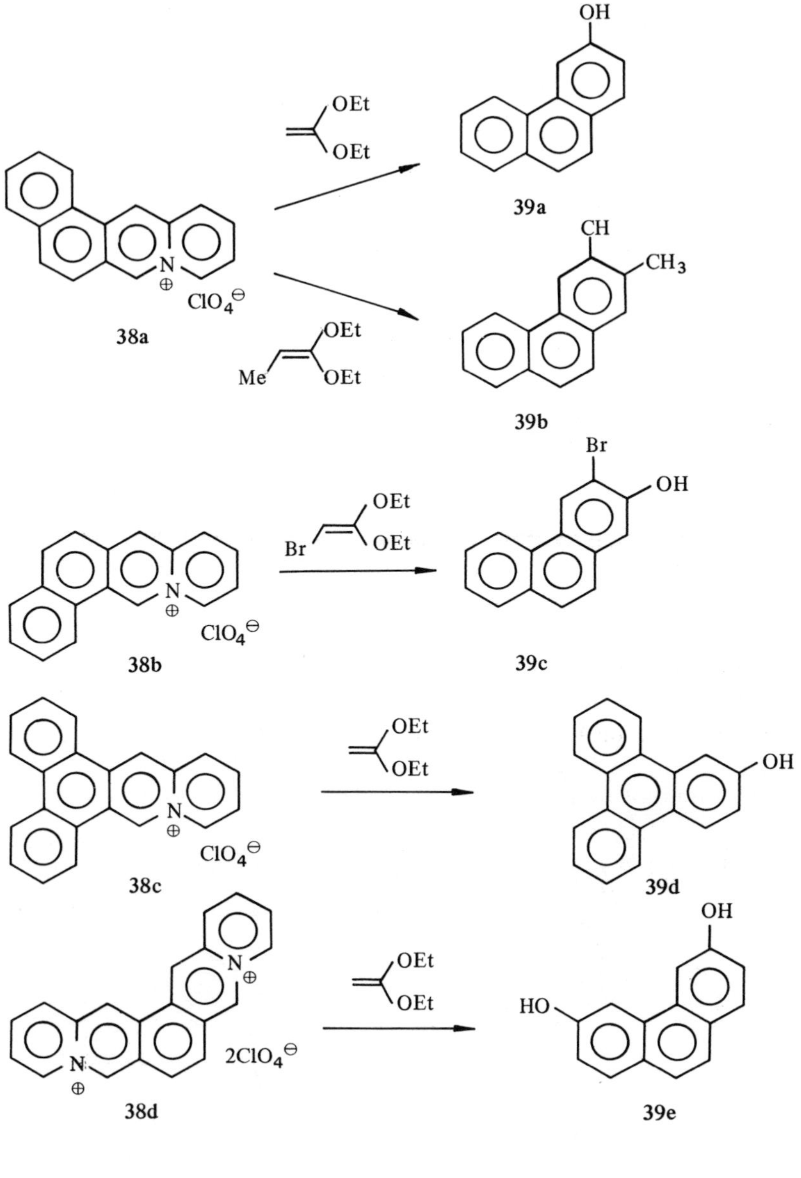

An anthracene synthesis was made possible via this route by utilizing benzyne cycloadditions to the quinolizinium salts.[9] The *in situ* diazotization of anthranilic acid[10,11] produces benzyne that cycloadds to the quinolizinium salt producing the adduct (40). Reduction with sodium borohydride gave the piperdine derivative (41) which fragmented to the anthracene (42) upon heating in acetic acid, acetic anhydride, or a toluene hydrochloric acid mixture.

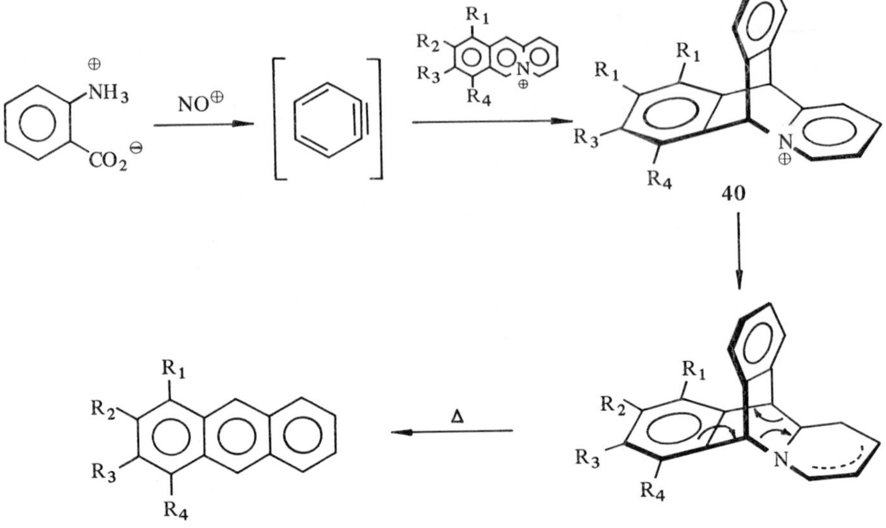

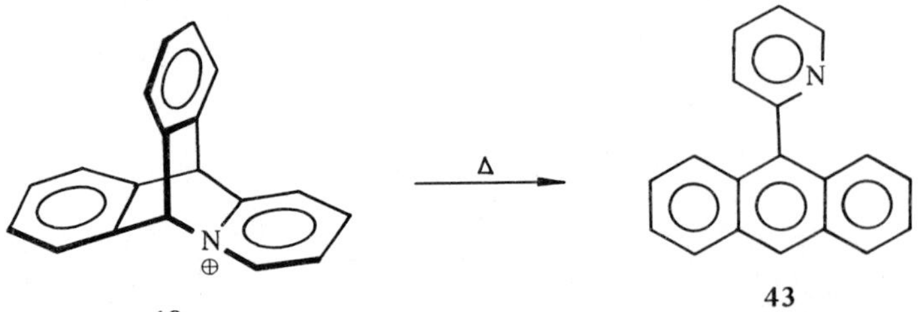

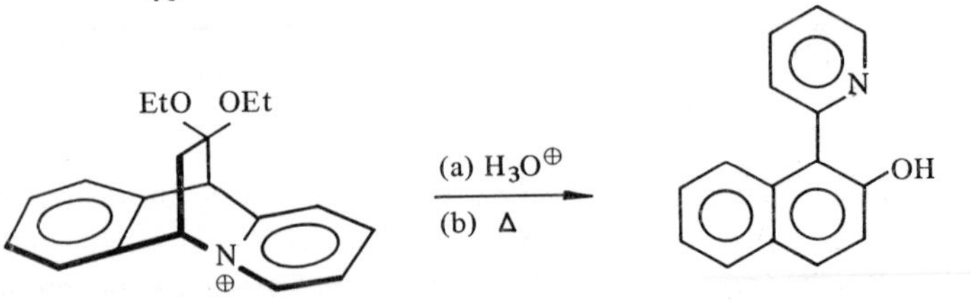

If the benzyne (40) or ketene acetal adducts (34) are thermolyzed prior to the reduction step, an efficient route is opened to pyridyl anthracenes (43)[11] and pyridyl naphthol (43a)[12] respectively. This method of obtaining polynuclear aromatic systems should find considerable utility when the ease of construction of various quinolizinium salts is taken into account.

E.   From Pyridazines

Highly substituted benzene derivatives (46) have been prepared[13] by an "inverse" Diels-Alder reaction where the dienophile (yne-amine) is electron rich and the diene [pyridazine (44)] is electron poor. The reaction proceeds by the 3,6-addition of the yne-amine to (44) leading to the bridged adduct (45) which extrudes nitrogen as it passes on to the benzene derivative. It is of interest that a different pathway is taken when the pyridazine ester (47) is employed. Cycloaddition occurs at the 2,5-position affording the adduct (48) containing nitrogen at the bridgehead position. The latter extrudes hydrogen cyanide producing the pyridine system, 49. The pyridazine precursors are readily available by similar reactions starting from tetrazine derivatives[14] (50 → 51).

F.   From s-Tetrazines

As mentioned above, the availability of pyridazines from s-tetrazines should allow the preparation of a variety of benzene derivatives. However, the tetrazine system itself has been shown to act as a precursor to aromatic compounds. Thus, by double cycloaddition of cyclobutadiene or benzocyclobutadiene to s-tetrazines (52), the bridged adducts 53 and 54 were formed. Thermolytic decomposition of 53 produced the dihydronaphthalene 55 in 78% yield along with a minor amount of an isomeric product. Photolytic or thermal decom-

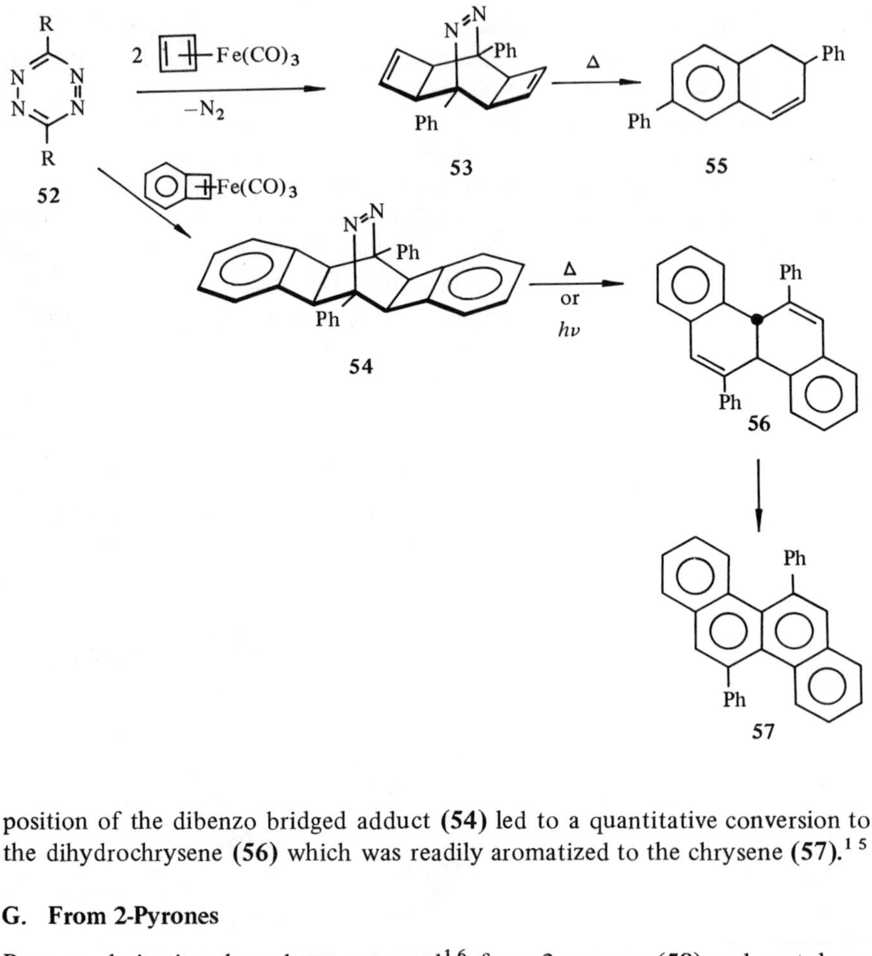

position of the dibenzo bridged adduct (54) led to a quantitative conversion to the dihydrochrysene (56) which was readily aromatized to the chrysene (57).[15]

G.   From 2-Pyrones

Benzene derivatives have been prepared[16] from 2-pyrones (58) and acetylenes by Alder and Rickert in 1937. The process is a typical Diels-Alder reaction which is followed by loss of carbon dioxide to provide the aromatic product. The reaction has been carried out with a wide range of substituents on both the pyrone and acetylene components.[17-19] Since there are two possible modes of cycloaddition, isomeric mixtures could result. However, if either the pyrone or the acetylene is symmetrically substituted, only a single product will result.

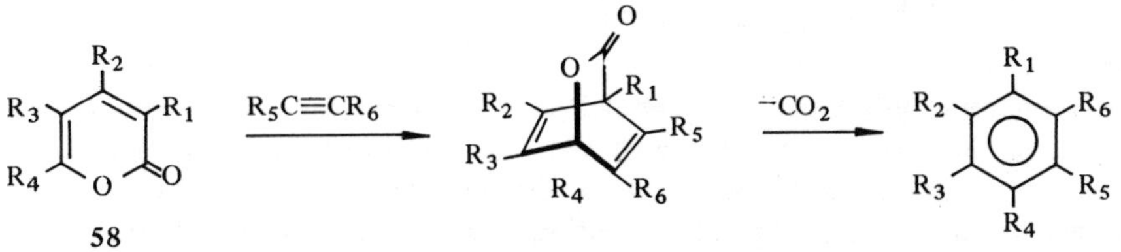

58

R = H, alkyl, aryl, MeO, CO_2Et, etc.

A study was recently performed[20] to assess the regioselectivity of unsymmetrically substituted acetylenes and unsymmetrically substituted pyrones. The results indicated that varying ratios of isomers are found that are mainly dependent on the position and nature of the substituents. For example, reaction of 4,6-diphenyl-2-pyrone **(59)** and triphenyl pyrone **(61)** with methyl propiolate gave only **(60)** and **(62)** respectively. The other possible isomers, **60a** and **62a**, were not found.

59 **60** (70%) **60a** (0%)

61 **62** (55%) **62a** (0%)

63 **64** (40%) **65** (60%)

66 **67** (84%) **68** (16%)

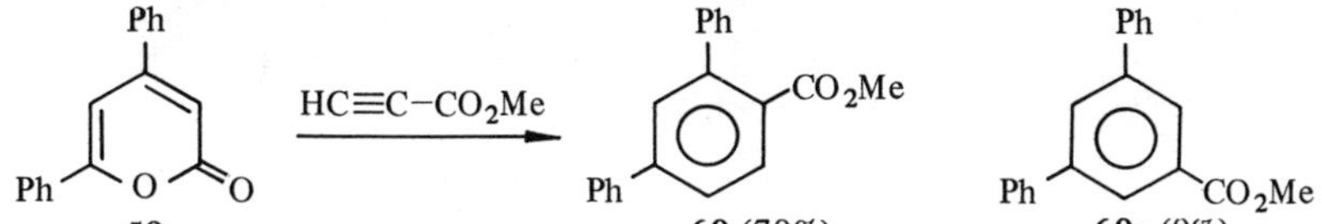

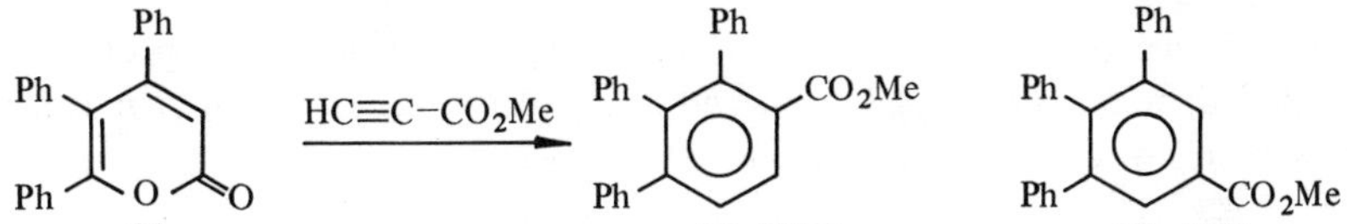

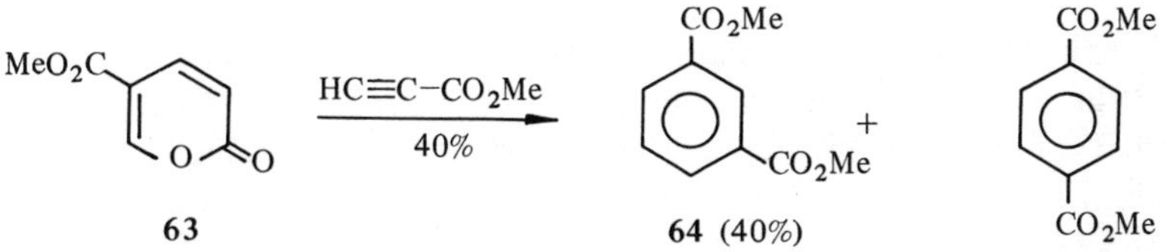

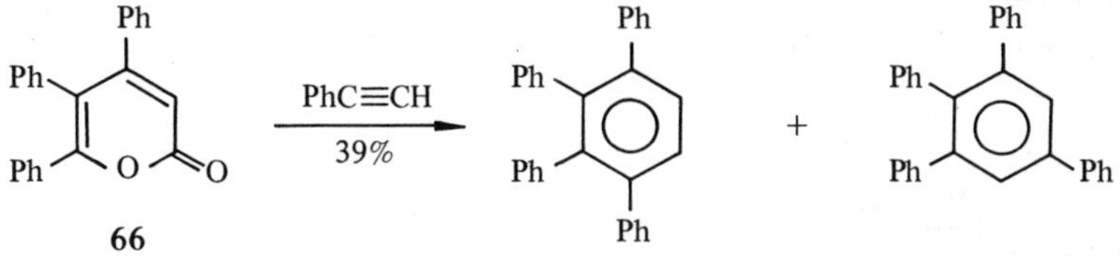

On the other hand, pyrones **63** and **66** gave mixtures **64**, **65** and **67**, **68** respectively. The use of Huckel MO calculations to predict the direction of addition were only partly successful because of the unknown values necessary to weigh the relative importance of steric and polar effects. Nevertheless, if a mode of addition had to be predicted, the similarities observed in this process to that of unsymmetrically substituted butadienes (the sites of highest electron density will cluster closest together) provide a useful rule of thumb.

Since any synthesis of benzene derivatives from pyrones will depend heavily on the availability of various derivatives of pyrones, a survey of synthetic routes reveals that procedures are indeed abundant.[16-20]

A versatile synthesis of pyrones (**69**) has been described by Markl[21] based on the condensation of cyclic ketones and β,β-dichloroacrolein producing the dichlorodieneone (**70**). Diels-Alder addition to the cycloalkanopyrones leads to good yields of cycloalkano benzenes (**71**).

H. From Thiophenes and Benzothiophenes

Thiophenes have already been shown to exhibit useful chemical properties related to the synthesis of hydrocarbons (Chapter 2). Recent studies have revealed that aromatic compounds are also possible from the thiophene nucleus by virtue of a cycloaddition-extrusion sequence.[22,23] Although thiophenes are said to be exceptions in the Diels-Alder reaction,[24] Wynberg has successfully implemented this process using dicyanoacetylene. The products formed are phthalonitriles (**72**) and the yields reflect the necessity for having electron

R_1	R_2	R_3	R_4	% Yield (72)
Me	t-Bu	H	Me	51
Me	H	H	Me	49
H	Ph	Ph	H	18
Me	H	H	H	9
H	H	H	H	8

donating groups present on the thiophene molecule. Heating the thiophene derivative in excess dicyanoacetylene furnished the phthalonitriles directly presumably by passing through the unstable bridged sulfur adduct (73). Attempts to substantiate the presence of the intermediate (73) have not yet been fruitful.

A recent study[25] has shown that cycloaddition of acetylene dicarboxylic ester to the thiophene (74) at −30° proceeds initially to the thiabicyclohepta-diene and then to the thiepin (75). Sulfur extrusion slowly occurs under these conditions leading finally to the aromatic amine (76). This scheme represents the first successful synthesis of the 8π-electron system (75) which has been

reported to possess antiaromatic character. The stability of the thiepin has been attributed to the two methoxycarbonyl substituents, which cause a decrease in the electron density and thus reduce the formal antiaromatic character. This mechanistic pathway cannot be invoked to explain the synthesis of phthalo-nitriles (72) since the substitution pattern in the latter would be different from that observed for 76.

In a related study involving benzothiophenes, photochemical cycloaddition of acetylenes produced the unexpected adduct (77)[26] which is thermally transformed, in good yield, to the naphthalene (78). The expected 2 + 2 cycloadduct (79) was not observed in this case. The authors are uncertain as to the pathway taken in this unusual reaction and postulate three possible routes. In view of the study mentioned above (74 → 76), the mechanism appears to involve the

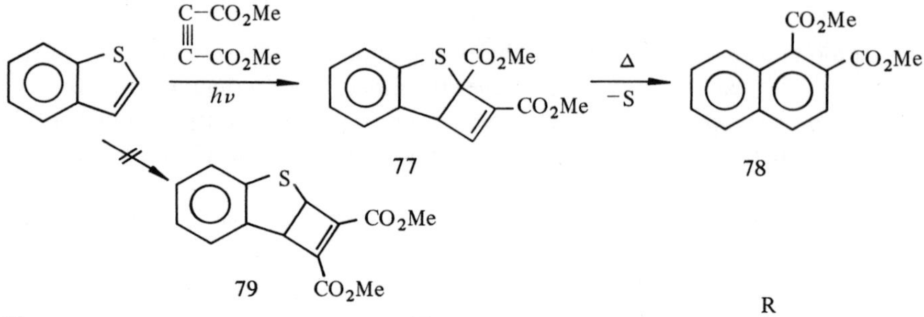

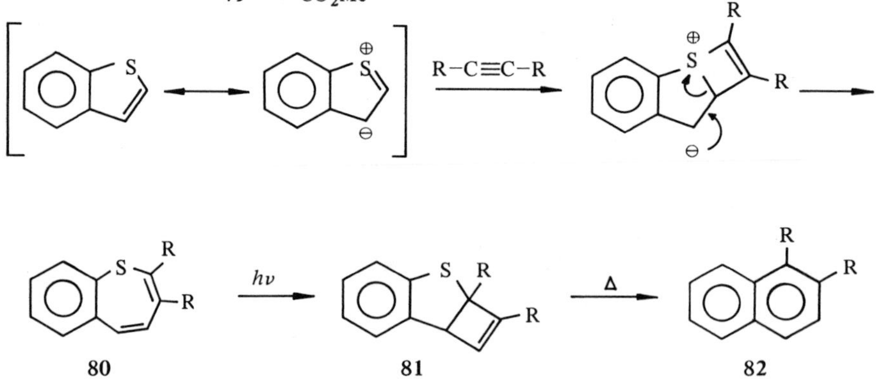

benzothiepin (80) which rearranges to the tricyclic system (81) and is ultimately converted to the naphthalene (82).

A variety of naphthalenes were prepared in this manner and are shown below in tabulated form:

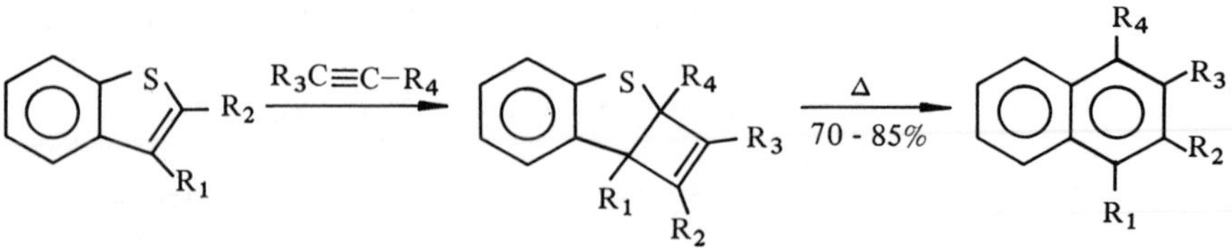

R_1	R_2	R_3	R_4
H	H	CO_2Me	CO_2Me
Me	H	CO_2Me	CO_2Me
Me	Me	CO_2Me	CO_2Me
H	H	H	CO_2Me
H	Me	H	CO_2Me
H	Me	CO_2Me	Ph
H	Cl	H	Me

This seemingly general thermal rearrangement provides an attractive route to highly substituted naphthalenes, some of which are difficult to obtain through conventional procedures.

I. From Furans

Just as cycloaddition of highly reactive dienophiles to thiophenes leads to aromatic systems, a similar process involving furans is also known.[27] The *in situ* generation of benzyne (83) in the presence of furan was reported[28] to produce the addition product (84). Inspection of the latter indicates that it has the further potential to serve as a dienophile. Treatment of 84 with 2,3-dimethyl-butadiene led to the Diels-Alder adduct (85) in good yield (80-90%) which was smoothly hydrolyzed in acidic methanol to the 1,4-dihydroanthracenes (86).

Aromatization using chloroanil provided the anthracenes (87). The versatility of this anthracene synthesis was demonstrated with substituted benzynes and furans leading to various polyalkylated derivatives. For example, the use of the dimethyl benzyne and 2,5-dimethylfuran furnished the tetramethyl adduct (88) which underwent the Diels-Alder reaction with 2,3-dimethylbutadiene affording the hexamethyl adduct (89). Hydrolysis and oxidation of this intermediate gave 2,3,6,7,9,10-hexamethyl anthracene (90). Acid catalyzed rearrangement of

Diels-Alder adducts derived from furans and acetylene dicarboxylates (91) provides a useful synthetic route to polysubstituted phenols (92).[29]

Once again it should be stated that substituted aromatic compounds prepared by cycloaddition reactions involving simple starting materials opens a regiospecific route to substituted products and avoids the isomeric mixtures usually associated with direct introduction techniques.

J. From Pyrroles

It would be expected, *a priori*, that pyrroles should react with dienophiles in the manner observed for furans and thiophenes. However, pyrroles behave poorly in the Diels-Alder reaction.[30] N-Acetylpyrroles (93), on the other hand, condense with acetylene dicarboxylic ester in the presence of Lewis acids

furnishing the adduct (94).[31] Further rearrangement leads to the highly substituted aromatic product (95) in 85-90% yields. The presence of aluminum
chloride in forming 94 reduced the Diels-Alder reaction time from 7-8 days to
30 minutes (at 0°). The bicyclic adduct (94) need not be isolated. By allowing
the temperature of the Diels-Alder reaction to warm from 0° to 40°, the

93 (R$_1$, R$_2$ = H, Me)   94   95

facile rearrangement to the substituted benzene derivative ensues, thus making
this method a single operation synthesis of aromatics from pyrroles.

K. From Isoxazoles

The synthetic utility of the isoxazole ring (96) with regard to ring annelations
has also been employed as a means for construction of aromatic rings.[32] The
chloromethyl isoxazole (96) is elaborated to the cyclohexanone derivative
(97) using methyl magnesium carbonate and converted to the quaternary salt
(98) with triethyloxonium fluoroborate. Treatment with aqueous sodium
hydroxide provides the acyl phenol (99) presumably via intermediates A-D.
In this fashion dihydrophenanthrene (100) and the octahydro derivative (101)
were similarly prepared. A recent variation of this method has led to the
corresponding pyridine derivative (102).[33] The use of isoxazoles as a vehicle for
the synthesis of polycyclic ketones is discussed in Chapter 9 (Section V.E).

96   97   98

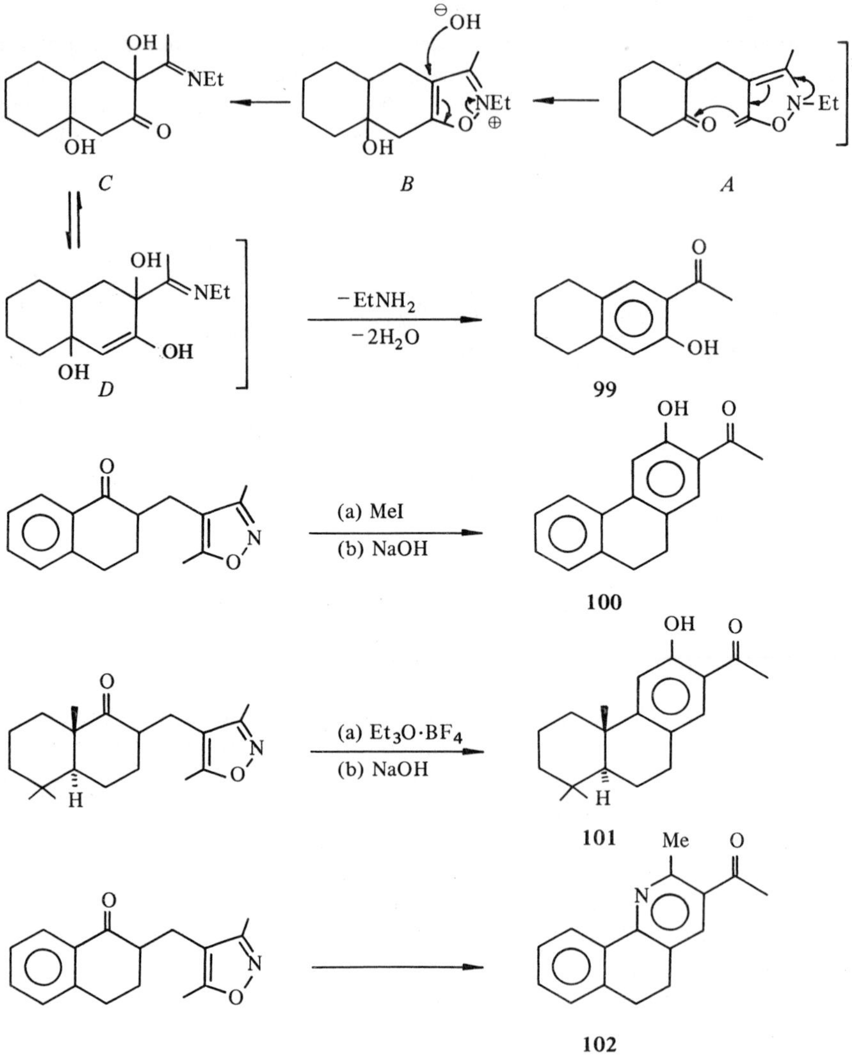

It is interesting to point out that the direction of ring closure in **98** may be altered by use of pyrrolidine as a base so that the proton on the methyl group adjacent to the iminium bond is removed. This results in the intermediate *E* which proceeds on to the aromatic amine **103**.

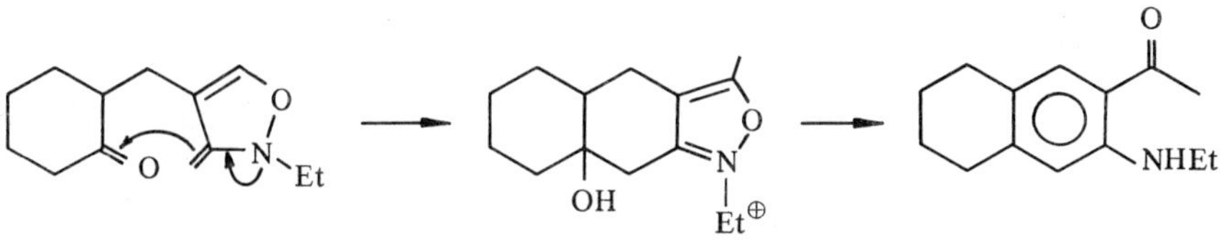

$$E \qquad\qquad 103$$

II. INTRODUCTION AND REMOVAL OF SUBSTITUENTS

A. Deoxygenation of Phenols with Tetrazoles

The use of a heterocycle as a reagent to perform a synthetic manipulation has been demonstrated by 1-phenyl-5-chlorotetrazole.[34] This system provides a general and facile method of directly removing a phenolic hydroxyl group. Thus, by first transforming the phenol hydroxyl into a heterocyclic ether

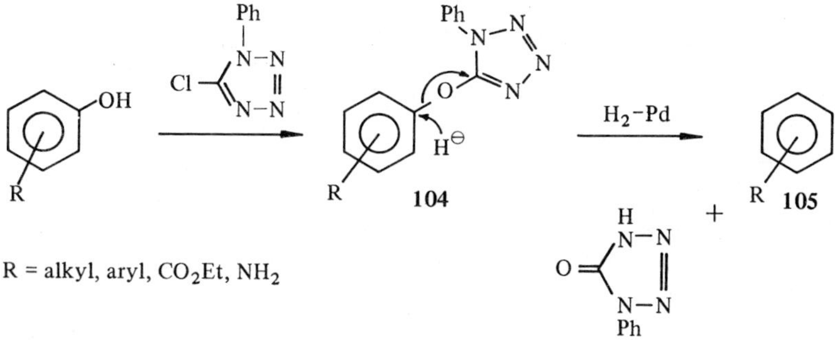

R = alkyl, aryl, CO$_2$Et, NH$_2$

(104) and then subjecting the latter to catalytic hydrogenolysis, the deoxygenated benzene derivative (105) is obtained. A variety of phenolic and naphthol derivatives were examined in this process and gave good yields of both the heterocyclic ether and the deoxygenated aromatic. This reagent has also been employed in the synthesis of complex molecules (106 → 107),[35] (108 → 109).[36] It is of interest to note that the formation of (109) represents a case where a hindered phenolic hydroxyl was cleanly removed by the tetrazole reagent.

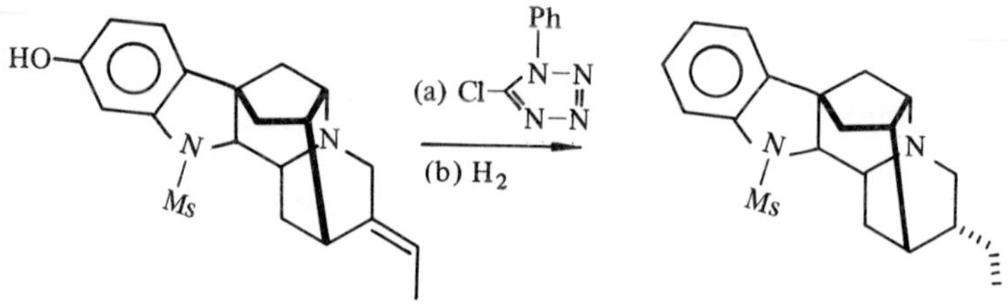

106   107

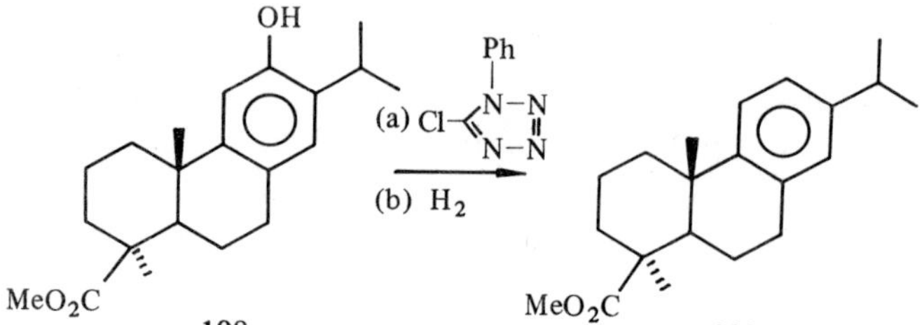

108   109

B.   Methoxylation of Aromatics with N-Methoxyphenanthridinium Salts

Direct alkoxylation of aromatic compounds is virtually unknown although hydroxylation has been achieved by a few reagents (peracids,[37] lead tetraacetate,[38] thallium trifluoroacetate,[39] peroxide-light,[40] and pyridine-N-oxides[41]). The photolytic cleavage of N-methoxyphenanthridinium perchlorate (110) in acetonitrile containing aromatic substrates (111) resulted in the formation of methoxy substituted aromatic systems (112).[42] Despite the poor yields, the method is potentially quite useful, and future work may increase the efficiency of the process. The isomer distribution in 112 is shown in the accompanying table. The results appear to support a radical mechanism where the methoxyl radical is the attacking species. Decomposition of 110 in acrylonitrile gave

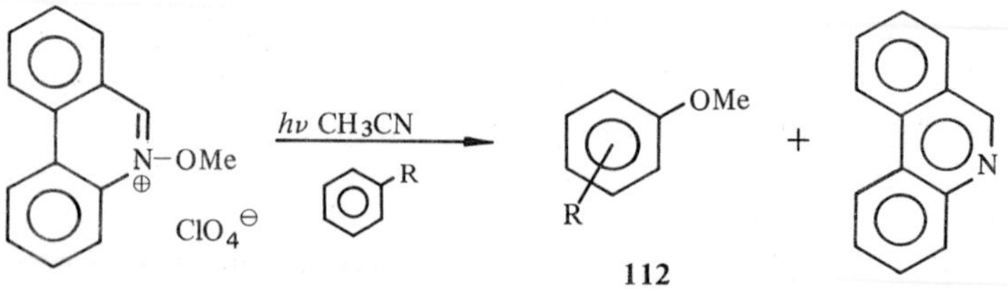

110   111   112

polyacrylonitrile, whereas in the presence of toluene, bibenzyl was isolated along with the methoxytoluenes. These results seem to lend credence to the radical process.

Aromatic Compound (111)	% Yield of 112	% Isomer Distribution in 112		
		Ortho	Meta	Para
Anisole	21.6	79	Trace	21
Toluene	4.9	70	13	17
Benzene	4.4			
Benzonitrile	7.6	73	5	22

C. Hydroxylation of Aromatic Compounds Using Pyridazine N-Oxides

Irradiation of a solution of pyridazine N-oxides (113) and aromatic substrates in dichloromethane has been reported[43-45] to lead directly to the hydroxylated derivatives (114) in yields up to 40%. The reaction proceeds by the direct transfer of oxygen from the N-oxide to the aromatic ring but the exact mechanism by which this occurs is still unknown.[46] A number of examples

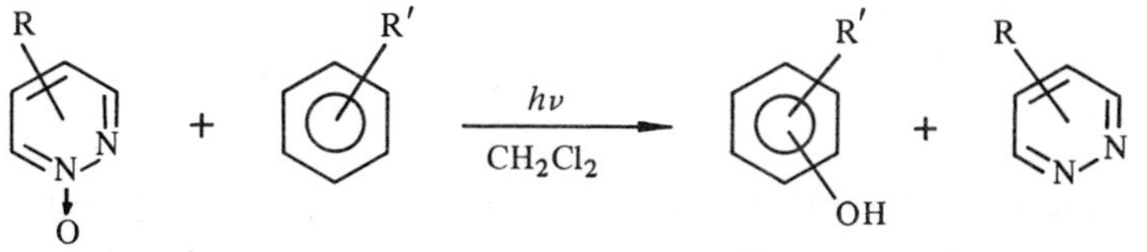

113   (R = Me, Cl)   114

have been offered to demonstrate this process and they are listed below:

1.    30%

2. 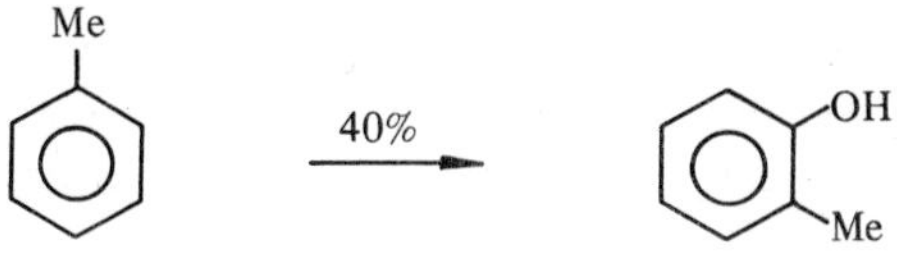   40%   + (meta and para)

3.

23%

4.

18%   +   3%

5.

10%   7%   4%

6.

14%   7%

In examples 4, 5, and 6 it is clear that one of the ring methyl groups has been replaced by oxygen which tends to support an addition-elimination mechanism. The lack of any methoxy-substituted benzenes would preclude any insertion process taking place, yet the isolation of the benzyl alcohol (example 6) may be regarded as a C–H insertion product. On the other hand, the latter may have arisen from a benzylic oxidation. Further details must be awaited before this process is completely understood. Nevertheless, its synthetic utility and further development need not await these results. A variety of pyridazine N-oxides bearing addition substituents have been employed and appear to give similar results. Oxidations of saturated and olefinic hydrocarbons have also been achieved producing alcohols and diols (see Chapter 7).

D. Formylation of Aromatic Nuclei and N-Methyloxazolinium Salts

The direct conversion of an aryl Grignard reagent to the aldehyde (117) has been reported[47] to take place by reaction with the highly electrophilic N-methyloxazolinium salt followed by hydrolysis of the intermediate adduct (116). Although formylation of Grignard reagents has been achieved using other electrophiles[48] (DMF, orthoformates), this method, by virtue of its simplicity, should provide a favorable alternative. In particular, the readily accessible 2-deuterooxazolinium salt 115 (A=D) allows a facile entry into C-1 deuterated aldehydes. The reaction is also characterized by the need for two equivalents of hexamethylphosphoramide (HMPA) to complex the Grignard reagent.[49]

Further application of the *N*-methyloxazolinium salt is discussed in a subsequent section (Chapter 9, Section II.A) with regard to a one-carbon homologation of Grignard reagents to aldehydes.

REFERENCES

1. For a detailed review of pyrylium salts and their application toward the synthesis of aromatic compounds, see K. Dimroth and K. H. Wolf, *Newer Methods of Preparative Organic Chemistry*, Vol. 3, Academic Press, New York, 1964, p. 357.

2. K. Hafner and H. Kaiser, *Annals*, **618**, 140 (1958).

2a. K. T. Potts, A. J. Elliot, and M. Sorm, *J. Org. Chem.*, **37**, 3838 (1972).

3. G. Suld and C. C. Price, *J. Am. Chem. Soc.*, **84**, 2090, 2094 (1962).

4. G. A. Reynolds and J. A. Van Allen, *J. Heterocycl. Chem.*, **8**, 301 (1971).

5. Z. Yoshida, S. Yoneda, H. Sugimoto, and T. Sugimoto, *Tetrahedron*, **27**, 6083 (1971).

6. A. R. Katritzky and Y. Takeuchi, *J. Am. Chem. Soc.*, **92**, 4134 (1970).

7. D. Bertelli, T. G. Andrews, and P. O. Crews, *ibid.*, **91**, 5286 (1969).

8. C. K. Bradsher and L. E. Beavers, *ibid.*, **77**, 4812 (1955); C. K. Bradsher and J. C. Parham, *J. Org. Chem.*, **28**, 83 (1963); C. K. Bradsher and T. W. G. Solomons, *J. Am. Chem. Soc.*, **82**, 1808 (1960).

9. D. L. Fields, *J. Org. Chem.*, **36**, 3002 (1971).

10. L. Friedman and F. M. Logullo, *J. Am. Chem. Soc.*, **85**, 1549 (1963).

11. D. L. Fields, T. H. Regan, and R. E. Graves, *J. Org. Chem.*, **36**, 2995 (1971).

12. D. L. Fields, T. H. Regan, and J. C. Dignan, *ibid.*, **33**, 390 (1968); D. L. Fields and T. H. Regan, *ibid.*, **35** (1970); **36**, 2986 (1971); **36**, 2991 (1971).

13. H. Neunhoeffer and G. Werner, *Tetrahedron Lett.*, 1517 (1972).

14. H. Neunhoeffer and H. Frühauf, *ibid.*, 3355 (1970); P. Roffey and J. P. Verge, *J. Heterocycl. Chem.*, **6**, 497 (1969), and references cited therein; H. Neunhoeffer and H. Frühau.., *retrahedron Lett.*, 3151 (1969).

15. L. A. Paquette, M. R. Short, and J. F. Kelly, *J. Am. Chem. Soc.*, **93**, 7179 (1971).

16. K. Alder and H. Rickert, *Berichte*, **70**, 1354 (1937).

17. E. Wenkert, D. B. R. Johnston, and K. G. Dave, *J. Org. Chem.*, **29**, 2534 (1964).

18. A. Padwa and R. Hartman, *J. Am. Chem. Soc.*, **88**, 1518 (1966).

19. J. Bulock and H. G. Smith, *J. Chem. Soc.*, 502 (1960).

20. J. K. Stille, J. A. Reed, C. L. Schilling, R. F. Tarvin, and T. A. Rettig, *J. Org. Chem.*, **34**, 2188 (1969).

21. G. Markl and R. Fuchs, *Tetrahedron Lett.*, 4691, 4695 (1972).

22. R. Helder and H. Wynberg, *Tetrahedron Lett.*, 605 (1972).

23. H. Wynberg and R. Helder, *ibid.*, 3647 (1972).

24. N. L. Allinger, M. P. Cava, D. C. De Jongh, C. R. Johnson, N. A. LeBel, and C. L. Stevens, *Organic Chemistry*, Wirth Publishers, New York, 1971, p. 122; L. A. Paquette, *Modern Heterocyclic Chemistry*, Benjamin, New York, 1968, p. 137.

25. D. N. Reinhoudt and C. G. Kouwenhoven, *Chem. Commun.*, 1232, 1233 (1972).

26. J. H. Dopper and D. C. Neckers, *J. Org. Chem.*, **36**, 3755 (1971).

27. E. Wolthius, *ibid.*, **26**, 2215 (1961).

28. G. Wittig and L. Pohmer, *Berichte*, **89**, 1334 (1956).

29. A. W. McCulloch, B. Stanovnik, D. G. Smith, and A. G. McInnes, *Can. J. Chem.*, **47**, 4319 (1969).

30. O. Diels and K. Alder, *Annals*, **490**, 267 (1931); R. M. Acheson, in *Adv. Heterocycl. Chem.*, A. R. Katritzky, ed., Academic Press, New York, **1**, 125 (1963).

31. R. C. Bansal, A. W. McCulloch, and A. G. McInnes, *Can. J. Chem.*, **48**, 1472 (1970).

32. M. Ohashi, T. Maruishi, and H. Kakisawa, *Tetrahedron Lett.*, 719 (1968).

33. G. Stork, M. Ohashi, H. Kamachi, and H. Kakisawa, *J. Org. Chem.*, **36**, 2784 (1971).

34. W. J. Musliner and J. W. Gates, *J. Am. Chem. Soc.*, **88**, 4271 (1966).

35. S. Sakai, T. Hamamoto, M. Wakabayashi, K. Takahashi, Y. Ohtani, and J. Haginiwa, *Tetrahedron Lett.*, 1489 (1969).

36. J. W. Huffman, *J. Org. Chem.*, **35**, 3154 (1970).

37. R. L. Dannley and G. E. Corbett, *ibid.*, **31**, 153 (1966).

38. R. E. Partch, *J. Am. Chem. Soc.*, **89**, 3662 (1967).

39. E. C. Taylor, H. W. Altland, R. H. Danforth, and G. McGillivray, *ibid.*, **92**, 3520 (1970).

40. K. Omura and T. Matsuura, *Tetrahedron*, **24**, 3475 (1968).

41. J. Streith and C. Sigwalt, *Bull. Soc. Chim.*, 1157 (1970).

42. J. D. Mee, D. W. Heseltin, and E. C. Taylor, *J. Am. Chem. Soc.*, **92**, 5814 (1970).

43. T. Tsuchiya, H. Arai, and H. Igeta, *Tetrahedron Lett.*, 2747 (1969).

44. H. Igeta, T. Tsuchiya, N. Yamada, and H. Arai, *Chem. Pharm. Bull.*, **16**, 167 (1968).

45. T. Tsuchiya, H. Arai, and H. Igeta, *Tetrahedron Lett.*, 2213 (1970).

46. Light induced decomposition of pyridine *N*-oxides in benzene has been reported to give a 15% yield of phenol with an oxepine intermediate a likely possibility; J. Streith, B. Danner, and C. Sigwalt, *Chem. Commun.*, 979 (1967).

47. A. I. Meyers and E. W. Collington, *J. Am. Chem. Soc.*, **92**, 6676 (1970).
48. J. Carnduff, *Quart. Rev. Chem. Soc.*, **20**, 169 (1966).
49. A. I. Meyers, E. W. Collington, R. Brimkmeyer, *Org. Synthesis*, in press.

6 HALIDES

The more prominent routes[1] to halides originate in transformations involving alcohols, alkenes, alkanes, and other halides via halogen addition, substitution, or displacement reactions. However, several useful methods stem from heterocyclic compounds serving as precursors, reagents, or vehicles.

I. FROM 2-ALKYLTHIO-2-THIAZOLINES

The synthetic utility of 2-alkylthio-2-thiazolines (1) as a "riveting" agent for hydrocarbons (2) has been described in an earlier section (p. 26). By invoking a slight modification in the procedure, this interesting heterocycle has been shown to serve as a useful vehicle in homologating alkyl halides by one or

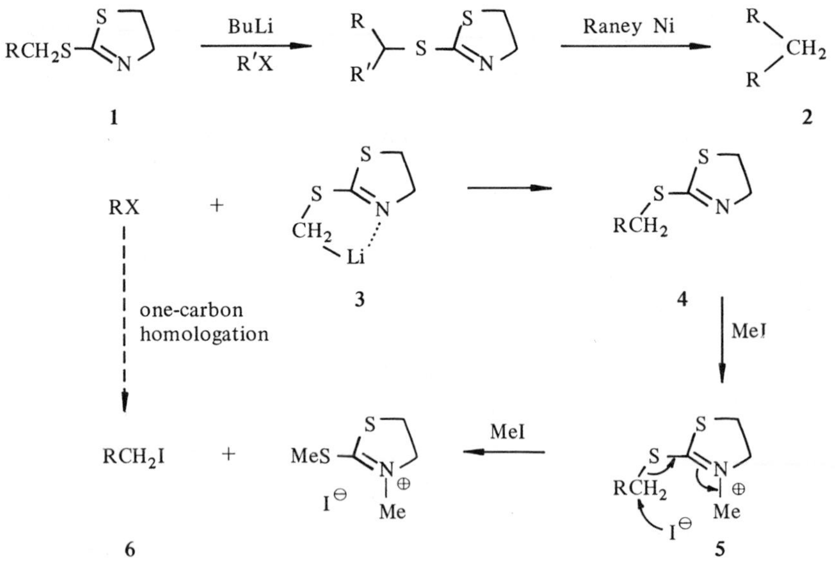

three carbons.[2] Alkylation of the lithio salt of 2-methylthio-2-thiazoline (3) with an alkyl halide led to the elaborated thiazoline (4) which was transformed into the corresponding methiodide salt (5) by heating in dimethylformamide.

The latter spontaneously underwent displacement of the thiazoline by iodide ion producing the homologated iodide (6). The excellent leaving group ability of the thiazolinium salt (5) represents the significant feature of this useful reaction. In this fashion the following alkyl halides were iodomethylated to their corresponding homologs:

$PhCH_2Br \longrightarrow 4 \longrightarrow PhCH_2CH_2I$		75% (from 4)
$PhCH_2CH_2Br \longrightarrow 4 \longrightarrow PhCH_2CH_2CH_2I$		70% (from 4)
$n\text{-}C_9H_{19}Cl \longrightarrow 4 \longrightarrow n\text{-}C_{10}H_{21}I$		68% (from 4)

The three-carbon homologation of alkyl halides was cleverly implemented by employing the lithio salt of 2-allylthio-2-thiazoline (7), prepared from allyl bromide and 2-mercaptothiazoline. Addition of an alkyl halide in THF-HMPA solvent resulted in the alkylated thiazoline 8, which was similarly treated with excess methyl iodide forming the labile heterocyclic intermediate (9). After removing the precipitated salt (10), the *trans*-allylic iodide (11) was isolated free of any *cis*-isomer. The facile SN_2' process that takes place on the intermediate methiodide salt (9) leads to an iodopropenylation of alkyl halides and pure *trans*-olefins—not in the least a trivial result. This method should find wide applicability both in the preparation of alkyl halides and in *trans*-olefins.

II. FROM *N*-NITROSOOXAZOLIDINONES

The versatile heterocycle, 5,5-dialkyl-*N*-nitrosooxazolidinone (12) has been previously discussed (Chapter 3, p. 55) with respect to its fragmentation to vinyl carbenes (13) and vinyl cations (14). By performing the base catalyzed ring opening in the presence of alkali metal halides, the vinyl cation is efficiently captured by the halide ion forming vinyl halides (15) in good yield.[3] The reaction appears to have considerable generality provided there are geminal substituents on the starting heterocycle. The latter is derived from Reformatsky products (β-hydroxy esters) and hydrazine (p. 58), and its accessibility,

therefore, should present little difficulty.

A study was undertaken by Newman to assess the degree of stereoselectivity, if any, that accompanies this synthesis of vinyl halides. It was found that there was, indeed, a high percentage of *trans*-vinyl halide formed (93-95%). It is, therefore, reasonable to conclude that the diazonium intermediate 16 decomposes by nitrogen departure from the most accessible side (*trans* to the *t*-butyl group). The resulting vinyl cation 17 which seems to retain its configuration is also attacked predominantly at the most accessible side giving rise to the *trans*-vinyl halide. A high degree of stereoselectivity would not be expected if

93 - 95%   5 - 7%

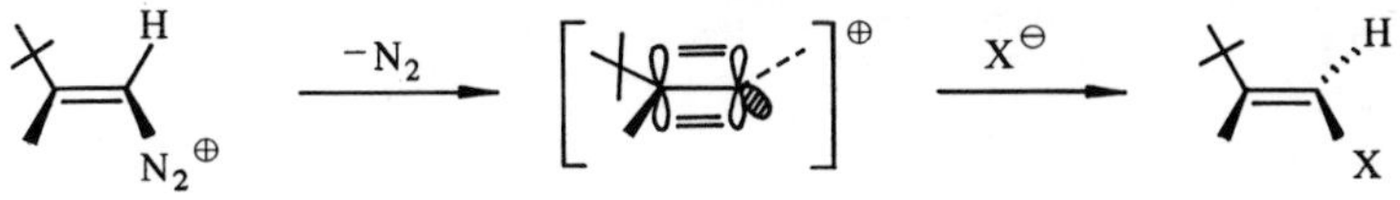

16   17

the alkyl groups on the oxazolidinone were of comparable steric bulk.

III. FROM 1,3,5-TRICHLORO-2,4,6-TRIOXOHEXAHYDRO-*s*-TRIAZINE (TRICHLOROISOCYANURIC ACID)

Direct chlorination of ethers,[4,5] alkenes,[6] and benzylic[7] and aromatic substrates has been accomplished using trichloroisocyanuric acid (**18**). It has proved to be a convenient alternative to chlorine in these transformations. The synthetic utility of this reagent may be demonstrated by the following:

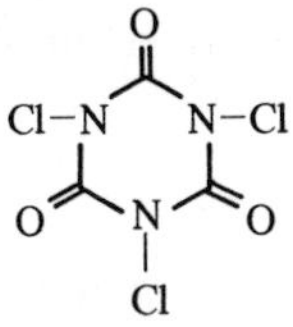

18

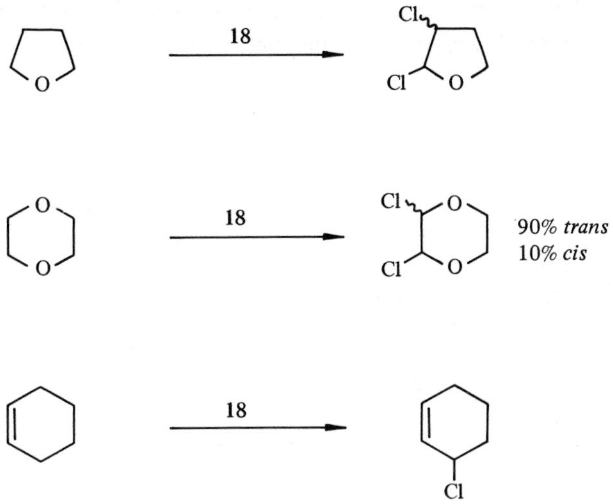

90% *trans*
10% *cis*

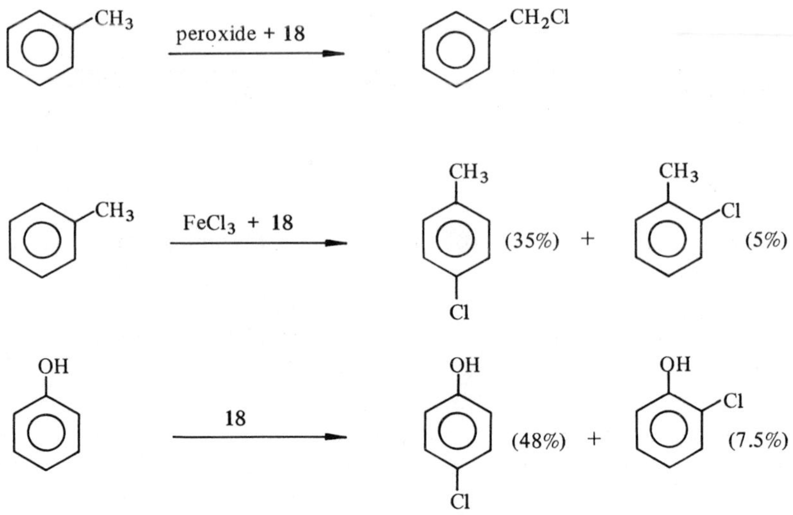

It is of interest to point out that reaction with phenol (or aniline) gives only monochlorinated derivatives in contrast to the polyalkylation normally observed with molecular chlorine.

IV.   FROM 2,2-TRICHLORO-1,3,2-BENZDIOXAPHOSPHOLENE

A useful reagent for transforming ketones into vinyl halides is the trichlorophospholene (19). Heating a carbonyl compound with 19 produces the vinyl halide directly and the following examples have been reported.[8]

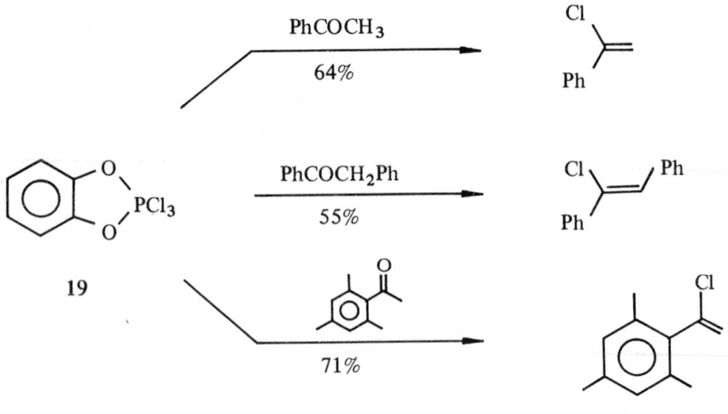

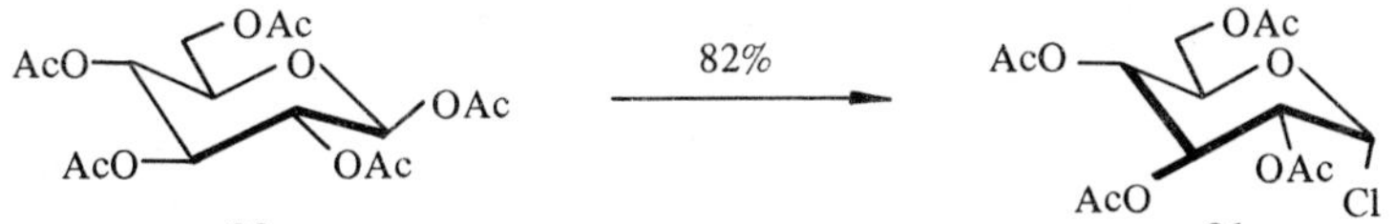

Reaction of the penta-acetylglucose derivative **20** with **19** also led to chlorination, presumably via the hemiacetal, and produced the inverted α-chloro-
glucose derivative **21**. N-Benzoylpiperidine (**22**) was transformed into 1,5-
dichloropentane (**23**) with the phospholene trichloride while ethylene carbonate
(**24**) gave the chloroethyl chloroformate (**25**). The ready availability of **19**
should render it as a widely useful chlorinating agent of a variety of molecules.

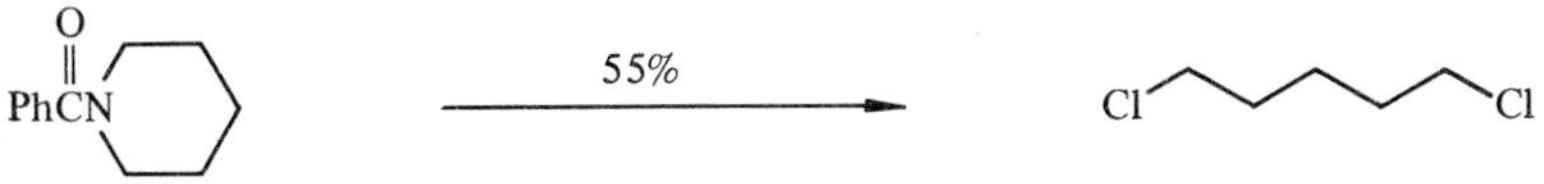

V. FROM 2-OXAZOLINES

In molecules containing vicinal amino and hydroxyl functions (**26**), the latter
may be stereospecifically transformed into halogen with retention of configuration by invoking the intermediacy of an oxazoline (**28**).[9] Thus, by treatment of
the corresponding N-acyl derivative (**27**) with thionyl chloride, the temporarily
constructed oxazoline (**28**) is formed through nucleophilic displacement of the
chlorosulfite by the amide carbonyl. Nucleophilic attack by chloride ion results
in regeneration of the open chain chloro amide (**29**) with retention of configuration.

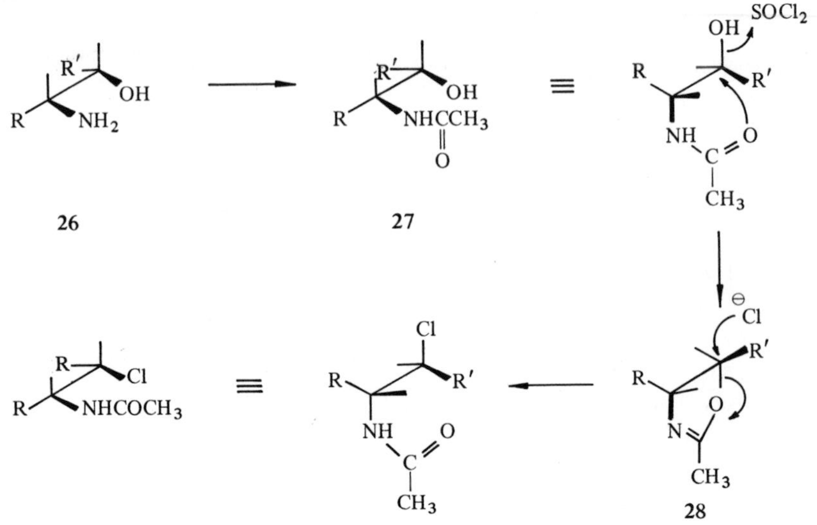

In this fashion, the *erythro* isomers of N-benzoyl-β-aryl serines (30) gave *erythro* β-chloro-β-aryl alaninates (31). It should be understood, of course, that the transformation of hydroxyl to chloride via the 2-oxazoline is limited specifically in this instance to β-amino alcohols; yet the concept of a heterocycle used as a vehicle is well illustrated in this process.

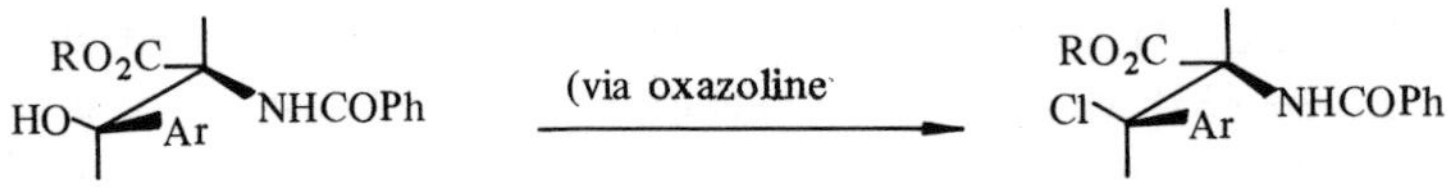

REFERENCES

1. (a) R. T. Morrison and R. N. Boyd, *Organic Chemistry*, Allyn and Bacon, New York, 3rd ed., 1973; (b) I. T. Harrison and S. Harrison, *Compendium of Organic Synthetic Methods*, Wiley-Interscience, 1971, p. 329; (c) H. O. House, *Modern Synthetic Reactions*, Benjamin, New York, 1972, p. 422.

2. K. Hirai and Y. Kishida, *Tetrahedron Lett.*, 2743 (1972).

3. M. S. Newman and C. D. Beard, *J. Am. Chem. Soc.*, **92**, 4309 (1970).

4. E. C. Juenge, P. L. Spangler, and W. P. Duncan, *J. Org. Chem.*, **31**, 3836 (1966).

5. W. P. Duncan, G. D. Strate, and B. G. Adcock, *Org. Prep. Proced.*, **3**, 149 (1971).

6. K. Zeigler, *Annalen.*, **551**, 80 (1942).
7. E. C. Juenge, D. A. Beal, and W. P. Duncan, *J. Org. Chem.*, **35**, 719 (1970).
8. H. Gross and J. Gloede, *Berichte,* **96**, 1387 (1963).
9. S. H. Pines and M. A. Kozlowski, *J. Org. Chem.*, **37**, 292 (1972).

7 ALCOHOLS, MERCAPTANS, AND THEIR ACYL AND ALKYL DERIVATIVES

When a heterocycle plays a role in the formation of one of the titled classes of compounds, the carbon skeleton that accompanies the functional group usually represents the more significant portion of the molecule. This aspect is, therefore, the major advantage in considering heterocycles as precursors or vehicles to alcohols, mercaptans, and their derivatives. Direct introduction, removal, and manipulations of oxygen and sulfur moieties are, on the other hand, mainly left to the time tested reagents known to all practicing organic chemists.

The material in this chapter is presented in four parts·

1. Alcohols.
2. Esters
3. Mercaptans.
4. Ethers

As usual, the preparation of these four groups of compounds is dependent on the synthetic utility of a specific heterocyclic system.

I. ALCOHOLS

A. From Isoxazolidines

The lability of the N—O linkage in isoxazolidines provides the basis for the synthesis of alcohols as demonstrated by Huisgen in 1960.[1] In his extensive studies on 1,3-dipolar additions involving nitrones (1) and olefins, the resulting isoxazolidines **2** and **3** were subjected to catalytic hydrogenation producing the corresponding alcohols **4** and **5** in high yield. The presence of the benzylamine moieties in these heterocycles resulted in concomitant hydrogenolysis of the amino group, thus affording the alcohols. In general, whenever a benzyl amine group resides on the isoxazolidine, loss of the amino group accompanies the reductive ring cleavage.

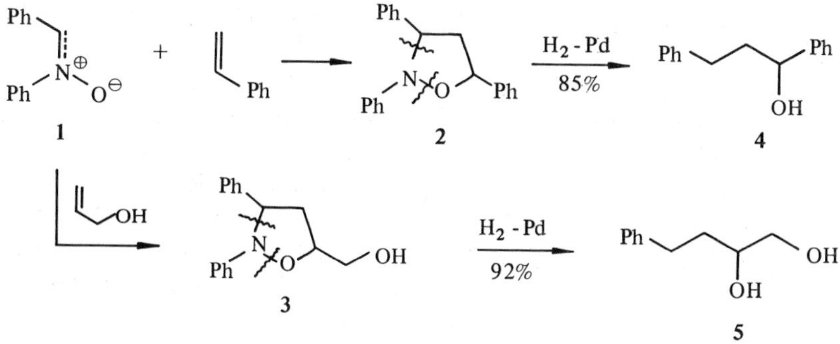

Another useful variation of this technique was reported in 1960 by LeBel[2] which involved the intramolecular cycloaddition of nitrones to olefins. The resulting bicyclic isoxazolidine (**6**) which had the *cis*-configuration could be readily reduced to the *cis*-amino alcohol **7**. In this series, the amine function is retained during hydrogenolysis because of the absence of a labile benzyl group.

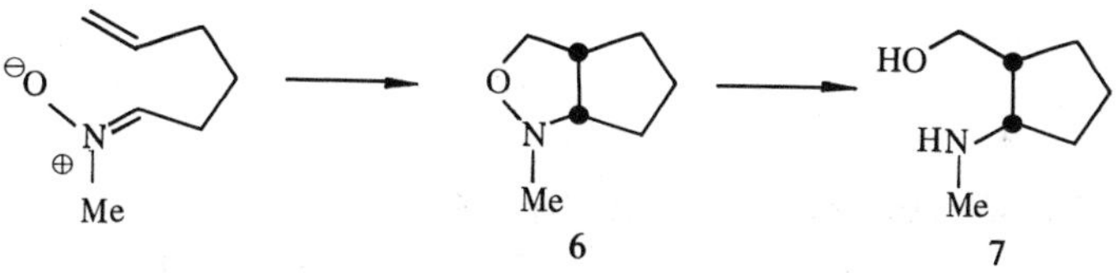

Of further interest is the fact that besides being an excellent route to amino-cyclopentyl carbinols, the stereochemistry of these systems is purely in the *cis*-series. However, in the cyclohexyl series, the amino cyclohexylcarbinols (**9**) derived from the fused isoxazolidine (**8**) were a mixture of *cis* and *trans* isomers.

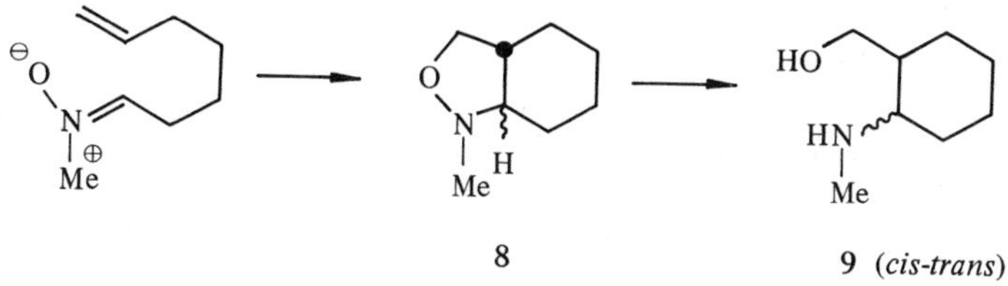

A study to determine the factors influencing the stereochemical results has been reported.[3]

Isoxazolidines derived from electrophilic olefins have been shown to provide hydroxy compounds bearing additional substituents.[4] Thus the cycloaddition of benzophenone oxime to methylvinyl ketone leads to the isoxazolidine **10** which after masking the carbonyl groups was reductively cleaved to the alcohol **11**. The loss of the amino group is expected in light of the factors mentioned earlier. Consideration of the skeletal structure of **11** and its substituents depicts the advantages of preparing hydroxy compounds from the versatile isoxazolidine ring.

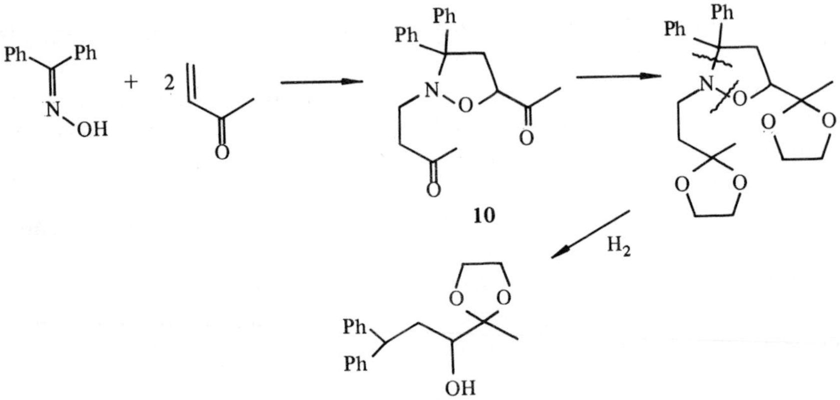

10

11

B.   From Tetrahydrofurans and Dihydropyrans

The employment of these rather prosaic heterocycles in the preparation of alcohols is generally not appreciated. In 1950 Crombie[5] showed that a useful extension of a halide to an alcohol by four or five carbons was quite feasible using tetrahydrofurans and tetrahydropyrans, respectively. The readily formed 2,3-dichloro derivatives **12** and **13**, when treated with a Grignard reagent,

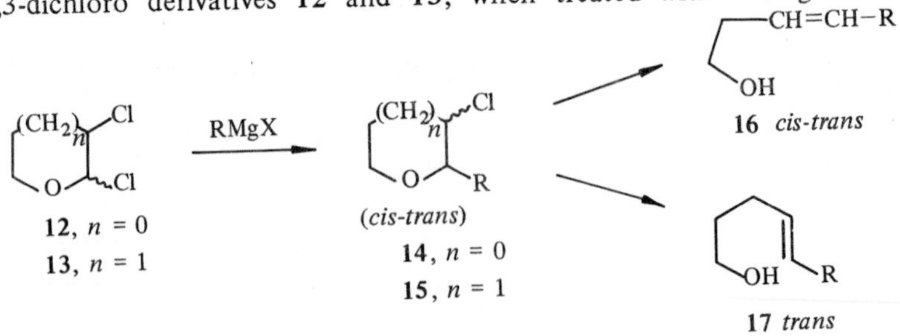

12, $n = 0$
13, $n = 1$

(*cis-trans*)
14, $n = 0$
15, $n = 1$

16 *cis-trans*

17 *trans*

led to a *cis-trans* mixture of 2-alkyl-3-chloro derivatives **14** and **15**. Reductive cleavage with sodium in ether provided the unsaturated alcohols extended by four (**16**) or five (**17**) carbons, respectively. Of particular interest is the fact that the *cis-trans* pyran mixture (**15**) gave only the *trans*-alcohol **17**, whereas the *trans*-furan **14** gave only the *trans*-alcohol (**16**). The *cis*-furan gave a mixture of *cis-trans*-olefinic alcohols **16**. Of course, simple catalytic hydrogenation con-verted these substances into the four- and five-carbon homologated alcohols. This method was used to prepare specific alkanols (e.g., **18**) in the synthesis of wool wax components[6] in 60% yield.

An elegant synthesis of (+)-6-methyloctanoic acid (**22**) was reported using the tetrahydrofuran approach.[7] By treating 2,3-dichlorotetrahydrofuran with the

Grignard reagent of optically active 1-bromo-2-methylbutane (from the readily available "fusel alcohol"), a *cis-trans* mixture of the 2-alkyl tetrahydrofuran (**19**) was obtained. Reductive cleavage then led to the corresponding *cis-trans* mixture of unsaturated alcohols **20** (recall that only pyrans gave pure *trans* olefinic alcohols). This mixture was of no consequence, since catalytic hydrogenation reduced the isomeric alcohols to a single substance **21** which was then oxidized to the desired optically active acid **22**.

In Chapter 4 (p. 76) the formation of acetylenic alcohols derived from tetrahydrofurans and tetrahydropyrans was discussed. It is noteworthy to reiterate that treatment of 2-alkyl-3-chloro tetrahydrofuran (**23**) with sodium amide rather than the reducing system, sodium-ether, gave acetylenic alcohols (**24**) instead of olefinic alcohols.

Tetrahydrofurans and -pyrans have been reported by Eliel[8] to serve as useful precursors to a variety of hydroxy ethers **25** and **26** which are difficult to prepare by other means. The cleavage of the C–O bond was accomplished

using the "mixed-hydride" reagents (LiAlH$_4$-AlCl$_3$ or LiAlH$_4$-BF$_3$). It is well known that the cyclic ethers are stable to lithium aluminum hydride in the absence of the Lewis acid and it is presumed that the latter initiates the process by formation of an oxonium complex (**27** and/or **28**). Hydride addition to the weakened C–O bond results in the observed products. It is to be noted that if the ring oxygen is complexed, the reaction leads to the hydroxyethers **25**, whereas tetrahydrofuran (or pyran) results if the exocyclic oxygen is complexed. The site of complexation and the resulting mode of ring opening has been found to mainly parallel the polar nature of the R group on the exocyclic oxygen. The bulk of R also appears to play a role in the direction of cleavage. Thus varying the nature of R from primary to secondary to tertiary alkyl results in increasing yields of the hydroxy ethers (entries 1-3).

Entry	R	%
1	n-Bu	10
2	i-Pr	43
3	t-Bu	72
4	PhCH$_2$	nil
5	PhCH$_2$CH$_2$	nil
6	CF$_3$C(CH$_3$)$_2$	nil
7	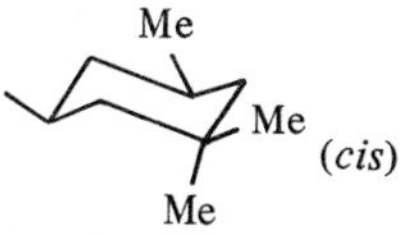 (cis)	38
8	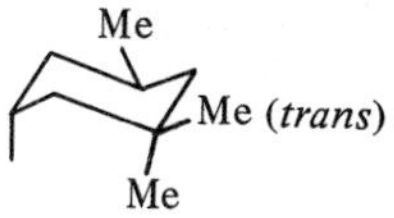 (trans)	62

The presence of electron withdrawing substituents on the pyran causes a drastic decrease in hydroxy ether formation (entries 4-6). The yields obtained by employing configurational isomers (entries 7, 8) in the cyclohexane series were also in accord with the steric effects noted. A similar reaction pattern was observed for the 2-alkoxytetrahydrofurans.

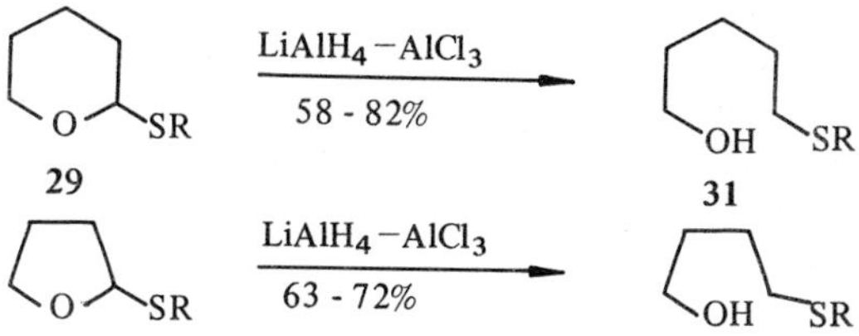

This study was extended[9] to 2-thioalkyltetrahydropyrans (29) and -furans (30) producing hydroxy thioethers 31 and 32, respectively. Unlike the oxygen analogs mentioned above, the sulfur derivatives underwent selective cleavage of the ring C–O bond. Amino alcohols (**34**) were also made available[10] from the

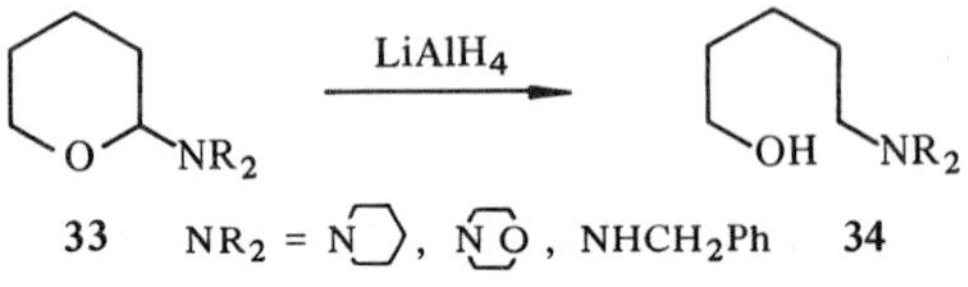

LiAlH4
NR2
OH
NR2
NR2 = N, N O, NHCH2Ph

33 **34**

appropriately substituted tetrahydropyranylamines on treatment with lithium hydride. Ring cleavage in this series proceeded smoothly without the need of a Lewis acid affording the 5-aminopentanols (**34**) in yields ranging from 67 to 92%. The availability of the lone pair on nitrogen for overlap is probably responsible for the weak C–O bond *(A, B)*, hence for the facile reductive

NR2
NR2
OH
R′
N–CH2R′
R
R

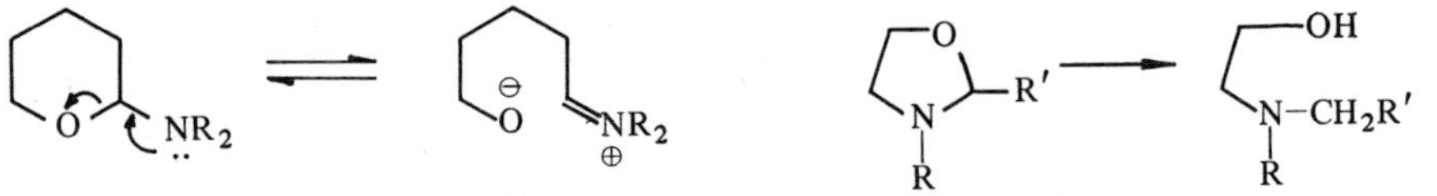

A *B* **35** **36**

cleavage. This process is analogous to the ready ring cleavage of oxazolidines (**35**) to amino alcohols (**36**) with lithium aluminum hydride.[11]

C. From Dihydrothiapyrans

A stereospecific total synthesis of racemic *Cecropia* juvenile hormone **37** was achieved[12] by use of the thiapyran **38** as a starting point. The versatile functionality present in this heterocycle allowed a series of unique elaborations leading, ultimately, to the trien-ol **39**. Epoxidation and oxidation gave the racemic C_{18}-juvenile hormone.

S
BuLi
DABCO
S
Li
Br
S
S
S
BuLi - DABCO

38 **40**

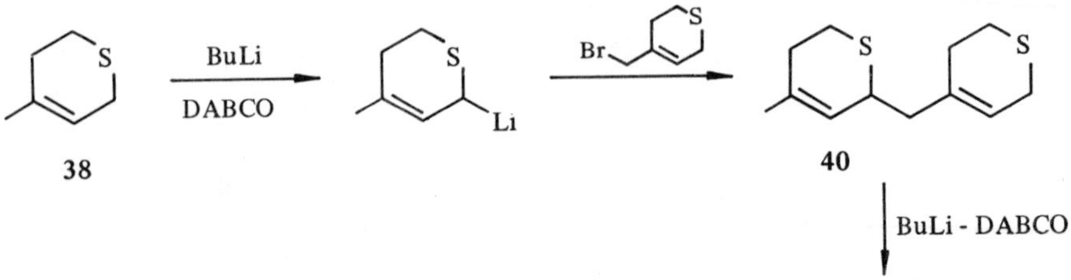

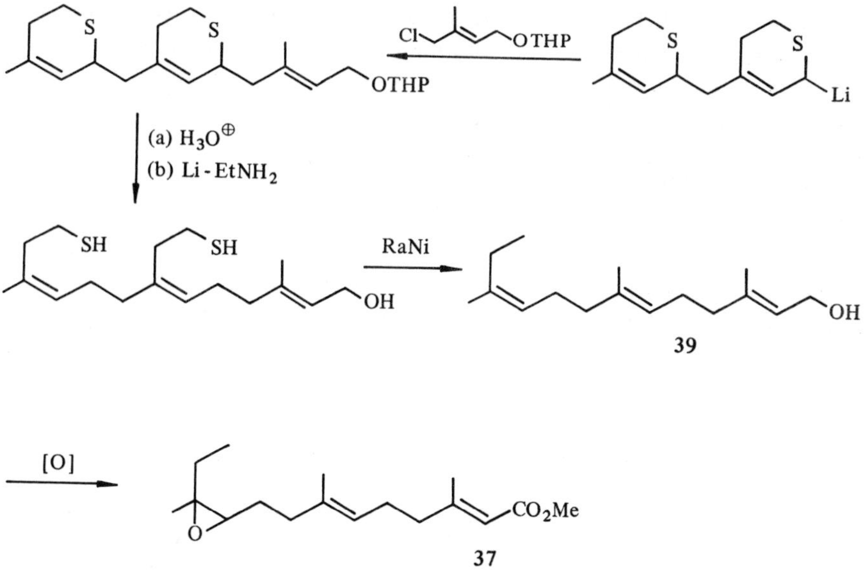

The key intermediate **40** was prepared from the lithio salt of the dihydrothia pyran and the bromomethyl dihydrothiapyran. This technique holds consider- able promise for the preparation of a variety of *trans*-substituted olefinic alcohols.

D. From Furans

The lithio salt of furan (**41**) reacts with organoboranes in an unusual manner leading to cyclic borates **42** in good yield. Oxidation with aqueous hydrogen peroxide affords the *cis*-diols **43** in nearly quantitative yields.[13] The reaction

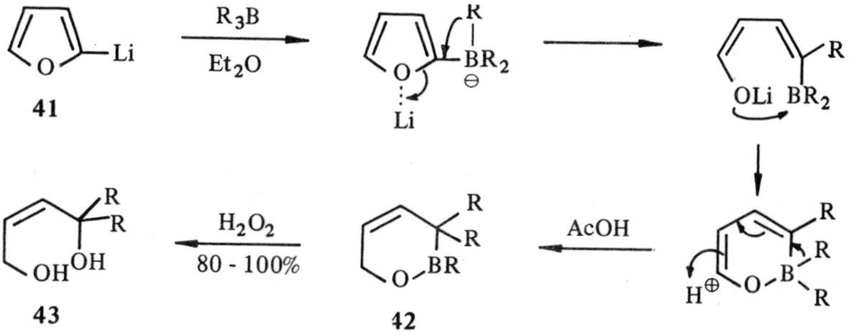

R = Et, *n*-Pr, *n*-Pent, *n*-Hex

failed in cases where the trialkylborane contained highly bulky groups (i.e., cyclopentyl or cyclohexyl). This technique of using a metallated heterocycle in reaction with an organoborane should provide some very interesting possibilities toward the synthesis of functionalized molecules.

E. From Boroxazolines

This novel ring system **44** has been implicated in the conversion of trialkylboranes to trialkylcarbinols **(45)**.[14] Although never isolated, it represents an example of a heterocycle being formed as a transient species whose properties are responsible for a useful transformation. It, therefore, becomes feasible to

R	%
n-Bu	73
n-Heptyl	79
n-Octyl	75

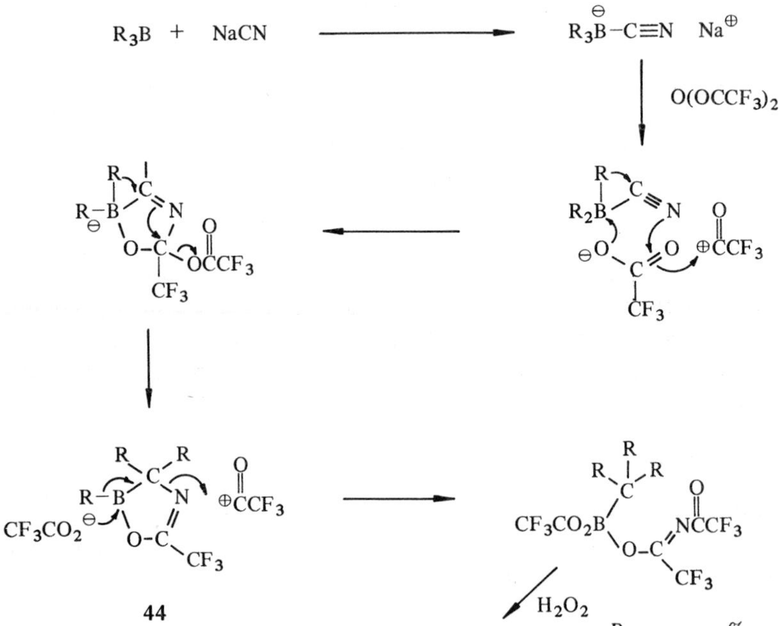

convert olefins to trialkylcarbinols using this simple two-step procedure.

F. From 1,3-Dioxolanes, 1,3-Dioxanes, 1,3-Oxathiolanes, and 1,3-Oxathianes

The mixed-hydride reductive cleavage of tetrahydrofuran and -pyrans discussed previously (Section X) has also been applied[15] to 1,3-dioxolanes (46) and -dioxanes (47). Excellent yields of the corresponding hydroxyethers 48 are obtained in this manner.

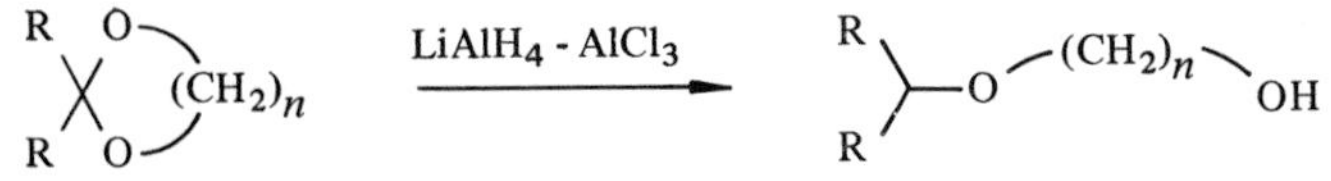

46, n = 2
47, n = 3

48 n = 2,3

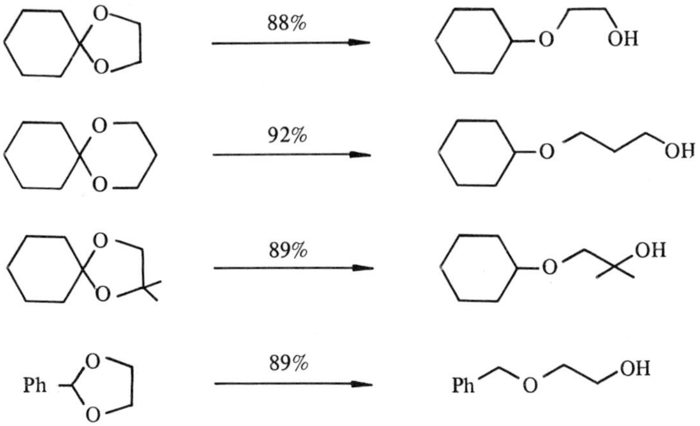

88%

92%

89%

89%

Specific examples are given to illustrate the utility of these heterocycles on treatment with the mixed hydride. The last two examples were reported to produce mixtures of hydroxy ethers due to the unsymmetrical nature of the starting materials. These results appear to depend on a delicate interplay of polar and steric effects.

In an analogous reaction[16] sequence, 1,3-oxathiolanes (49) and -oxathianes (50) have been transformed into their respective hydroxyalkylthioethers (51).

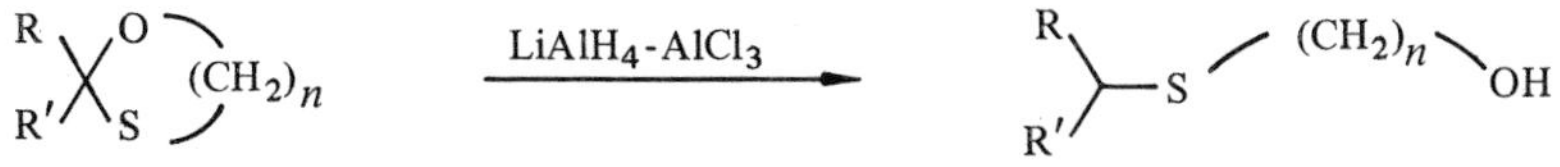

49, n = 2
50, n = 3

51, n = 2,3

R	R'	n	%51
Me	Me	2	83
n-Hexyl	H	2	74
n-Hexyl	Me	2	79
Ph	Me	2	79
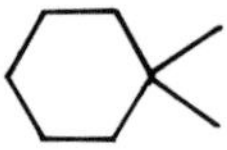		2	91
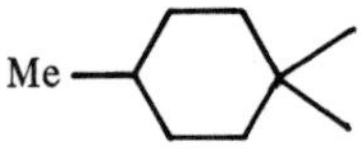		2	80
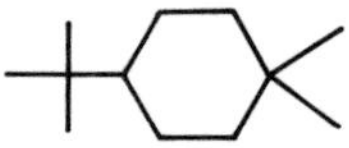		2	80
3-Cholestanylidene		2	80
2-Bornylidene		2	14
Ph	H	3	98
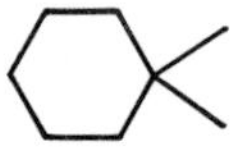		3	91

II.   ESTERS

A.   From 1,3-Dioxolanes and 1,3-Dioxanes

The oxidative cleavage of the titled heterocycles (52, 53) has been reported[17] to lead to bromoethyl (54) and bromopropyl (55) esters. This transformation is effected by treating the dioxolane or dioxane with N-bromosuccinimide in benzene or carbon tetrachloride in the presence of a small quantity of

52, $n = 2$
53, $n = 3$

54, $n = 0$
55, $n = 1$

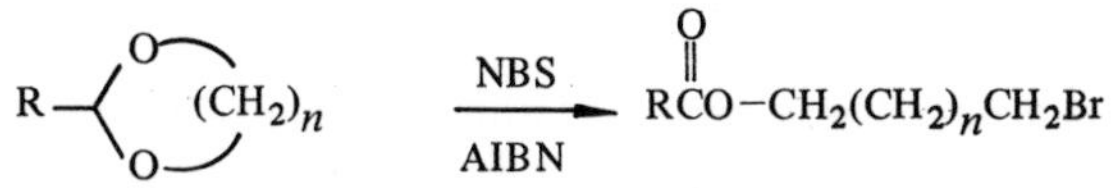

R	n	%54,55
Me	0	66
H	0	30
n-Pr	0	67
Ph	0	76
Ph	1	95

56

57

NBS

54,55

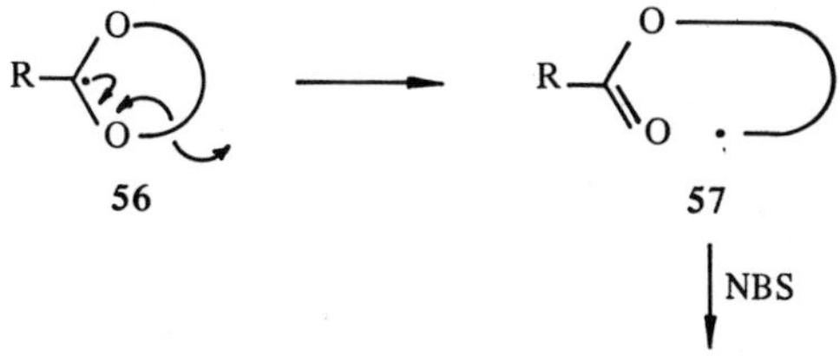

2,2′-azo-*bis*-isobutyronitrile (AIBN). The reaction conditions are rather mild (30-45°) and the yields of products are generally quite satisfactory. The process is suggested to proceed via the radical **56** while the subsequent step (**57** → **54, 55**) is still open to question.

B. From Trichloroisocyanuric Acid

A reagent that has been utilized in the transformation of ethers to esters is the well-known trichloroisocyanuric acid **58**.[18] By stirring a cold mixture of the

R = n-Bu, Et
R′ = n-Bu, Et

58

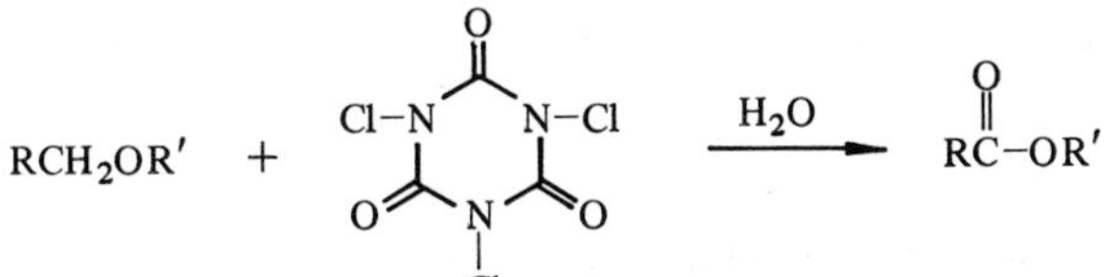

ether and water (∿3 molar excess) with **58**, the esters were produced directly. Thus, n-butylether gave n-butyl acetate in 100% yield while a 49% yield of ethyl acetate was obtained by the oxidation of diethyl ether. This important synthetic transformation, although still in an early developmental stage, should be quite useful if its generality holds in other systems.

C.   From Hydantoins

A neutral, selective reagent for acylating phenolic hydroxyl groups in the presence of alcoholic groups was shown[19] to be a hydantoin derivative (59). Treatment of *p*-hydroxybenzyl alcohol with 3-acetyl-1,5,5-trimethyl hydantoin

59 in anhydrous acetonitrile produced a 91% yield of *p*-acetoxybenzyl alcohol **(60)**. Similar results were obtained with 17β-estradiol which led to the 3-acetoxy derivative **(61)**. This selectivity has been proposed to be the result of steric factors present in the acylating agent. However, the greater acidity of the phenolic hydroxyl should also be taken into account. The crystalline reagent **(59)** was readily prepared by treating 1,5,5-trimethyl hydantoin[20] with acetic anhydride.

D.   From Pyridine and Quinolines

Although many esters are prepared from acid chlorides, alcohols, and pyridine, the role of the latter is not specifically related to its heterocyclic nature. Amines other than pyridines have been successfully employed as a "proton sponge," and only in special cases does the heterocycle play a unique role. One such case is the *O*-acylation of β-keto esters in which pyridine is suggested to serve as a template for the reaction.[21] For example, the reaction of benzoyl chloride with ethyl benzoylacetate **62** in pyridine gives, via the adduct **63**, an 83% yield of ethyl-benzoyloxycinnamate **64**. The intermediacy of **63** has been confirmed

PhCOCl + PhCCH$_2$CO$_2$Et

62

63

64

in the quinoline series by isolation of **65** and its ultimate conversion to quinoline and the benzoyloxycinnamate **64**.

$$\left(\begin{array}{c} \text{PhCOCl} \\ + \\ \text{PhCOCH}_2\text{CO}_2\text{Et} \end{array} \right) \longrightarrow \ \ \overset{\Delta}{\longrightarrow} \ \ \textbf{64}$$

65

The reorganization of the intermediate **65**, which takes place upon heating, was found to be essentially concerted in nature. Using a variety of substituents on both phenyl groups, the effect on the rate of formation of **64** was small. A negligible dependence on the rate with solvent polarity was also noted.[22] The synthetic role of pyridine and quinoline in this process is, therefore, unique, since other tertiary amines give only poor yields of O-acylation.

E. From Pyridazones (Cleavage of Esters)

As mentioned in Chapter 1, the driving force to form a heterocycle is sometimes useful in performing a synthetic transformation. Letsinger[23] found that esters of β-benzoylpropionic acid **66** are cleaved quantitatively by dilute solutions of hydrazine hydrate in pyridine and acetic acid (4:1). The facile formation of the dihydropyridazone **67** is responsible for this rapid release of the alcohol. The conditions for this deblocking of nucleosides are extremely mild with the solution behaving as an essentially neutral medium. Esters of β-cyanoethylphosphoric acid and ethers of *p*-monomethyltriphenyl carbinol are stable to this reagent. Thus the facile condensation of hydrazine and the β-benzoyl propionates serves to liberate a variety of alcohols possessing various sensitive groups.

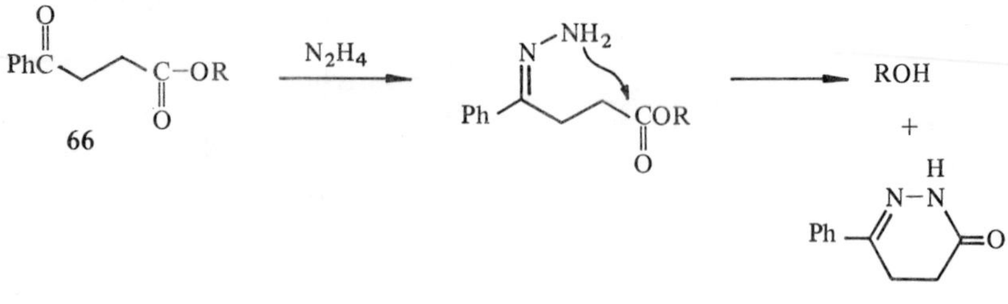

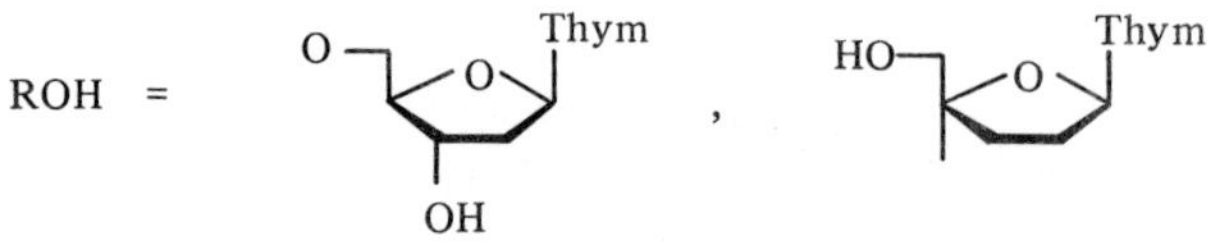

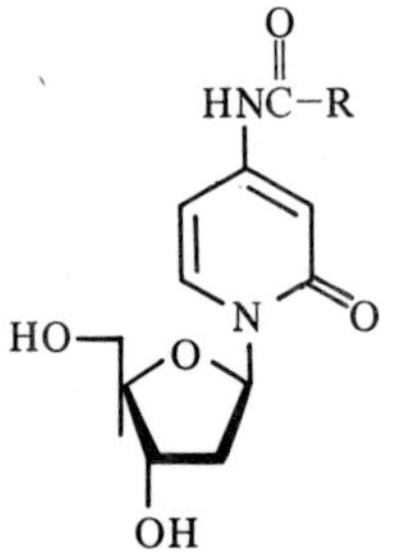

68, R = Ph
69, R = OC—Bu^i

Extension of this study to other common deoxyribonucleosides **68** resulted in cleavage of the *N*-benzoyl group in addition to the β-benzoylpropionic acid blocking group. However, replacing **68** with the isobutyloxycarbonyl group **69** allowed the hydrazine to pyridazone cleavage to occur without affecting the amide function.

III.   MERCAPTANS

A.   From Oxathiolanes and Oxathianes

In his extensive study of reductive cleavage of oxathiolanes (70) and oxathianes (71), Eliel showed that hydroxythio ethers (72) are formed when these hetero-

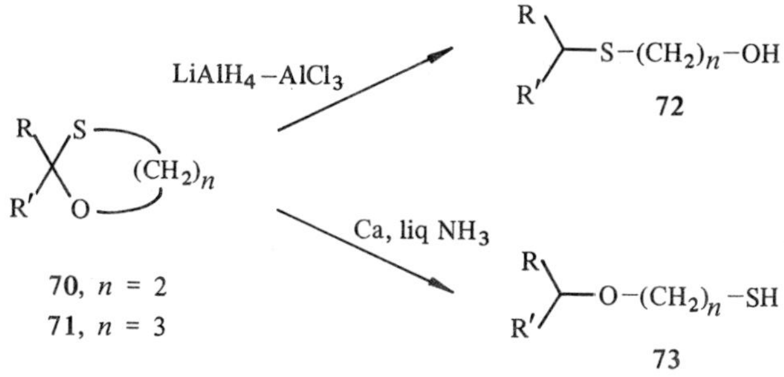

cycles are treated with the mixed hydride reagent (LiAlH$_4$ - AlCl$_3$; sec IF). On the other hand,[24] the path of the reductive cleavage may be reversed by employing calcium in liquid ammonia producing the mercapto ethers 73. Thus alcohols or thiols may be produced from 70 and 71 by choice of the appropriate reducing agent.

The process is reported to proceed by the transfer of two electrons from the metal-ammonia solution to the heterocycle leading to intermediates 74 and 75. The latter is protonated by the ammonia solvent and, upon quenching, produces

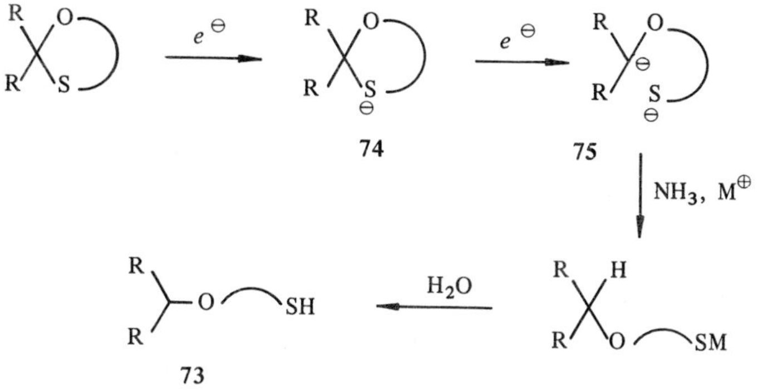

the mercapto ether 73. A variety of mercapto ethers have been prepared via this route and a sampling of these is given below.

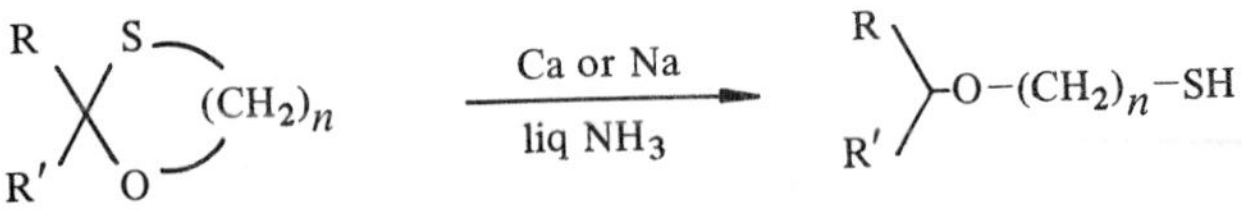

$$\underset{R'}{\overset{R}{\diagdown}}\hspace{-0.5em}\underset{O}{\overset{S}{\diagup}}\hspace{-0.5em}(CH_2)_n \xrightarrow[\text{liq NH}_3]{\text{Ca or Na}} \underset{R'}{\overset{R}{\diagdown}}\hspace{-0.5em}O-(CH_2)_n-SH$$

R	R′	n	%
i-Bu	H	2	7
Me$_3$C	H	2	22
PhCH$_2$	H	2	73
i-Pr	Me	2	72
n-Hex	Me	2	41
	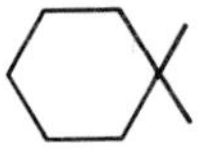	2	49
	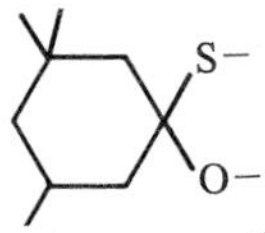	2	61
	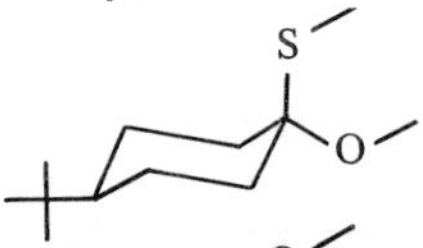	2	80
	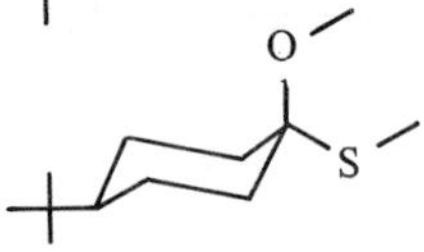	2	61
i-Bu	H	3	66
Me$_3$C	Me	3	89
		3	85
	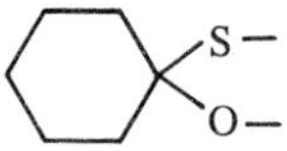	3	70
	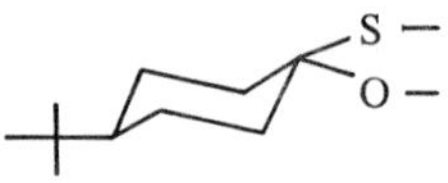	3	85
		3	∼30

144

B. From 1,3-Dithiolanes and 1,3-Dithianes

Just as the oxathianes were reductively cleaved to mercapto ethers with dissolved metal ammonia solutions, the related dithiolanes (76) and dithianes (77) were transformed into mercaptothio ethers (78).[25] A number of these products were prepared in good yield, and this method represents a distinct improvement over earlier methods. With the availability of a wide variety of dithiane

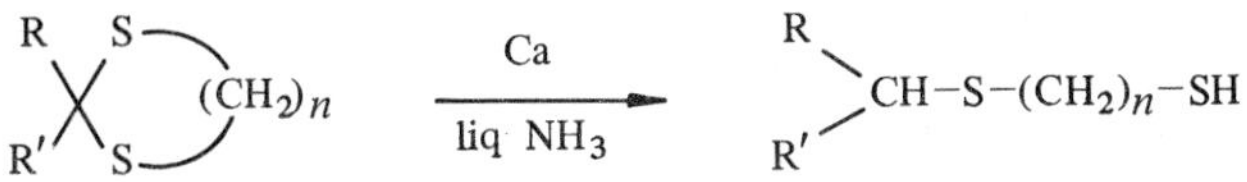

76, $n = 2$
77, $n = 3$

R	R'	n	%
n-Pr	H	2	91
Et	Et	2	90
PhCH$_2$	H	2	94
PhCH$_2$	Me	2	76
PhCH$_2$CH$_2$	H	2	94
i-Bu	H	3	85
HOCH$_2$CH$_2$	H	3	83

derivatives (77) via metallation and alkylation of the parent heterocycle,[26] this approach to mercaptans takes on added importance.

C. From 1,3-Thiazolidines and 1,3-Thiazanes

Metal hydrides reduce thiazolidines (79)[27] and thiazanes (80)[28] to the β-amino mercaptans (81) and the γ-amino mercaptans (82). The reactions are carried out using lithium aluminum hydride in ether and the products are obtained in good yields. In the case of the thiazolidines (79), the use of excess lithium aluminum

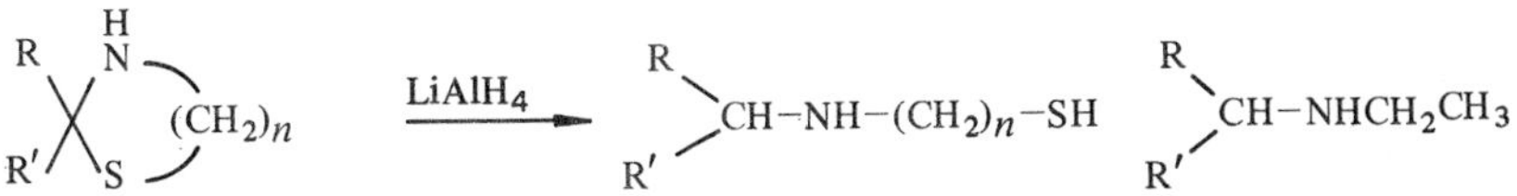

79, $n = 2$
80, $n = 3$

81, $n = 2$
82, $n = 3$

83

R	R'	n	% 81, 82
n-Bu	H	2	60
Ph	H	2	70
Ph	Me	2	56
			57
	3-Cholestanylidene	2	75
Et	H	3	53
Ph	H	3	56
Ph	Me	3	66
	Cyclohexylidene	3	55
	4-t-Butylcyclohexylidene	3	59

hydride must be avoided since hydrogenolysis of the thiol group occurs leading to **83**. This sequential reaction was not observed in the thiazane **(80)** series. Sodium borohydride was also found to be effective in transforming **79** and **80** into their corresponding alkylaminomercaptans.[29] It is noteworthy that the reductive cleavage of **79** and **80** appears to proceed by direct hydride attack on C-2 rather than through its proposed[30] open chain tautomer **84**.

79, 80   84   81, 82   85

This was verified[28] by showing that the N,2-diethyl thiazane **(85)** reduced as smoothly as the NH derivative **(80)**. Furthermore, the absence of the C=N link was noted in the starting thiazanes and thiazolidines when examined by infrared and ultraviolet techniques. However, this does not exclude the presence of 84 in small concentrations in a rapidly equilibrating mixture. Since the N-ethyl thiazane **(85)** cannot exist in tautomeric forms of comparable energies, direct hydride attack is probably the correct path to ring cleavage.

IV.   ETHERS

A.   From Aryloxypyridinium Salts

The readily available N-aryloxypyridinium salts **86**, derived from addition of aryldiazonium salts to pyridine N-oxide,[31] were recently reported[32] to decompose, on heating, to diaryl ethers **88** in moderate yields (20-47%). The reactive intermediate is suggested to be the aryloxenium ion **87** based on (a) ortho-para ratios obtained for **88**, (b) product distribution when compared to known radical precursors, and (c) reactions with a typical cation trap producing the benzoxazole **90**. It is important to point out that the decomposition of N-methoxy pyridinium salts (Chapter 5, Section B-2) proceeds through methoxyl radicals presumably because they lack the ability to delocalize the electron deficiency on oxygen (e.g., **87**). Although diarylethers **88** could be reached via

the classical approach using p-nitrophenyl halides and sodium phenolates, the direct aryloxylation of aromatic substrates opens new avenues to this class of compounds. The generation of substituted aryloxenium ions **87** containing trifluoromethyl groups has also been successfully implemented, and further studies should widen the scope of this interesting method.

REFERENCES

1. R. Grashey, R. Huisgen, and H. Leitermann, *Tetrahedron Lett.*, No. 12, 9 (1960).
2. N. A. LeBel and J. J. Whang, *J. Am. Chem. Soc.*, **81**, 6334 (1959); N. A. LeBel, M. E. Post, and J. J. Whang, *ibid.*, **86**, 3759 (1964).

3. N. A. LeBel and E. G. Banucci, *J. Org. Chem.*, **36**, 2440 (1971).

4. A. Lablache-Combier and M. L. Villaume, *Tetrahedron*, **24**, 6951 (1968).

5. L. Crombie and S. H. Harper, *J. Chem. Soc.*, 1707, 1714 (1950).

6. F. W. Hougan, D. Ilse, D. A. Sutton, and J. P. de Villiers, *ibid.*, 98 (1953).

7. L. Crombie and S. H. Harper, *ibid.*, 2685 (1950).

8. E. L. Eliel, B. E. Nowak, R. A. Daignault, and V. G. Badding, *J. Org. Chem.*, **30**, 2441 (1965).

9. E. L. Eliel, B. E. Nowak, and R. A. Daignault, *ibid.*, **30**, 2448 (1965).

10. E. L. Eliel and R. A. Daignault, *ibid.*, **30**, 2450 (1965).

11. N. G. Gaylord, *Experientia*, **10**, 351 (1954).

12. K. Kondo, A. Negishi, K. Matsui, D. Tunemoto, and S. Masamune, *Chem. Commun.*, 1311 (1972).

13. A. Suzuki, N. Miyaura, and M. Itoh, *Tetrahedron*, **27**, 2775 (1971).

14. A. Pelter, M. G. Hutchings, and K. Smith, *Chem. Commun.*, 1048 (1971).

15. E. L. Eliel, V. G. Badding, and M. N. Rerick, *J. Am. Chem. Soc.*, **84**, 2371 (1962).

16. E. L. Eliel, L. A. Pilato, and V. G. Badding, *ibid.*, **84**, 2377 (1962); E. L. Eliel, E. W. Della, and M. Rogic, *J. Org. Chem.*, **30**, 855 (1965).

17. J. D. Prugh and W. C. McCarthy, *Tetrahedron Lett.*, 1351 (1966).

18. E. C. Juenge and D. A. Beal, *ibid.*, 5819 (1968).

19. O. O. Orazi and R. A. Corral, *J. Am. Chem. Soc.*, **91**, 2162 (1969).

20. O. O. Orazi and R. A. Corral, *Experientia*, **21**, 508 (1965).

21. R. L. Stutz, C. A. Reynolds, and W. E. McEwen, *J. Org. Chem.*, **26**, 1684 (1961); P. E. Wright and W. E. McEwen, *J. Am. Chem. Soc.*, **76**, 4540 (1954); W. R. Gilkerson, W. J. Argersinger, and W. E. McEwen, *ibid.*, **76**, 41 (1954).

22. W. E. McEwen and G. F. Pollard, personal communication.

23. R. L. Letsinger and P. S. Miller, *J. Am. Chem. Soc.*, **91**, 3356 (1969).

24. E. L. Eliel and T. W. Doyle, *J. Org. Chem.*, **35**, 2716 (1970).

25. B. C. Newman and E. L. Eliel, *ibid.*, **35**, 3641 (1970).

26. D. Seebach, *Synthesis*, **1**, 17 (1969).

27. E. L. Eliel, E. W. Della, and M. M. Rogic, *J. Org. Chem.*, **27**, 4712 (1962).

28. E. L. Eliel and J. Roy, *ibid.*, **30**, 3092 (1965).

29. J. C. Getson, J. M. Greene, and A. I. Meyers, *J. Heterocycl. Chem.*, **1**, 300 (1964).

30. E. D. Bergmann and A. Kaluszyner, *Rec. Trav. Chim.*, **78**, 327 (1959).

31. R. A. Abramovitch, S. Kato, and G. M. Singer, *J. Am. Chem. Soc.*, **93**, 3074 (1971).

32. R. A. Abramovitch, M. Inbasekaran, and S. Kato, *ibid.*, **95**, 5428 (1973).

8 AMINES

I. FROM AZIRIDINIUM SALTS

New synthetic methods for obtaining stable aziridinium salts developed by Leonard and co-workers[1] now make this interesting heterocyclic system available for a variety of transformations. The most general route to stable aziridinium salts **2** is by the reaction of diazomethane with an imminium perchlorate or tetrafluoroborate **1** in dichloromethane at 0°.

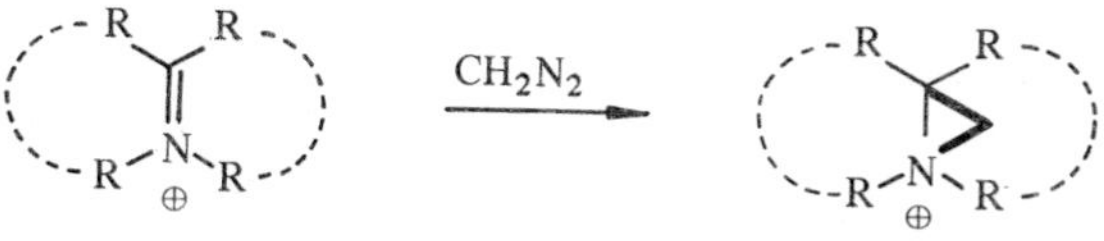

Other approaches to aziridinium salts were also developed from β-haloethylamines (**3**, **4**) and by quaternization of aziridines **5**. The success of these methods lies mainly in the fact that the aziridinium gegenions ($BF_4^{\ominus}$, $ClO_4^{\ominus}$) are poorly nucleophilic, thus inhibiting ring opening of the highly strained heterocycle.

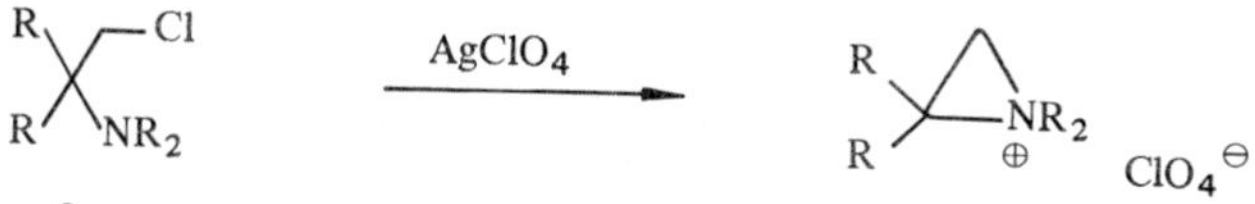

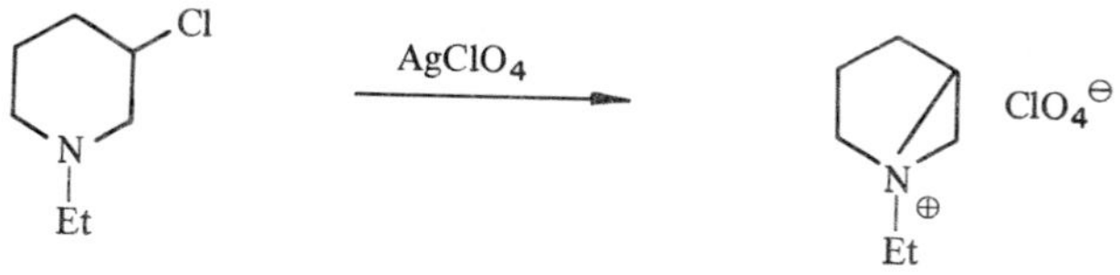

The utility of aziridinium salts in heterocyclic syntheses has been amply demonstrated. Their ability to undergo dipolar additions with aldehydes, ketones, nitriles, nitrones, and isocyanates has already been reviewed[1] and is outside of the scope of this work. With regard to the preparation of amines, the aziridinium salts are rather useful in introducing the aminoethyl group ($-CH_2CH_2NH_2$) into various nucleophilic reagents.

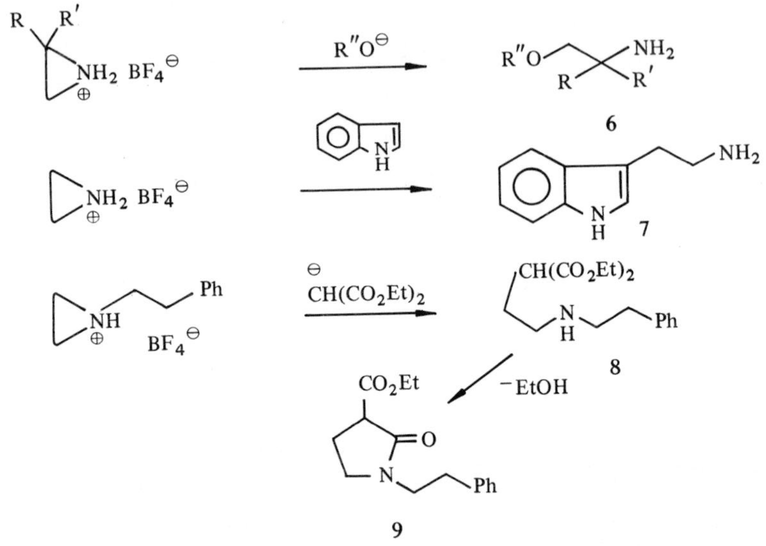

Since it is now feasible to prepare and utilize stable aziridinium salts, their reaction with alkoxides leads to β-alkoxy amines (**6**). The nucleophilic 3-position of indole may be aminoethylated with aziridinium salts affording tryptamine **7** in 80-90% yield. Reaction with diethyl malonate carbanion leads to the amino-ester **8** which rapidly cyclizes to the lactam **9**. Hydrogenation of the $C-N^{\oplus}$ linkage in aziridinium salts has provided a route to a,a-dimethylamines **10**.

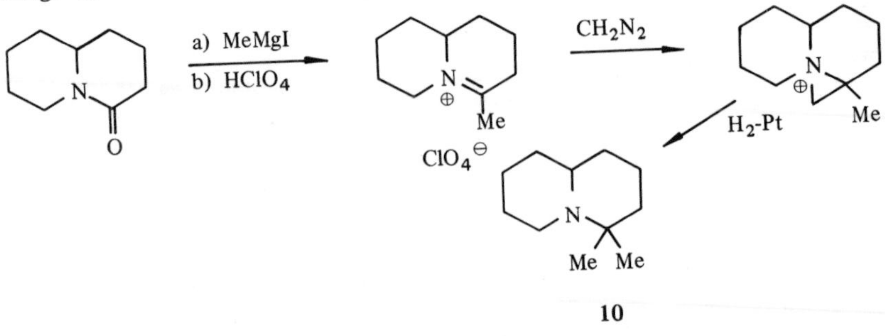

Aziridinium salts (**11**) have been formed intramolecularly and are used to prepare, in a stereospecific manner, the pyrrolidine **12** and the piperidine **13**

11 **12** **13**

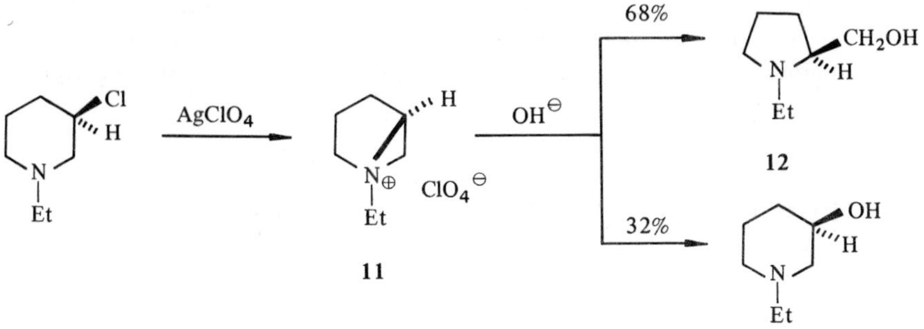

alcohols.[2] Since the latter are 95 ± 5% configurationally pure products, the aziridinium ions must all be opened by a concerted process.

II. FROM THIAZOLIDINES

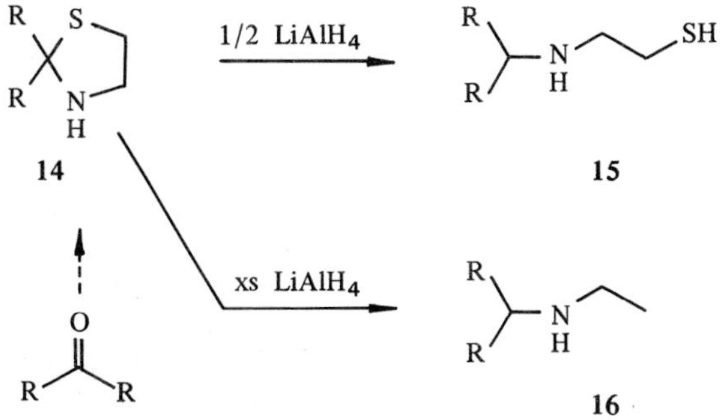

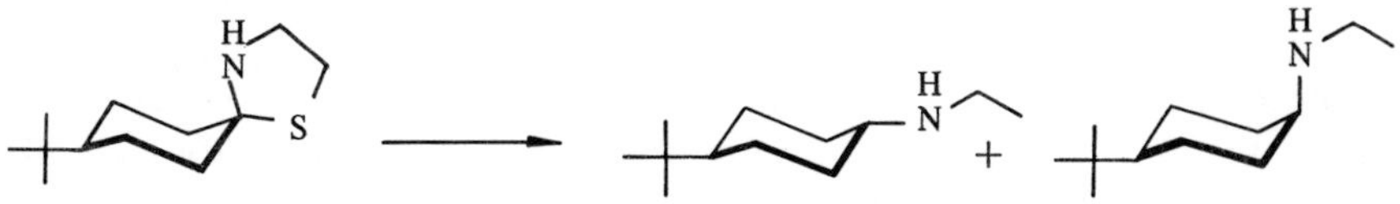

In the previous chapter (p. 145), the reaction of thiazolidines **14** with a stoichiometric quantity of lithium aluminum hydride led to good yields of

14 **15** **16**

17 **18** (85%) **19** (15%)

amino mercaptans **15**. If an excess of the hydride reagent is used, the major product is the dialkyl amine **16** resulting from hydrogenolysis of the thiol function.[3] This behavior was demonstrated by the efficient transformation of the thiazolidine **17** into an 85:15 mixture of the cyclohexyl amines **18** and **19**. Since thiazolidines are readily prepared from carbonyl compounds and 2-mercaptoethyl amine, this method represents a useful conversion of the carbonyl group to the amino group.

III. FROM 1,3,2,4-DIOXATHIAZOLE S-OXIDES

An efficient and convenient synthesis of isocyanates (**20**), which are excellent precursors to amines (**21**), has been achieved[4] by passing through the titled heterocycle (**22**). The latter is formed from hydroxamic acids **23** and thionyl

chloride or by dipolar addition of sulfur dioxide to nitrile oxides **24**. By heating the intermediate heterocycle **22** above its melting point or in an inert solvent (n-octane, 100°), a smooth rearrangement ensues, producing the isocyanates in quantitative yield. Simple hydrolysis of isocyanates leads to the primary amines, also in excellent yields. The utility of this interesting process has been exemplified by the following transformations which took place in 70-80% overall yields. It should be noted that this process provides an alternative to other carbon degradative methods (Hofmann, Wolfe, Curtius, etc.).

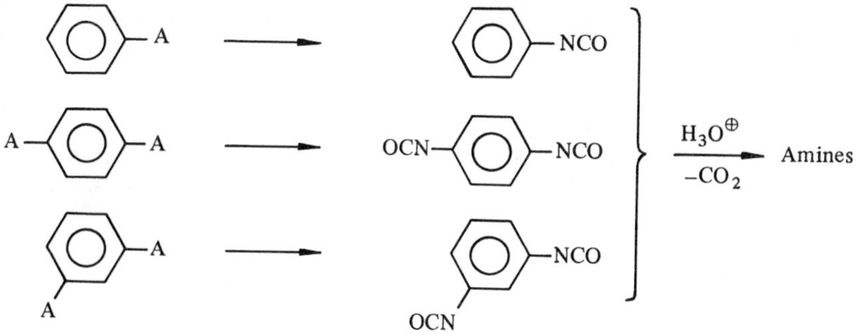

$$A = (C\equiv N \rightarrow O \quad \text{or} \quad \overset{O}{\overset{\|}{C}}-NHOH)$$

IV.   FROM 4,5-DIPHENYL-4-OXAZOLIN-2-ONES

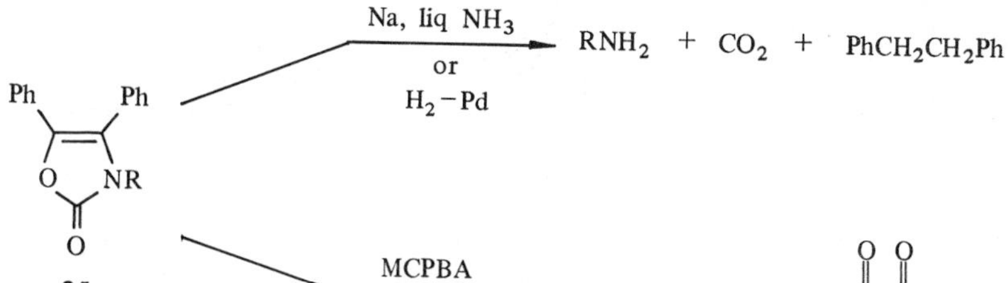

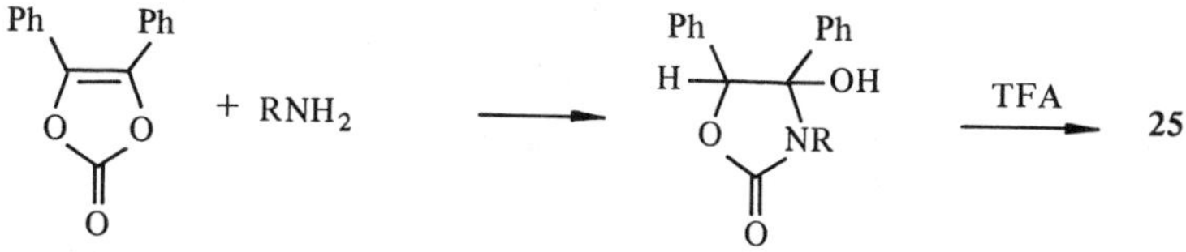

Sheehan[5] recently found that the oxazolin-2-ones **25** may be readily cleaved under reductive (sodium in liquid ammonia or catalytic hydrogenation) or oxidative (*m*-chloroperbenzoic acid) conditions to primary amines. This behavior was capitalized upon by utilizing **25** as a protecting group, since it was stable to a variety of conditions usually employed to remove amine protecting groups. The oxazolin-2-one derivative is prepared by treating a primary amine with the cyclic carbonate of benzoin **26** in dimethyl formamide and dehydration of the initially formed carbinol **27**. Regeneration of the amine under the oxidative or reductive conditions occurs in good yield. If the amine is optically active, no racemization was observed either in the protecting step or the de-blocking reaction. The protecting group abilities of **25** were exhibited with phenethyl amine, L-alanine, L-valine, and the dipeptide, L-Ala-GlyOEt.

The oxazolin-2-one **25** also has the potential of serving as a source of primary amines from alkyl halides, in a manner analogous to the Gabriel synthesis. Thus the parent heterocycle **25a** may be alkylated[6] to the N-alkyl derivative **25**

and the primary amine then released by the oxidative or reductive cleavage. This would avoid the necessity of generating the amine under acidic or alkaline conditions and may circumvent racemization in optically active compounds.

V. FROM IMIDAZOLIUM SALTS

Treatment of imidazolium salts **28** with excess sodium borohydride has been shown to lead to unsymmetrically substituted ethylene diamines **29** and **30**.[7] The results described in the table show that sodium borohydride cleaves the imidazolium ring predominantly between C-2 and nitrogen bearing the benzyl group. In order to assess the role of the N-benzyl substituent, the N-phenethyl nidazolium salt (last entry in the table) was also subjected to borohydride

R	R'	%	%
		29	30
Me	H	85 (71)	8
Et	H	81 (50)	15
n-Pr	H	80 (53)	16
Me	Me	93 (65)	6
Et	Me	77 (38)	23
Me	H	91 (61)	3
(N-Phenethyl)			

reduction. In this case cleavage occurred to a higher degree of regioselectivity (91:3) than that observed for the N-benzyl derivatives. These results suggest that the approach of hydride is controlled by steric factors and ring cleavage takes place from the more accessible side of the imidazolium salt. It is almost certain that more equal mixtures of **29** and **30** would result if R on **28** were varied to groups larger than ethyl or n-propyl. Nevertheless, the diamines **29** were formed with reasonable efficiency and could be isolated in yields given in parentheses.

VI. FROM 2-HYDROXYPYRIMIDINES

The finding[8] that 4,6-dimethyl-2-hydroxypyrimidine **31** can be stereospecifically reduced to the cis-dimethyl cyclic urea **32** and ultimately cleaved to meso-1,3-diaminopentane represents a novel approach to diamines. The fact that the pyrimidine was formed smoothly by the reaction of acetylacetone **34** with urea extends this method to the preparation of diamines from 1,3-diketones. Reduction of oxime derivatives of diketones gives mixtures of diastereoisomers and is, therefore, unsuitable as a technique for stereospecific syntheses. The aforementioned approach to diamines achieves more significance by a recent report[9] that elaboration of 4-methyl-2-hydroxypyrimidines **35** is now readily accomplished. Thus treatment of the latter with butyllithium or sodium amide gives rise to the dianion **36** which is efficiently alkylated by various electrophiles to **37**. This behavior, coupled with the fact that 2-hydroxypyrimidines are reduced and cleaved to 1,3-diamines, now provides a synthetic sequence of considerable interest and scope (Scheme 1). One may visualize

R$_1$	R$_2$	R$_3$
Me	Ph	$-CH_2\underset{\underset{OH}{\mid}}{CH}Ph_2$
Me	Ph	$-CH_2CH_2Ph$
Ph	H	$-CH_2CH_2Ph$
Ph	H	(HO-cyclohexyl)$-CH_2$
Ph	H	$-CH=CPh_2$

Scheme 1

4,6-dimethyl-2-hydroxypyrimidine **31** as a "masked 1,3-diaminopentane" **38** capable of being homologated via reactions **35** → **36** → **37** and then reduced and hydrolyzed to the diamines **39**.

VII. FROM QUINAZOLINES

In Chapter 5, a discussion concerning the *construction* of aromatic nuclei from heterocycles was presented. The *direct* introduction or removal of substituents in aromatic compounds was also mentioned. However, substitution of aromatic substituents has been left to the particular class of functional group in question.

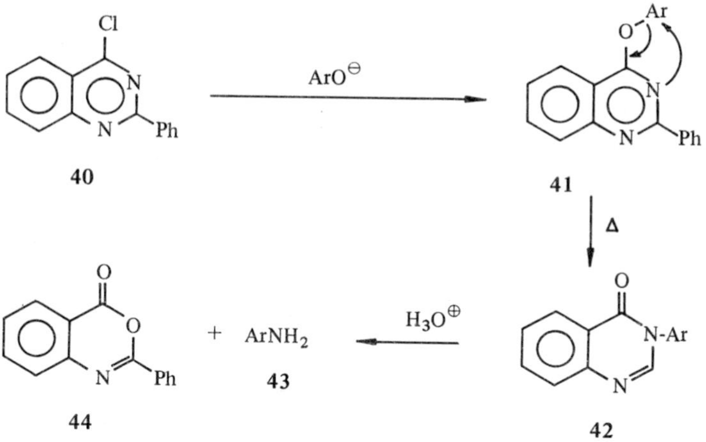

Ar	Yield, % 43
Ph	71
2,4-Cl$_2$Ph	64
2,3,6-Me$_3$Ph	70
4-NO$_2$Ph	42

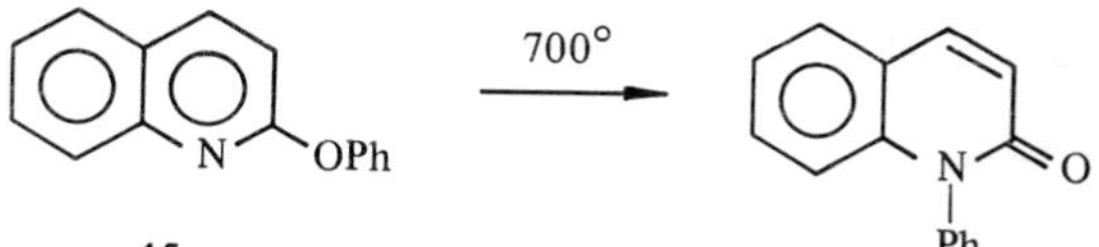

While most direct substitution of groups bound to an aromatic nucleus is limited to diazonium salts or aromatics containing strong electron withdrawing substituents, the use of the quinazoline **40** has opened new routes to this type of transformation.[10] A general procedure for the conversion of aromatic hydroxy compounds to the corresponding amine has been noticeably absent from organic synthesis. By condensation of a sodium phenoxide with 4-chloro-2-phenylquinazoline **40** in dimethylformamide, a high yield (69-82%) of the 4-aryloxyquinazoline **41** is obtained. The second step in this sequence relies on a thermal O to N aryl migration to **42** which occurs at 295-325° in mineral oil as a solvent. The process has been suggested to take place via an intramolecular migration. This is based on kinetic and substituent studies although simple crossover experiments were not performed. This type of rearrangement was first observed by Chichibabin[11] in 1924 in the quinoline (**45**) and pyridine (**46**) series and has gone virtually unnoticed although it appears to be quite general.[12] Furthermore, there is a formal resemblance of this rearrangement to that reported in 1926 by Chapman[13] for the formation of N,N-diarylbenzamides (47) from N-arylbenzimidates.[14]

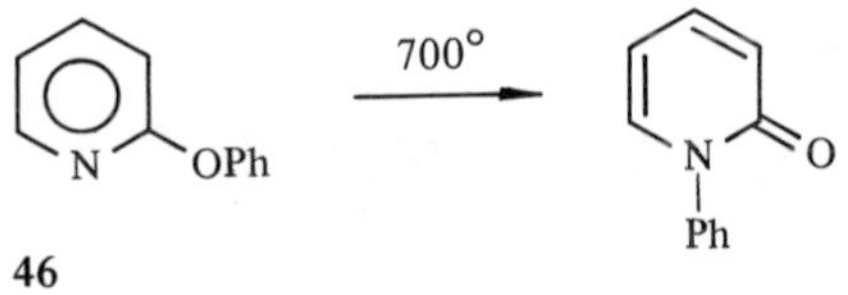

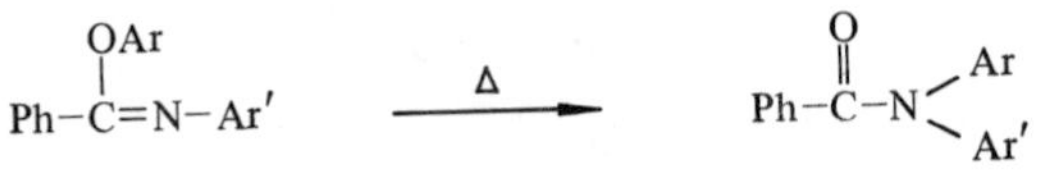

Finally, liberation of the aromatic amine 43 is accomplished by mild alkaline or acidic hydrolysis of the rearranged quinazolone 42 giving rise also to the oxazinone 44. This aniline synthesis should lend itself to the preparation of a variety of aryl amines as well as N^{15} labeled derivatives. The latter would be prepared by initially forming the quinazoline 40 (N^{15}) by reaction of benzoxazinone 44 and N^{15}-ammonium hydroxide followed by phosphorus pentachloride.

Application of this method to the steroid field has been reported.[15,16] By utilizing 40 in the manner described above, the estrone derivatives 48 and 49 have been converted to their amino analogs 50 and 51.

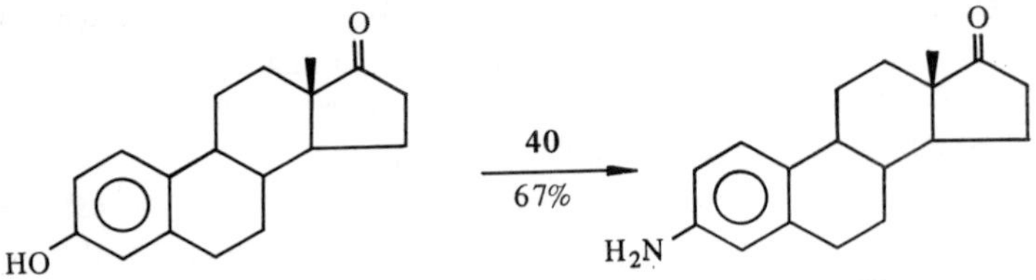

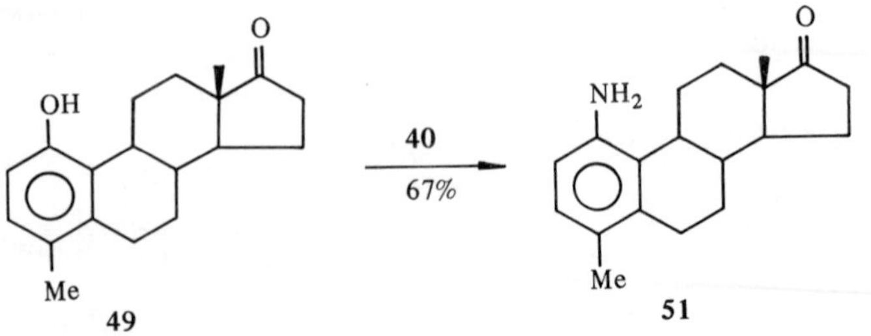

REFERENCES

1. D. R. Crist and N. J. Leonard, *Angew. Chem. Int. Ed.,* **8,** 962 (1969).
2. C. H. Hammer and S. R. Heller, *Chem. Commun.,* 919 (1966).
3. E. L. Eliel, E. W. Della, and M. M. Rogic, *J. Org. Chem.,* **27,** 4712 (1962).
4. E. H. Burke and D. D. Carlos, *J. Heterocycl Chem.,* **7,** 177 (1970).
5. J. C. Sheehan and F. S. Guziec, *J. Am. Chem. Soc.,* **94,** 6561 (1972); *J. Org. Chem.,* **38,** 3034 (1973).
6. R. Gompper, *Berichte,* **89,** 1748 (1956).
7. E. F. Godefroi, *J. Org. Chem.,* **33,** 860 (1968).
8. R. O. Hutchins and B. E. Maryanoff, *ibid.,* **37,** 1829 (1972).
9. J. F. Wolfe and T. P. Murray, *Chem. Commun.,* 336 (1970).
10. R. A. Scherrer and H. R. Beatty, *J. Org. Chem.,* **37,** 1681 (1972).
11. A. E. Chichibabin and N. P. Jeletsky, *Chem. Ber.,* **57,** 1158 (1924).
12. R. A. Scherrer, U. S. Patent 3,238,201 (1966); *CA.,* **64,** 17614b (1966); Germ. Patent 1,190,951 (1965); *CA.,* **63,** 4209d (1965).
13. A. W. Chapman, *J. Chem. Soc.,* 1992 (1925).
14. A. W. Chapman, *J. Chem. Soc.,* 1743 (1927); K. B. Wiberg and B. I. Rowland, *J. Am. Chem. Soc.,* **77,** 2205 (1955); J. W. Schulenberg and S. Archer, *Org. React.,* **14,** 1 (1965).
15. D. F. Morrow and M. E. Butler, *J. Org. Chem.,* **29,** 1893 (1964); D. F. Morrow and R. M. Hofer, *J. Med. Chem.,* **9,** 249 (1966).
16. R. B. Conrow and S. Bernstein, *Steroids,* **11,** 151 (1968).

9 CARBONYL COMPOUNDS —ALDEHYDES, KETONES, AND THEIR FUNCTIONALIZED DERIVATIVES

Nowhere is the synthetic utility of heterocyclic compounds more dramatically evident than in the preparation of carbonyl compounds. In fact, it was this particular route to carbonyl compounds that spurred the author to examine the general synthetic utility of heterocycles. There are so many excellent examples to relate in this area that this chapter is organized into five sections dealing with the formation of carbonyl derivatives as a function of changes in carbon-chain length. The following sections deal with the preparation of carbonyl compounds derived via heterocyclic precursors, intermediates, or reagents:

I. No change in carbon length (including carbonyl transpositions).

$$(R-C) \longrightarrow R\overset{|}{C}{=}O$$

$$R-\overset{\overset{\displaystyle O}{\|}}{C}-A \longrightarrow R-\overset{\overset{\displaystyle O}{\|}}{C}-B$$

II. One-carbon change in length.

(R−C) + (C=O) ⟶ R−C−C=O (one carbon increase)

(R−C−C=O) − (C) ⟶ R−C=O (one carbon decrease)

III. Two-carbon change in length.

(R−C) + (C−C=O) ⟶ R−C−C−C=O

IV. Three-carbon change in length.

(R−C) + (C−C−C=O) ⟶ R−C−C−C−C=O

V. Four-carbon change in length.

(R−C−) + (−C−C−C−C=O) ⟶ R−C−C−C−C−C=O

160

In each case the carbonyl compound (aldehyde or ketone) will have undergone the indicated homologation or degradation under the influence of a heterocycle. In Section II there are several examples of a one-carbon degradation leading to aldehydes, but degradative transformations involving two, three, or four carbons are not known except with nonheterocyclic reagents (e.g., alkenes + ozone or peracids).

I. CARBONYL COMPOUNDS WITH NO CHANGE IN CARBON CHAIN LENGTH

$$(R-C) \longrightarrow R-C{=}O$$

$$\underset{\overset{\|}{O}}{R-C-A} \longrightarrow \underset{\overset{\|}{O}}{R-C-B}$$

A. From Pyridines

The direct conversion of alkyl halides containing the CH_2X group to aldehydes via the pyridinium salt 1 and the nitrone 2 was reported by Kröhnke[1] in 1936.

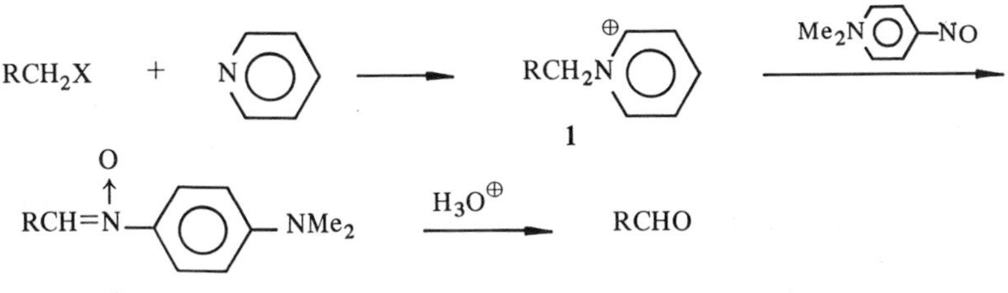

Since then, the method has been widely used in the preparation of a variety of aldehydes in moderate-to-good yields. Alkyl, benzyl, and allyl halides or tosylates with additional functionality have been employed as starting materials.[2] A representative list of examples is given:

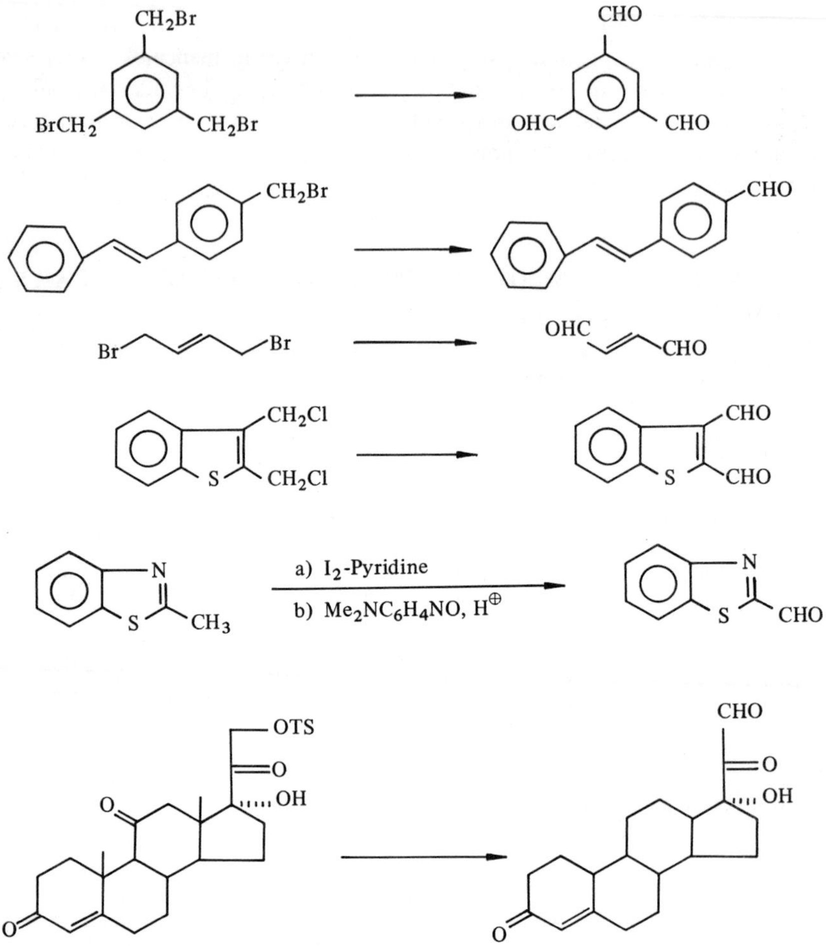

a) I_2-Pyridine

b) $Me_2NC_6H_4NO$, $H^{\oplus}$

Alkyl halides are also converted to aldehydes when treated with pyridine N-oxides (3).[3] Electrophilic attack on the oxygen of the N-oxide is well known,[4] producing the N-alkoxypyridinium salts 4 which are fragmented, in base or by heating, to the aldehydic product.

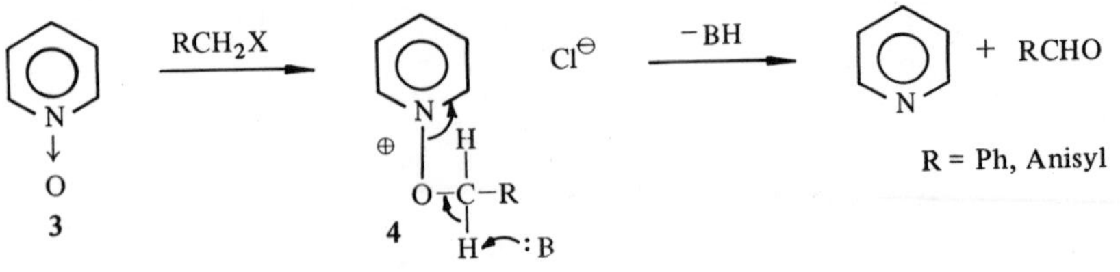

R = Ph, Anisyl

When styrene oxide was allowed to react with 2,6-lutidine *N*-oxide in the presence of perchloric or trifluoroacetic acid, the alkoxyammonium salt **5** was isolated in good yield.[5] This was offered as support for the suggested intermediacy of an alkoxypyridinium salt in the conversion of styrene oxide to phenacyl alcohol **6** on heating with pyridine *N*-oxide.

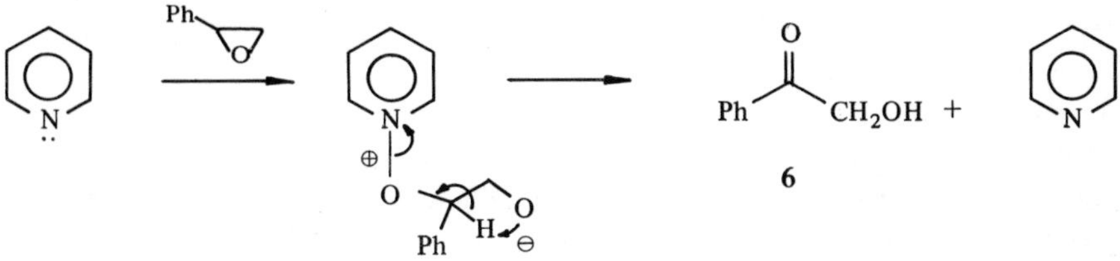

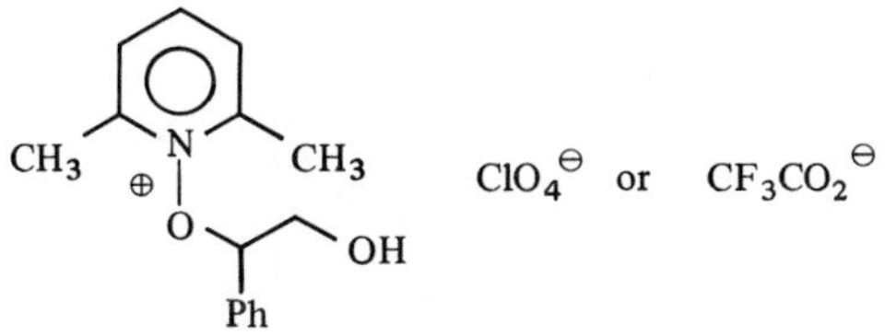

Danishefsky[6] has utilized the pyridine ring as a starting point for a novel annelation of cyclic ketones. Thus 6-vinyl-*a*-picoline **7**, which contains an eight-carbon unit, is transformed via the series of reactions **7 → 8 → 9 → 10 → 11** into the tricyclic ketone **12**. Note that there has been no change in the number

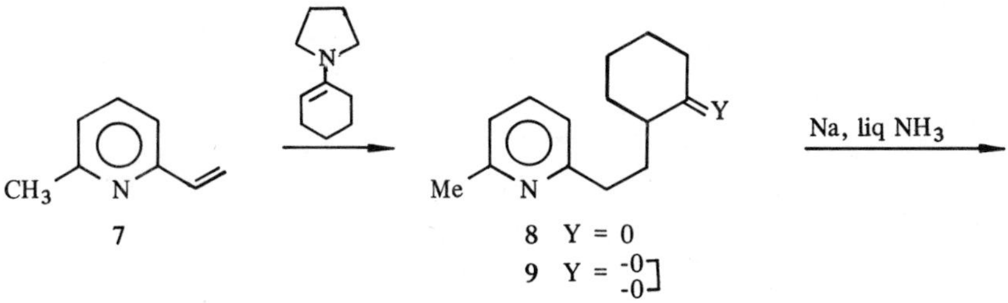

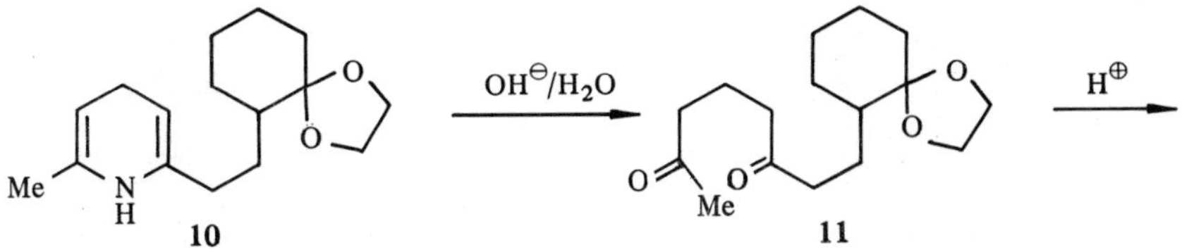

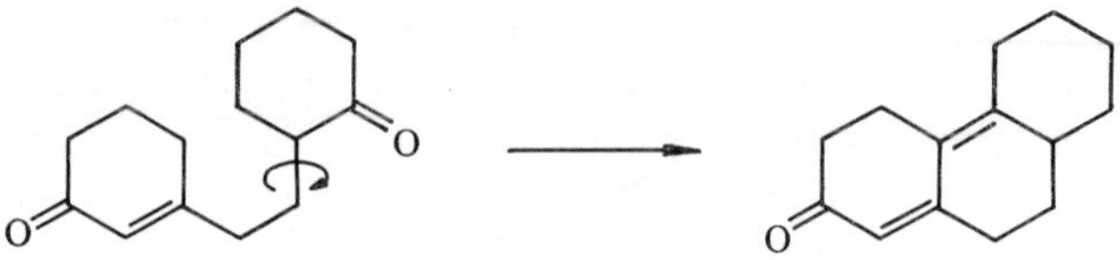

12

of carbon atoms in going from **8** to **12**. This interesting method has recently been applied to a ready synthesis of D-homo-oestrone **14**[7] beginning with the bicyclic ketone **13**. The possibility of obtaining isomeric ketones in this sequence was also examined and some conclusions were reached that appeared to

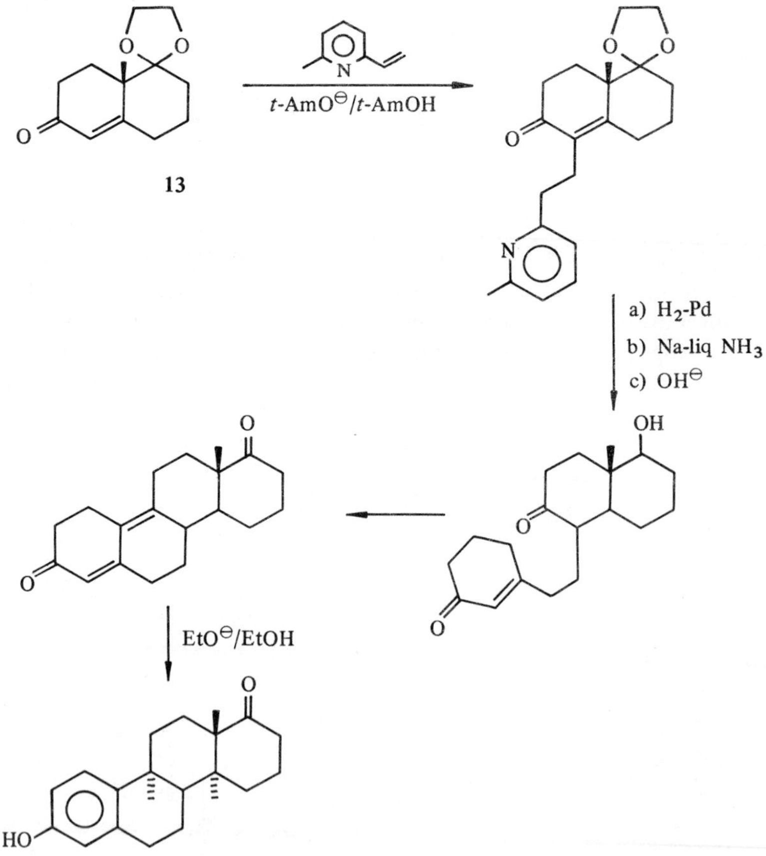

13

a) H_2-Pd
b) Na-liq NH_3
c) $OH^{\ominus}$

EtO$^{\ominus}$/EtOH

14

be dependent on the nature of cyclic ketone produced.[8] In most cases, however, the desired route is favored (i.e., annelated cyclohexanones). Finally, pyridines have been transformed into the open chain conjugate aldehydes 18 by photohydration.[9] The intermediate Dewar pyridine 15 was implicated as an inter-

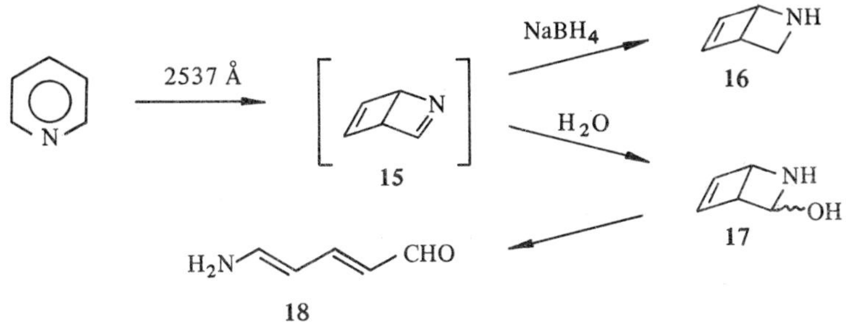

mediate by the actual trapping of its reduction product 16. The amino dienal 18, although not formed in yields sufficient to make this route synthetically attractive, nevertheless is the product of a remarkable transformation. The hydrated 2-azabicyclo(2,2,0) hexene 17 is a likely precursor to the aldehyde. In another interesting study, Olofson[10] showed that the bis-pyridinium salt 19, readily prepared from benzal chloride and pyridine, undergoes facile deuterium

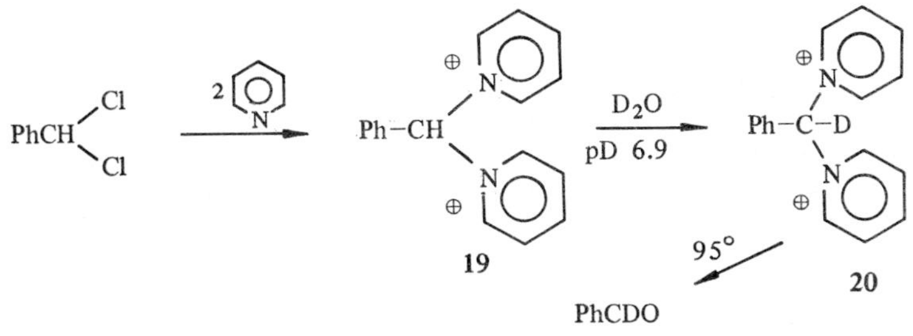

exchange to 20 (via the ylide) and leads to deuteriobenzaldehyde after hydrolysis. The isotopic purity of the deuteriobenzaldehyde thus prepared is 95 ± 1%. However, if hydroxide ion is introduced into an aqueous solution of 19, a product appears that was isolated and characterized as 21. The latter is

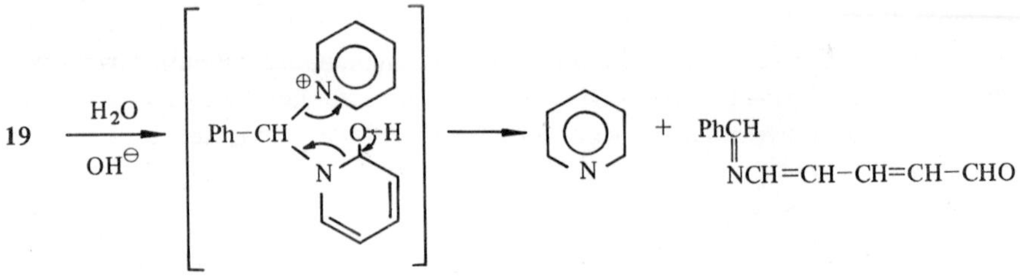

21

readily recognized as the benzaldehyde Schiff base of **18**. Further treatment in water converts **21** into benzaldehyde and pyridine, presumably by base catalyzed cyclization of **18**. These two reports, which have shown that the aminopentadienal **18** (or **21**) may be generated from a pyridine nucleus, should be further explored with regard to enhancing its efficiency and its potential synthetic utility.

B.   From Benzothiazoles

A useful transformation[11] of primary amines to aldehydes and ketones is based on the rapid reaction of benzothiazole-2-carboxaldehyde **22** to the imine **23**. The latter is treated with a catalytic amount of sodium methoxide or 1,4-diazabicyclo(2,2,2) octane (DABCO) in hexamethylphosphoramide which

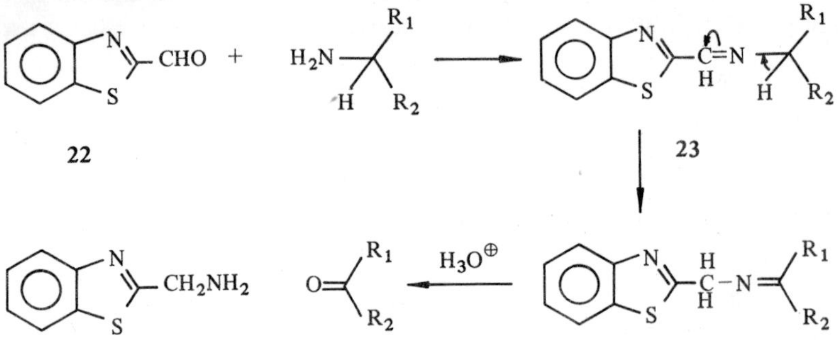

R_1	R_2	%
Ph	H	75
Ph	CH_3	97
Ph	Ph	94
Cyclopentyl		86
Cyclohexyl		69

effects a prototropic shift leading to the carbonyl precursor **24**. Hydrolysis leads to aldehyde or ketone in good yields. This method should provide a useful alternative to other methods[12] that perform the same task and, in particular, those that involve oxidation of primary amines.[13]

C. From 2-Oxazolines

Carboxylic acids have been reduced[14] to their corresponding aldehydes by

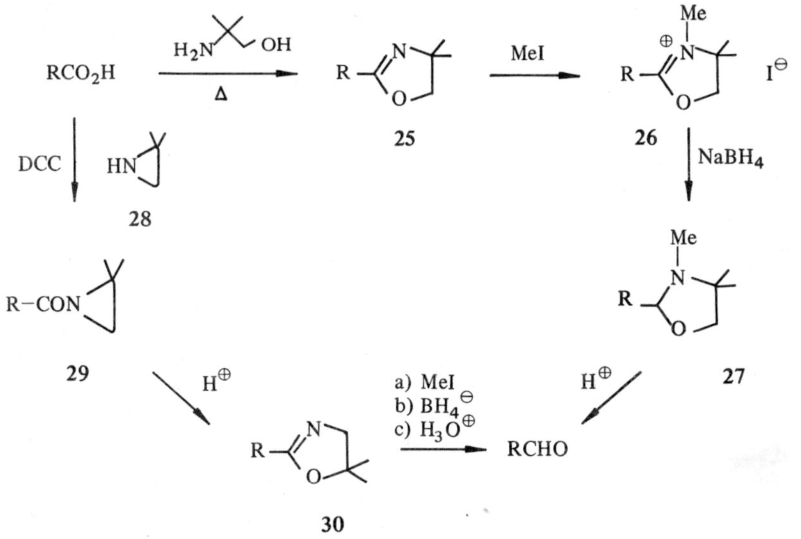

R	%
Phenyl	86
Cyclohexyl	78
t-Butyl	95
2-Phenethyl	72

converting the former to their 2-oxazoline derivatives **25**. Since **25** is resistant to borohydride reduction in alcoholic solvents, the highly electrophilic N-methyl salt **26** was prepared and readily gave the oxazolidine **27** when treated with borohydride. Acid hydrolysis generated the aldehydes in good yield. This method suffers from one major drawback, namely the harsh conditions necessary to form the oxazoline from the carboxylic acid. The method requires neat heating of a mixture of the amino alcohol and the acid at 150-200°, and even at these temperatures the oxazoline will only form if it can be distilled out of the mixture. A recent report[15] which describes a very mild conversion of carboxylic acids to oxazolines may greatly enhance the appeal for this aldehyde preparation. Oxazolines **30** are formed in high yield by treating the carboxylic acid with dimethylaziridine **28** in the presence of a dehydrating agent (dicyclohexylcarbodiimide) to form the acylaziridine **29**. The latter is smoothly rearranged to **30** by action of a catalytic quantity of acid.

D.   From Aziridines

Another technique for preparing aldehydes from their carboxyl derivatives involve the reduction of the acyl aziridines with lithium aluminum hydride.[16] This unique behavior is based on the assumption that, unlike other N,N-dialkyl amides, the acylaziridine shows negligible C–N p-π overlap (i.e., **31**) and thus renders the aziridine moiety vulnerable to nucleophilic displacement. A series of acylaziridines have been reduced by this method and the yields are presented.

$$RCO_2H \longrightarrow RCOCl \longrightarrow RC\overset{O}{\underset{N}{}} \xrightarrow{\text{LiAlH}_4} RCHO \qquad R-C\overset{O^{\ominus}}{\underset{\oplus N}{}}$$

31

R	%
n-Propyl	75
n-Hexyl	81
2-Pentyl	77
t-Butyl	80
i-Propenyl	40
Cyclopropyl	60

It is important to emphasize that the 60% yield of cyclopropane carboxaldehyde obtained represents the best approach, to date, to this molecule. Other methods which involve *any* type of acidic reagents result in extensive decomposition of cyclopropane carboxaldehyde.

E.   From N,N'-Carbonyldiimidazole

The important discovery by Staab[17] and co-workers that N,N-carbonyldiimidazole **32**, prepared from phosgene and imidazole, readily reacts with carboxylic acids to give acylimidazolides **33** provides the basis for a powerful synthetic tool. The lability of **33** to a host of nucleophiles (amines, alkoxides, hydride, carbanions, etc.) has opened routes to amides, esters, aldehydes, and ketones. The latter two classes are discussed here, while the former are

presented in the chapters dealing with amides and esters. The formation of the acylimidazolide (azolides) **33** is considered to pass through the adduct **34** with loss of one of the imidazole rings followed by rearrangement of the resulting anhydride **35** to the unstable system **36**. The anhydride **35** is also unstable and could not be isolated even at −50°. The reaction of carboxylic acids with carbonyldiimidazole **32** is highly efficient, proceeding in a variety of inert solvents at room temperature affording nearly quantitative yields of the acyl-imidazole **33**.

Reaction of **33** with 0.25 mole of lithium aluminum hydride in ether or tetrahydrofuran produces the corresponding aldehydes. The azolides of di-carboxylic acids **37** gave the dialdehydes **38**, whereas azolides derived from *N*-acylamino acids **39** were selectively reduced to *N*-acylamino aldehydes **40**.

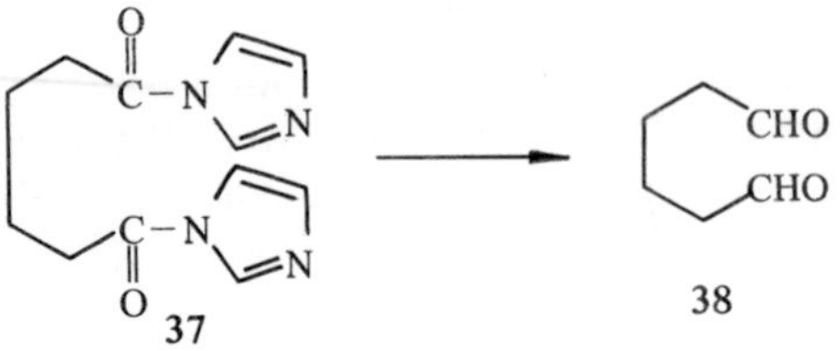

R	%
Phenyl	78
p-Tolyl	80
p-Nitrophenyl	88
Styryl	70
4-Pyridyl	82
n-Butyl	60
p-Acetamidophenyl	53
4-Carbomethoxybutyl	71

The process has been made even more efficient by eliminating the isolation of the acylimidazole. By simply treating the carboxylic acid with 32 (1:1 molar ratio) and immediately adding lithium aluminum hydride to the imidazolide solution, the aldehyde is formed directly. However, in this modification an additional 0.25 mole of hydride agent must be added to neutralize the imidazole which is liberated. The yields of aldehydes produced in this sequential process are comparable to the two-step method and requires considerably less effort.

It is an easy matter to envision extending this conversion of acids to aldehydes to the analogous conversion to ketones. This was successfully accomplished by reacting the acylimidazoles with a nucleophile other than hydride, namely Grignard reagents. A variety of ketones 41 were prepared in this manner.

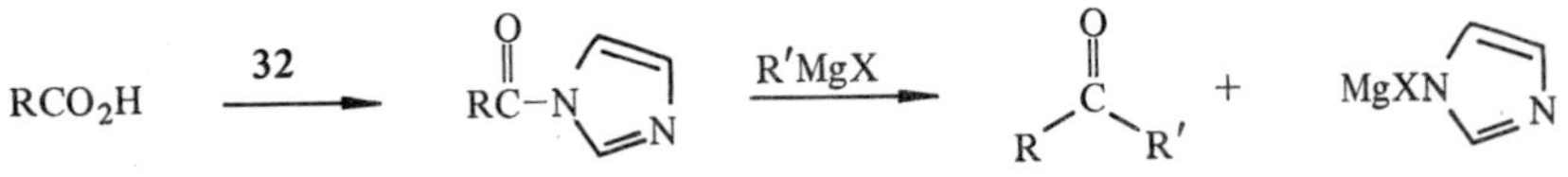

R	R'	%
Ph	Ph	72
3-Cl-Ph	Ph	94
4-Cl-Ph	Ph	91
3-Br-Ph	Ph	87
4-Br-Ph	Ph	88
Me	Ph	50
n-Pent	Ph	69
n-Pent	Et	64
n-Pent	Me	73

F. From Dihydro-1,3-oxazines

The report by Ritter and Tillmanns[18,19] that nitriles undergo reaction with certain glycols in the presence of cold concentrated sulfuric acid to give dihydro-1,3-oxazines **42** provides a route for the conversion of nitriles to aldehydes.[20] The oxazine intermediate may be smoothly reduced with sodium borohydride at low temperatures to the tetrahydrooxazine **43** which is readily cleaved in aqueous oxalic acid to the aldehyde **44**. By use of sodium borodeuteride, the C-1 deuteroaldehydes are also made available. The transformation of 1-cyano-1-carboethoxy cyclopropane to 1-formyl-1-carboethoxycyclopropane by this route[21] is exemplary of this technique. The oxazine route to aldehydes

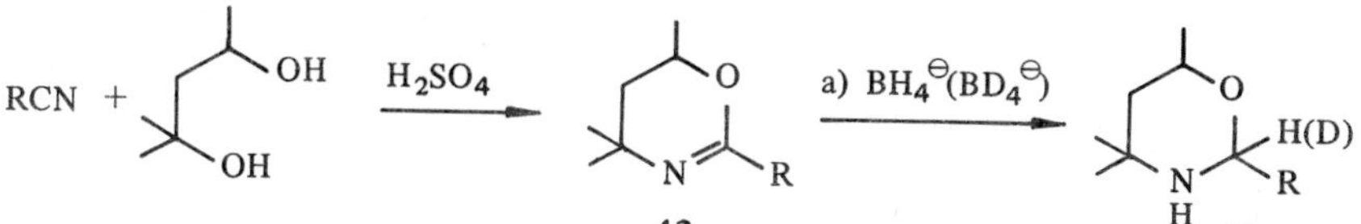

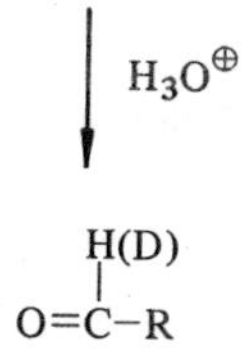

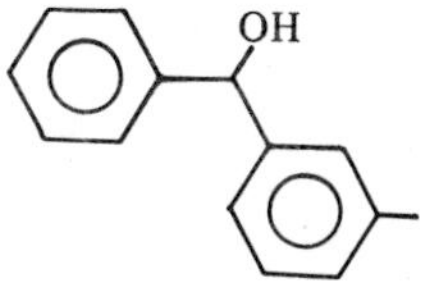

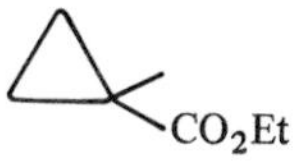

has considerably more versatility than simple conversion of nitriles to aldehydes, and this is discussed in a later section of this chapter.

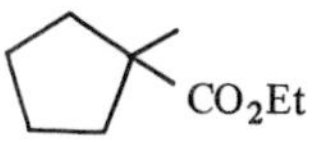

G. From 1,2,4-Triazoline-3,5-diones

A mild method for oxidizing primary and secondary alcohols to their respective carbonyl derivatives involves the use of the simple heterocycle, 4-phenyl-1,2,4-triazoline-3,5-dione **45**.[22]

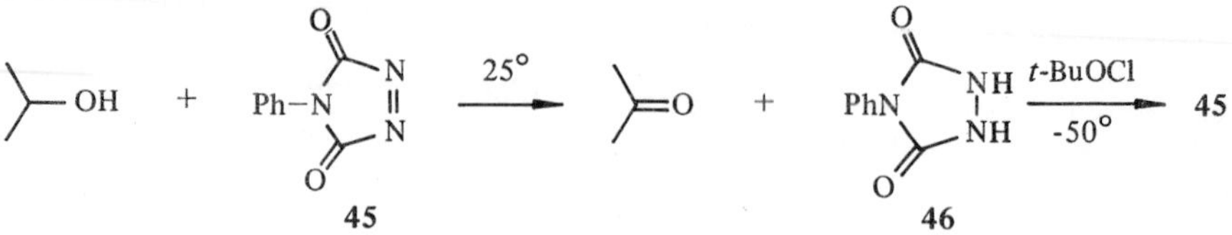

Alcohol	% Carbonyl
Benzyl alcohol	78
Benzhydrol	90
Cyclopentanol	62
Cyclohexanol	84
4-t-Butylcyclohexanol	75

Alcohols are oxidized in good yield by reaction with **45** in benzene at room temperature for a few hours. The reduced heterocycle **46** conveniently separates out during the reaction and may be used again after oxidation to **45** by t-butyl-hypochlorite at −50°.[23]

H. From 1-Chlorobenzotriazole

The mild oxidative properties of 1-chlorobenzotriazole **47** was reported by Rees[24] and demonstrated with a variety of substrates. Besides being useful in oxidizing the hydrazo (−NH−NH−) group to azo (−N=N−) compounds, it oxidizes primary and secondary alcohols to aldehydes and ketones, respectively.

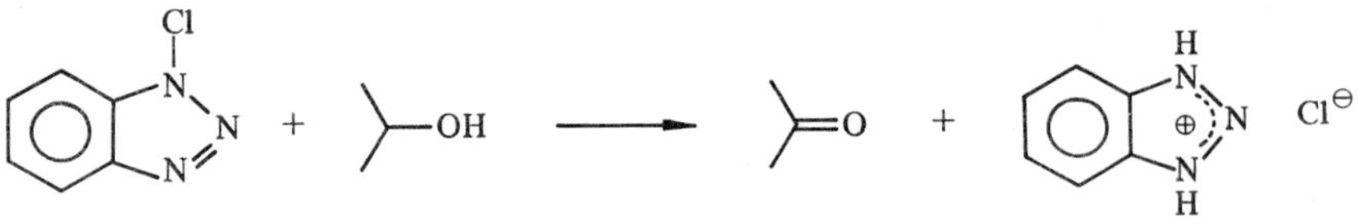

Alcohol	% Carbonyl
Benzyl alcohol	70
i-Propanol	70
Cyclohexanol	70
1-Phenylethanol	65
Benzhydrol	65

This method, like that described above (Section G) offers the convenience of clean separation of products and reagents. The benzotriazolinium salt **48** crystallizes from the solution (dichloromethane, benzene, or carbon tetrachloride) leaving the pure carbonyl product in solution. This method is, therefore, a useful alternative to the Cookson technique[22] and only further use in synthesis will determine the relative value of either heterocyclic reagent.

I. From Triazolines

The finding by the DuPont group[25] that it is possible to harness the versatile chemical properties of cyanogen azide has led to a new and simple approach to

cyclic ketones via a ring enlargement technique. By treating cyclopentene with cyanogen azide a good yield of the cyanamide **50** results, presumably, by passing through the dipolar cycloaddition product **49**. Hydrolysis of **50** leads to the ketone. Although this observation by the DuPont group, in itself, provides a method of converting olefins to ketones via the temporarily constructed heterocycle **49**, an extension by McMurry[26] is worthy of note. The latter author envisioned that a migration of an alkyl residue rather than hydrogen (**49**) would lead to ring expanded ketones. Toward this end, the olefin **51** was treated with cyanogen azide and, indeed, produced the ring expanded ketone **52**. Other olefins, **53-55**, which were similarly subjected to cyanogen azide, led to the corresponding ketones in variable yields.

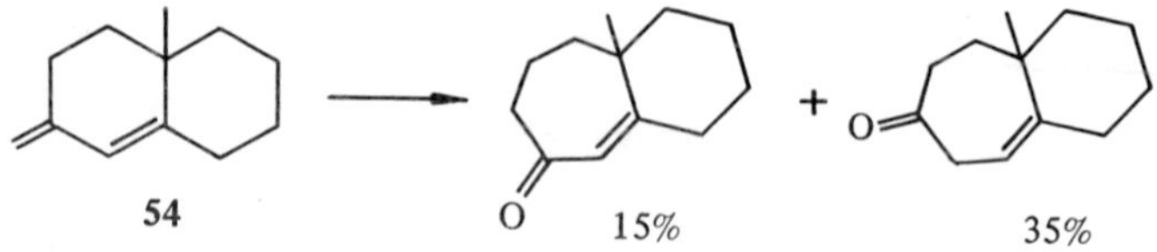

54   O   15%   +   O=   35%

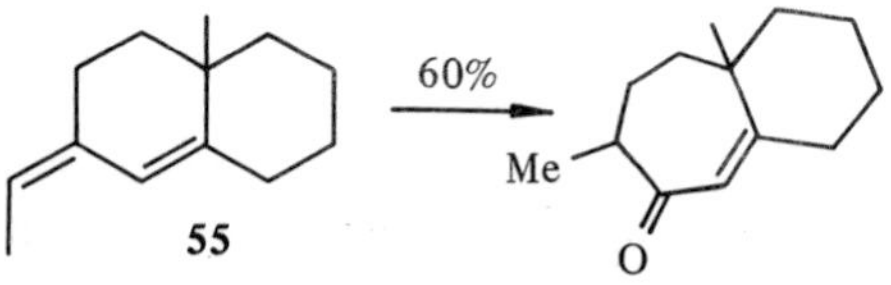

55   60%
            ──────▶
            Me

J. From Quinolines

Aldehydes are formed in excellent yields[27,28] by the acid catalyzed hydrolysis of Reissert compounds **56**. The latter, in turn, are prepared by reaction of acid chlorides with quinoline in the presence of potassium cyanide in a variety of solvents. Accompanying the aldehydic product is quinoline 2-carboxylic acid **57** (X=OH) or its amide **57** (X=NH$_2$). The reaction has been shown[29] to proceed via the intermediates **58** and **59**. Although this method of preparing

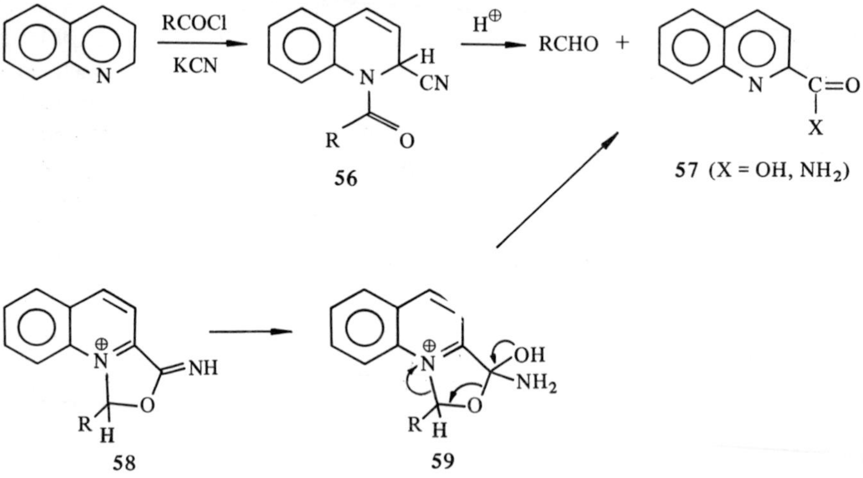

RCOCl
──────▶
KCN

H$^\oplus$
──────▶   RCHO   +

56

57 (X = OH, NH$_2$)

58   59

aldehydes from carboxylic acids has been largely displaced by newer techniques, its utility is still evident. Thus the aldehydes **60**,[30] **61**,[31] and **62**[31] were prepared via the Reissert intermediates in good overall yields starting from their

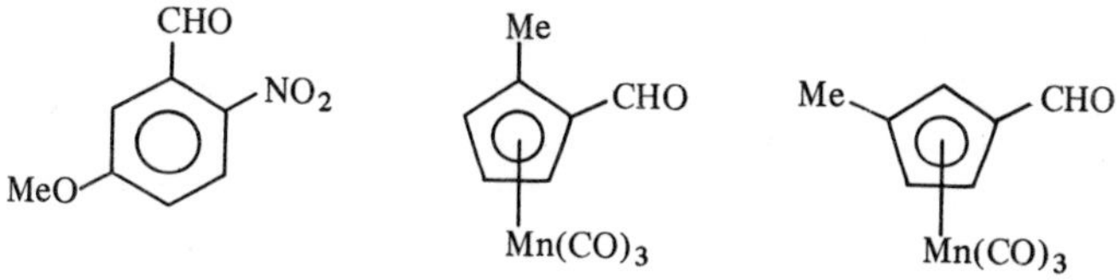

60   61   62

acid chlorides. The Reissert method has also employed isoquinolines and phenanthradines which appear to behave in a manner comparable to their quinoline analogs.[28] In fact, a report[32] described the photolysis of isoquinoline-derived Reissert intermediates **63** which led, in this case, to the cyanoisoquinoline **64** and the aldehyde. The yield of the latter was rather poor (5-14%) despite higher yields of the cyanoisoquinoline (30-70%). It is doubtful that this sequence

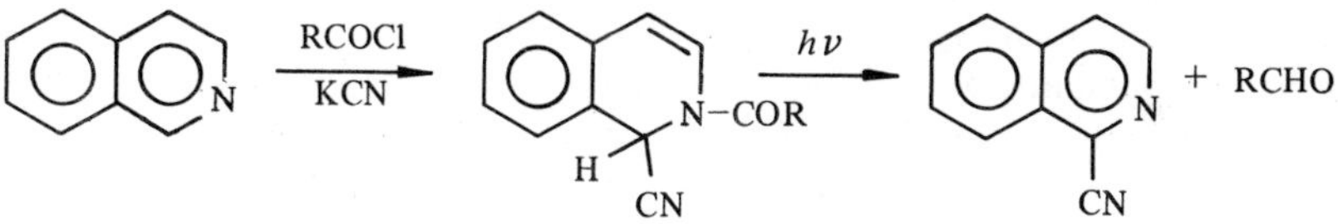

63   64

will exhibit any synthetic utility unless further detailed studies are performed.

K. From Oxiranes

Stork[33] reported a method for converting 1-alkynes **65** to their corresponding aldehydes **66** via the oxirane **67**. Thus, by addition of triethylsilane to the

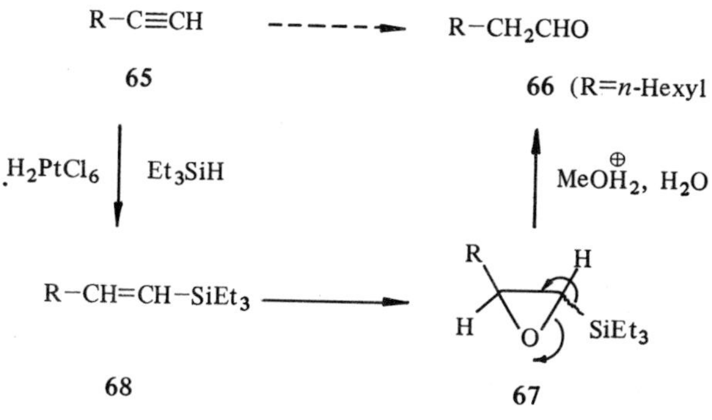

terminal acetylene in the presence of chloroplatinic acid, the vinylsilane **68** is produced which is readily oxidized (*m*-chloroperbenzoic acid) to the triethyl-silyl oxirane **67**. The vulnerability of β-alkoxyethylsilanes toward acid catalyzed Si—C cleavage was verified when **67** was treated with methanolic sulfuric acid. The expected mode of bond rupture occurred, producing the aldehyde. An extension of this method leads to a two-carbon homologation and is discussed in the section dealing with such transformations (Chapter 9, Section III.G).

L. From Azirines

Hassner and co-workers[34] reported a new approach to α-amino ketones **72** from olefins by way of the azirine derivative **71**. First, a general synthesis of vinyl azides was developed[35] which involved addition of iodine azide to olefins producing the iodoazide **69**. Base catalyzed elimination led to the vinyl azides **70**. If the latter are photolyzed in an inert solvent (pentane or cyclohexane) good yields of the azirines **71** are obtained. Hydrolysis of azirines is well known[36] to lead to α-aminoketones and this behavior forms the basis of the Neber rearrangement (**75** → **76**). By direct photolysis of vinyl azides **70** in methanol containing a catalytic quantity of sodium methoxide, it was possible to generate the azirine **71** *in situ* and proceed on to the amino ketone in a single operation. Thus it is necessary only to construct the synthetically useful azirine as a transient species to effect the transformation of alkenes to amino ketones. This method complements the Neber reaction by forming amino

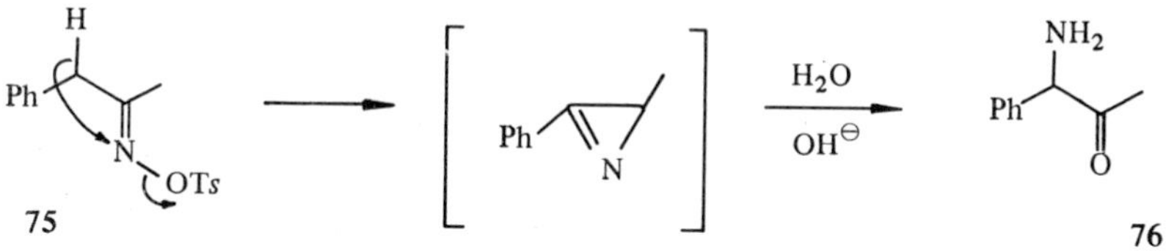

75   **76**

ketones that are isomeric to the Neber products. For example, the vinyl azide **73** affords the amino ketone **74**, whereas the oximino tosylate **75** undergoes the Neber reaction leading to aminoketone **76**.

M. From *N*-Aminoaziridine

In an earlier section (Chapter 3, Section III) the synthetic utility of the titled heterocyclic system was exhibited by stereospecific deamination leading to *cis* or *trans* alkenes. Eschenmoser[37] has recently demonstrated further utility of *N*-aminoaziridines (**77**) by transforming α-epoxyketones (**78**) to acetylenic aldehydes (**80**) of the same carbon content. The key step involves thermal

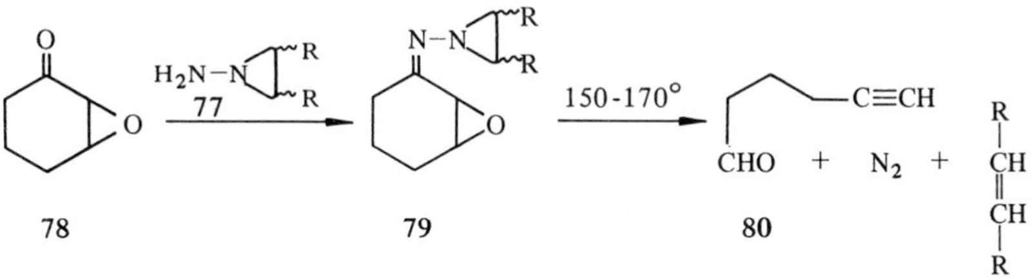

78   **79**   **80**

decomposition of the epoxy hydrazone **79** which produces during the fragmentation, along with the aldehyde, an olefin and nitrogen. Acetylenic ketones **82** are also generated by the thermal fragmentation of the hydrazone derived from the epoxyketone **81**. The mechanism of this fragmentation is postulated as

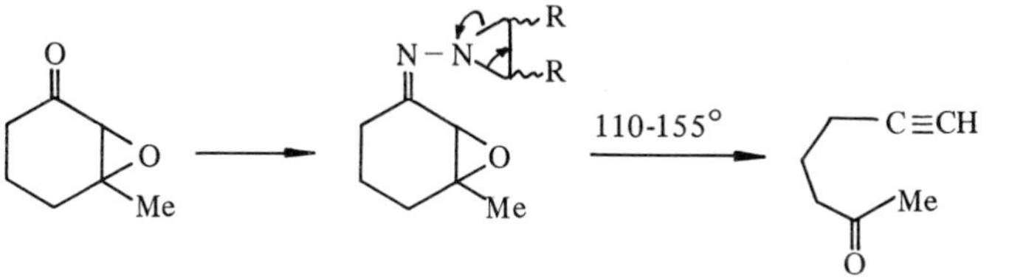

81   **81a**   **82**

initially proceeding through the diazo compound **83** which is analogous to the deamination of *N*-aminoaziridines leading to alkenes.[38] Ring cleavage of the epoxide group in **83** follows affording the diazonium ion **84** which fragments to the acetylene ketone **82**. This reaction has proved to be rather general and allows the aziridine to serve as a useful vehicle to arrive at various acetylenic carbonyl compounds. Further examples are given below.

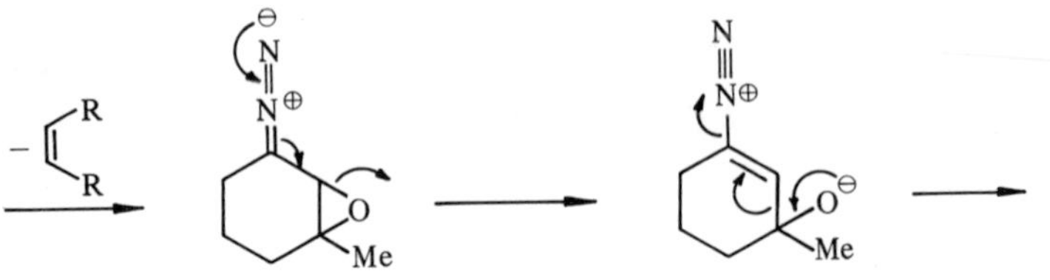

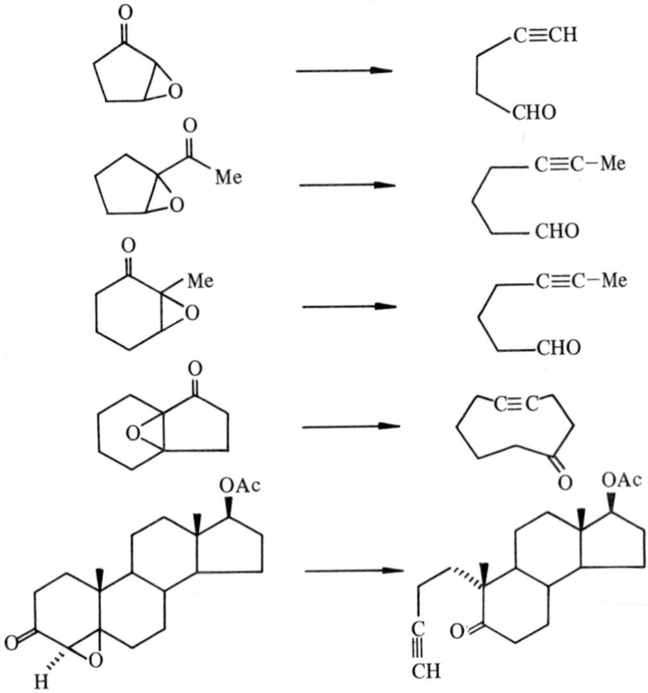

If the epoxyketone is not part of a cyclic ketone (e.g., **85**), fragmentation leads to acetylenes and carbonyl compounds (e.g., **86**). Thus the method allows a two-carbon degradation of an α,β-unsaturated ketone (the precursor to the epoxyketone). This may be demonstrated by the transformation of benzal-

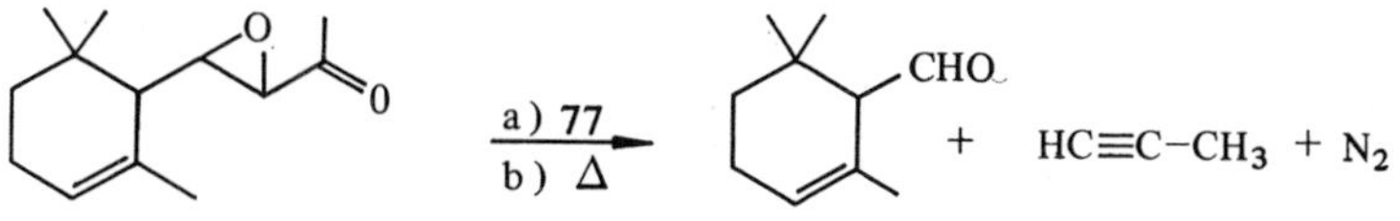

$$\text{85} \xrightarrow[\text{b) } \Delta]{\text{a) } 77} \text{86} + HC\equiv C-CH_3 + N_2$$

acetophenone 87 to phenylacetylene and benzaldehyde. The N-aminoaziridines 77 utilized in this reaction were obtained by several routes. The preferred

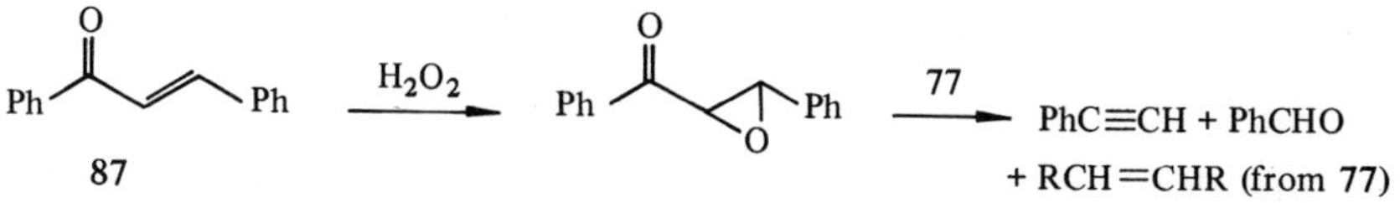

$$\text{87} \xrightarrow{H_2O_2} \quad \xrightarrow{77} PhC\equiv CH + PhCHO$$
$$+ RCH=CHR \text{ (from 77)}$$

method seems to be reaction of vicinal sulfonate esters with hydrazine.

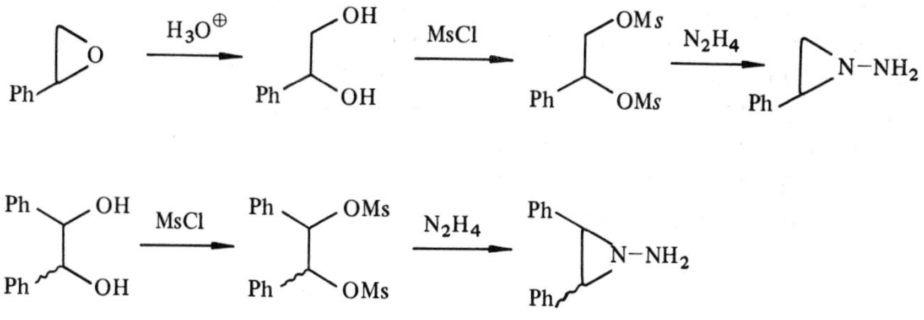

88, *meso*
89, *dl*

90, *cis*
91, *trans*

Scheme 1

According to Scheme 1, if meso-diols 88 are utilized as starting materials, the cis-disubstituted aziridine 90 results, whereas trans-aziridines 91 are formed from racemic diols 89.

N. From Dihydropyrans

The thermal reorganization of dihydropyrans 92 to cyclohexenyl carbonyl derivatives **93** was discussed earlier (Chapter 3, Section X). Although the preparation of cycloalkenes was the subject in focus, this method may also be utilized as a route to several carbonyl compounds.

O. Transposition of Carbonyl Groups

This section deals with the utility of heterocycles in effecting a carbonyl transposition with no change in carbon content in the molecule.

1. From Isoxazoles

The interchange of a carbonyl function in α,β-unsaturated ketones **94** to **95** was cleverly performed by Buchi[39] using the isoxazole system **97** as a vehicle. The sequence required that oxime **96** be readily transformed via mild oxidation (iodine-potassium iodide) to the isoxazole **97**. The process, which occurred in high yield, is a new method for preparing isoxazoles. Reductive cleavage of the latter using catalytic hydrogenation gave the enaminoketone **99** which can be hydrolyzed to the β-diketone **100** or further reduced to the β-amino-ketone **98**. More efficient, however, was the direct reduction of the isoxazole to the β-amino ketone **98** by action of sodium in liquid ammonia containing t-butanol. Distillation of **98** with a trace of acid present gave the transposed α,β-unsaturated ketone **95** in 70-75% yield. Examples of the compounds used in the conversion of **94** to **95** are shown in tabulated form.

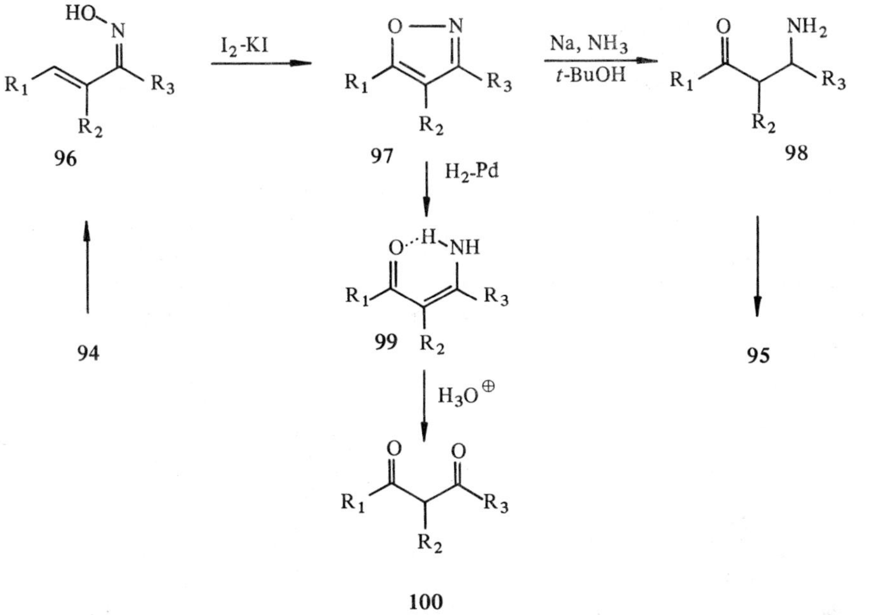

R$_1$	R$_2$	R$_3$
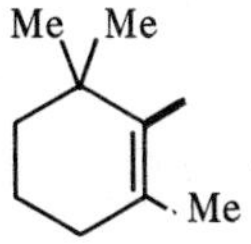	H	Me
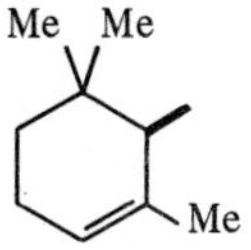	H	Me
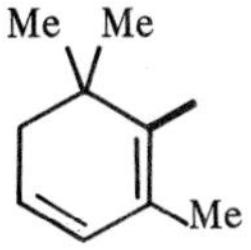	H	Me
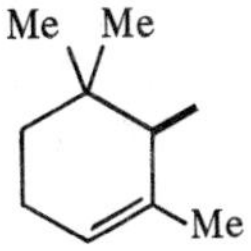	Me	Me
H	Me	Et

2. From 1,3-Dithianes

Transposition of the 3-carbonyl group in the bicyclic ketone **101** to the 2-position in **102** was performed[40] by constructing the dithiane intermediate **103**. Reduction to the alcohol followed by transformation into the dithiane acetate **104** provided the necessary precursor **105** after mercuric chloride deblocking. The ketoacetate was reduced to the bicyclic ketone **105** by treatment with calcium in liquid ammonia. The tedium of many steps necessary to effect the carbonyl transposition is tempered by the good overall yields (48-58%).

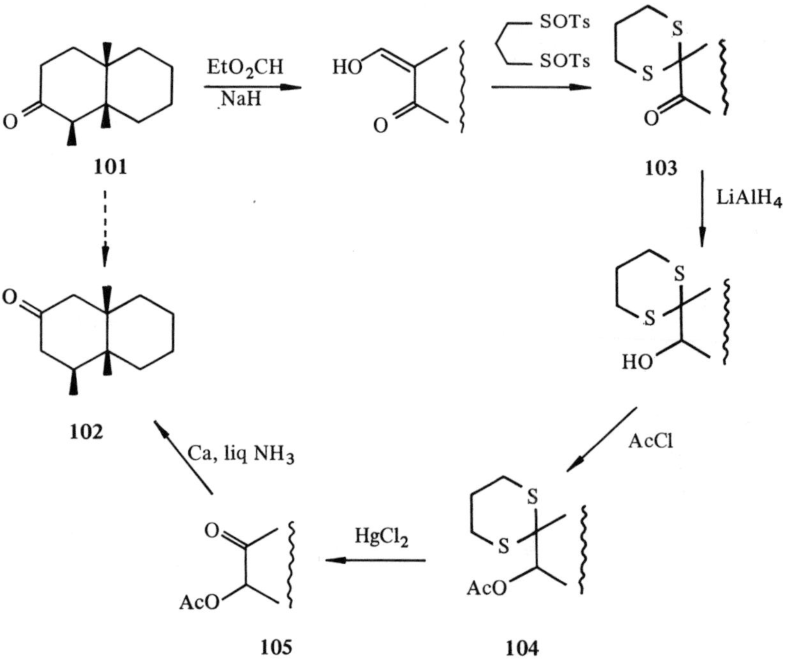

3. From Isoxazolidines

Bridged polycyclic homoconjugated ketones **107** were prepared[41a] from certain acetaldehyde derivatives **106** by invoking an isoxazolidine **109** formed by an intramolecular 1,3-dipolar cycloaddition of the nitrone **108**. The resulting isoxazolidine **109** was reductively fragmented to the amino alcohol **110** (cf. Chapter 7, Section I.A) and then, through a series of standard manipulations, gave the homoconjugated ketone **107** in 16% overall yield. The scheme is applicable to large scale work and appears to be quite general. For instance.

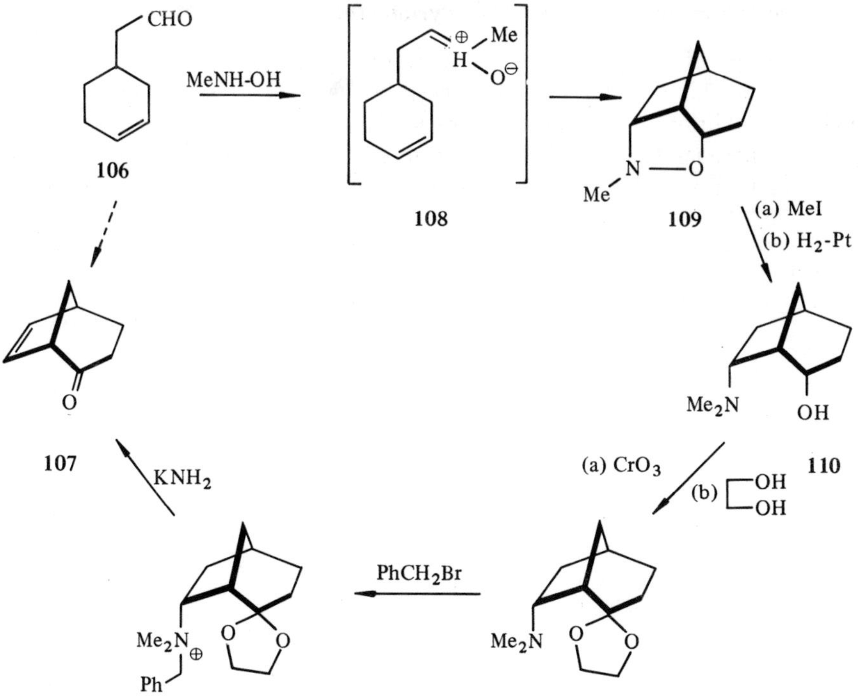

another report[41b] relates this sequence in an intermolecular manner involving the nitrone and terminal olefins producing α,β-unsaturated ketones:

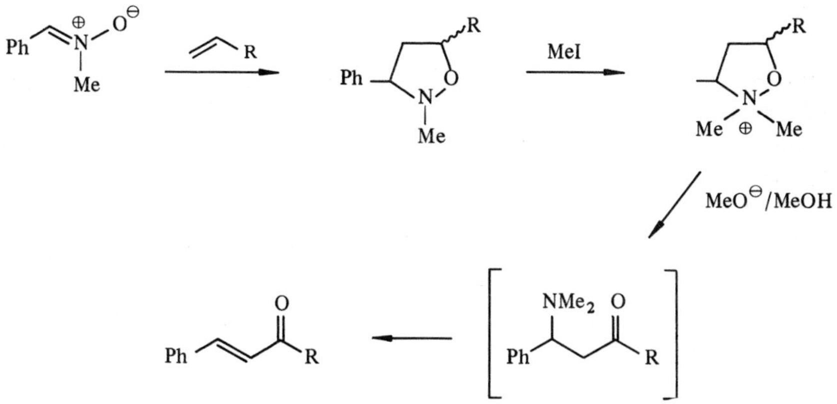

P. From 2-Mercaptopyridine [2(1H)-Pyridinethione]

The concept of using a labile acyl derivative 33 as seen in the technique of Staab[17] (Section E, p. 168) has been recently[42] extended to 2-acylthiopyridines 111. The addition of Grignard reagents to the latter gives, after hydrolysis, high yields of ketones 113. The fact that no tertiary alcohols are produced is due to the fact that the ketone is not released until after the addition of water to

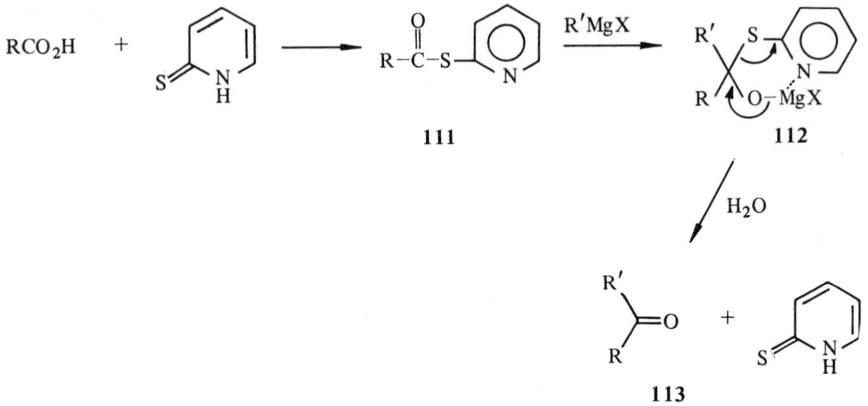

111   112

113

R	R'	% 113
Ph	Ph	94
$PhCH_2CH_2$	n-Bu	97
$PhCH_2CH_2$	Cyclohexyl	95
$PhCH_2CH_2$	sec-Bu	83
Me	Ph	91
n-Amyl	Ph	81
$-(CH_2)_4-$	Ph	92

decompose the complex 112. This method, based on the preliminary results, should provide a versatile and convenient synthesis of ketones from the corresponding acids. The active acyl derivative 111 is prepared by treating 2(1H)-pyridinethione with the acid chloride, ester, or free acid by standard coupling conditions. It is of interest to note that benzophenone has been prepared[42] in 98% yield when phenylmagnesium bromide was added to another active

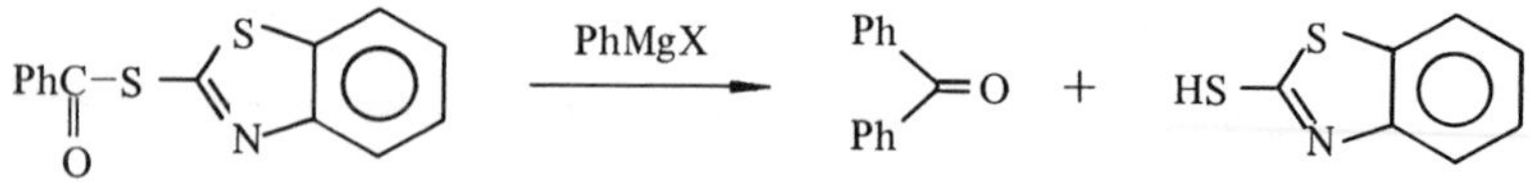

114   115

acyl derivative **114**. Thus 2-mercaptobenzothiazole **115** may also be acylated to give the acylthio-heterocycle whose vulnerability to nucleophilic displacement allows the formation of ketones.

Q. From Furans

Furfural alcohol, via its benzoate ester **115a**, has been pyrolyzed (640-700°) to methylenecyclobutanone **115b** in 40% yield.[42a] This method represents a practical approach to reasonable quantities of this interesting carbonyl derivative. Although no further details were reported, one may speculate that the

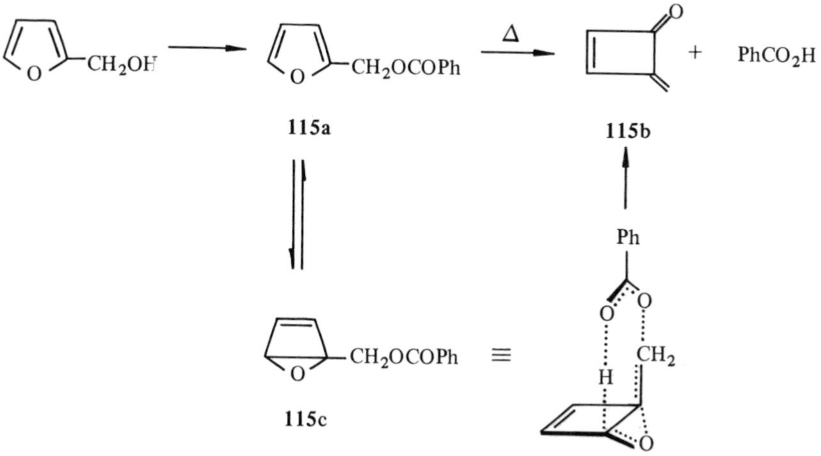

formation of **115b** proceeds through the valence tautomer **115c** whose geometry allows for a concerted elimination of benzoic acid in a manner analogous to the well-known pyrolysis of esters to olefins.

II. CARBONYL COMPOUNDS WITH ONE-CARBON CHANGE

$$(R-C) + (C=O) \longrightarrow R-C-\overset{|}{C}=O \quad \text{(increase in chain length)}$$

$$(R-C-C=O) - (C) \longrightarrow R-\overset{|}{C}=O \quad \text{(decrease in chain length)}$$

All of the following heterocycles (Sections A-G) except the last (Section H) are involved in adding one carbon to the final carbonyl product. The last system (pyridines) effects a one-carbon degradation to a carbonyl derivative.

A. From 2-Oxazolines

Grignard reagents may be formylated[43] to their corresponding aldehydes **121** by making use of the simple oxazoline **116**. The latter may be readily prepared by heating formic acid with 2-amino-2-methylpropanol and distilling the product directly from the mixture. Treatment of **116** with butyllithium followed by addition of deuterium oxide also provides the 2-deuterio-oxazoline **116a**. Conversion of **116** or **116a** to its methiodide salt **117** now provides the necessary formylating reagent. Addition of Grignard reagents, complexed by two equivalents of hexamethylphosphoramide (HMPA), to a suspension of **117** in THF leads to the adduct **118** in high yield. Hydrolysis (oxalic acid) then

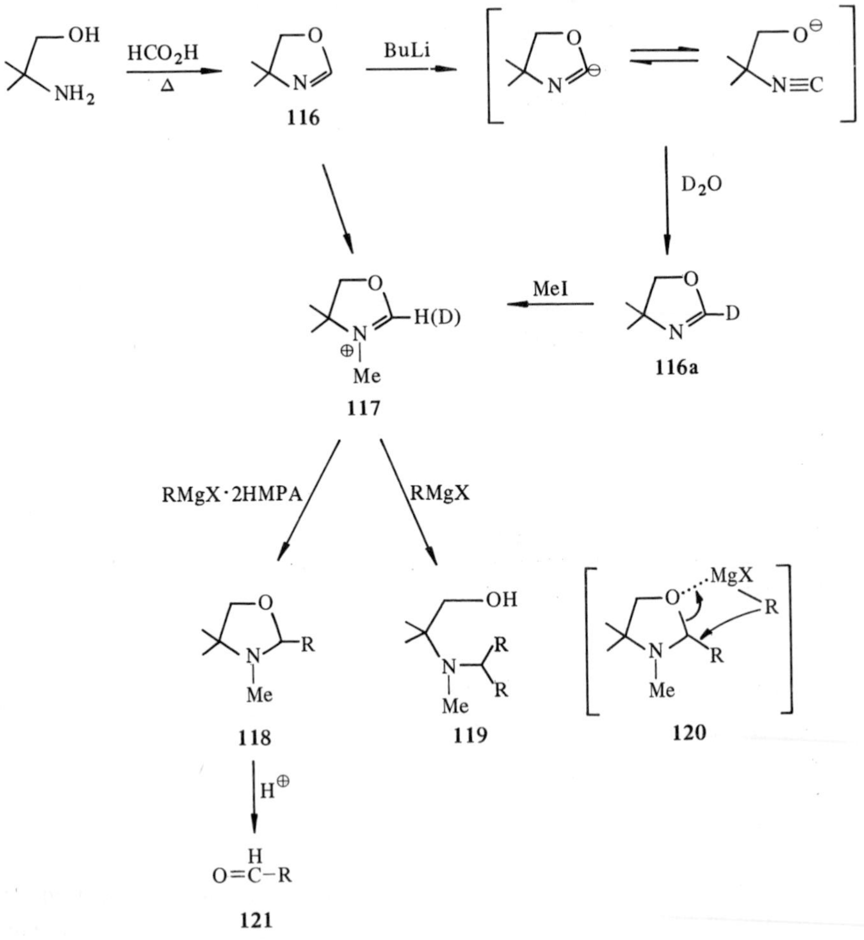

RMgX	Aldehyde	% 121
$PhCH_2-$	$PhCH_2CHO$	87
$PhCH=CH-$	$PhCH=CHCHO$	64
$PhC{\equiv}C-$	$PhC{\equiv}C-CHO$	51
o-(MeO)—Ph	o-(MeO)—PhCHO	90 (70, CDO)

produces the aldehyde **121**. If the Grignard reagent is added in the absence of HMPA, an excellent yield of the amino alcohol **119** is obtained. The latter presumably arises by complexation **(120)** of the initially formed oxazolidine **118** with Grignard reagent. The presence of the HMPA, therefore, appears to minimize the interaction of RMgX and the oxygen atom in the oxazolidine inhibiting the subsequent reaction leading to the amino alcohol **119**. Examina-

tion of the list of Grignard reagents employed indicates good yields of the corresponding aldehydes including C-1 deuterio aldehydes prepared from the 2-deuterio oxazoline **116a**. Noticeably absent from the list of aldehydes prepared are those derived from aliphatic Grignard reagents. This is due mainly to major difference in behavior when the latter are treated with the oxazolinium salt **117**. For example, when n-butylmagnesium bromide is added to **117**, very little (<20%) of the n-valeraldehyde is produced. Similar results were obtained with other n-alkyl Grignard reagents. A study involving the 2-deuterio oxazoline **116a** and 3-phenylpropyl magnesium bromide sheds light on this anomalous behavior. When the Grignard reagent, complexed with two equivalents of HMPA, was added to **117**, the products obtained were the 1-deuterio-3-phenyl propane **122** and phenyl propane **123**. The ratios (30:70) of the hydrocarbons indicated that proton abstraction, rather than addition, was occurring generating the ylides **124** (*A* and *B*). It is, therefore, clear that the Grignard complex with HMPA causes a significant enhancement in the base strength of the Grignard reagent when the latter is derived from an sp^3 (aliphatic) carbon. This formylat-ing method thus appears to be limited to Grignard reagents possessing sp^2 or sp hybridized carbanions.

B.   From Quinazolines

A conceptually similar process to the above mentioned formylation of Grignard reagents was also reported by Fales in 1955.[44] The addition of a variety of Grignard reagents in ether at room temperature to the quinazoline methiodide **125** gave the adduct **126** which, upon acidic hydrolysis, afforded the aldehydes **127**. In this method, the aliphatic Grignard reagent appears to be as efficient

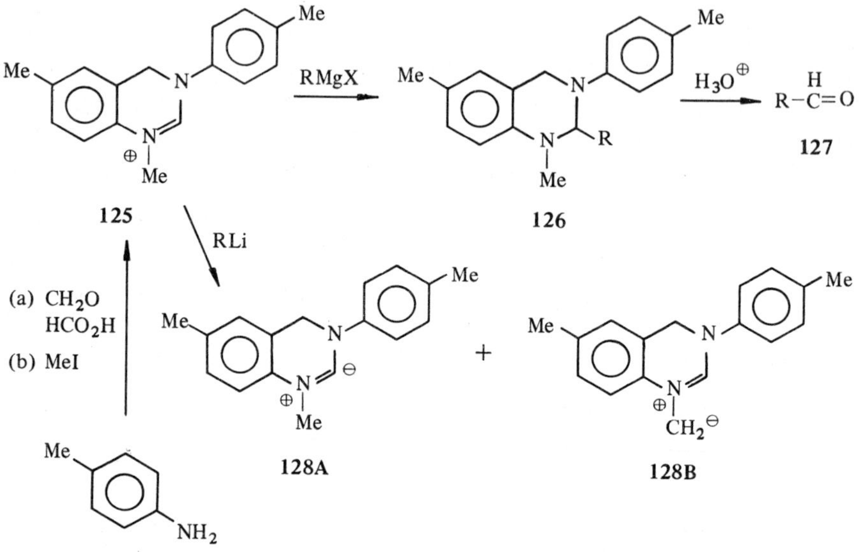

RMgX	Aldehyde	% **127***
Me	MeCHO	78
n-Bu	BuCHO	87
PhCH$_2$	PhCH$_2$CHO	74
n-Dodecyl	*n*-DodecylCHO	73
i-Pr	*i*-PrCHO	45
Ph	PhCHO	95
p-MeO-Ph	*p*-MeO-PhCHO	80

*Isolated as 2,4-dinitrophenylhydrazone derivatives.

in its conversion to the aldehyde as its aromatic counterparts. This is most likely due to the fact that the quinazoline **126** is not as susceptible to ring cleavage as the oxazolidine **118**. Note that the absence of an oxygen atom in **126** attests to the weaker complexation of nitrogen to magnesium salts. Steric effects in **126** may also be responsible for its inertness to the Grignard reagent. Furthermore, since this method did not employ HMPA, the base strength of the aliphatic Grignard reagent was not enhanced and added to **125** in the expected fashion. Support for this explanation may be gathered by the fact that when alkyl or aryl lithium reagents were added to **125**, very low yields of **126** (15-20%) were obtained. This must have been caused by proton abstraction generating the ylides **128** (*A* and *B*) as was observed with the oxazolinium salt **117**. The seemingly cumbersome structure of the formylating agent **125** is misleading. It is easily obtained by heating *p*-toluidine, formalin solution, and formic acid followed by quaternization with methyl iodide.

C. From 1,3-Dioxanes

A further technique for formylating Grignard reagents using a heterocyclic substrate was reported by Eliel[45] in his study of conformational stabilities of 1,3-dioxanes. The readily available 2-methoxy-1,3-dioxanes **129** (from the diol and trimethyl orthoformate) possessing an axial 2-methoxy group reacted very smoothly with Grignard reagents at room temperature to give the 2-substituted

RMgX	% **130**
Me	70
Et	75
i-Pr	63
Ph	95
p-F-Ph	94
p-Br-Ph	55
p-CF$_3$-Ph	89

1,3-dioxanes **130** in good yield. Irrelevant to the synthesis of aldehydes from the latter via acidic hydrolysis was the fact that the 2-substituent was 90-95% axially substituted. Thus the displacement of the methoxyl substituent took place with a high degree of stereoselectivity. The process was envisioned as a departure of the methoxyl substituent assisted by the nonbonded oxygen pair leading to the oxocarbonium ion **132**. Addition of the Grignard reagent then occurs from the topside (least hindered) affording the 2-alkyl-1,3-dioxane **130**.

Of further interest was the fact that the 2-methoxy dioxane **131**, containing an equatorial methoxy group, was completely inert to the Grignard reagent. Based on the rationale offered for this reaction, this would be consistent with the inability of the oxygen lone pair to displace the 2-methoxyl group. Since this study focused, primarily, on the mode of addition of RMgX to 2-methoxy-1,3-dioxane, aldehydes were not generated by acidic hydrolysis. However, the last step is sufficiently well known[46,47,48] that its success is virtually assured. While the reaction of Grignard reagents with simple orthoformates usually require heating in high boiling solvents and other forcing conditions, the reaction with the axial 2-methoxy-1,3-dioxanes **129** proceeds rapidly and exothermically.

D. From Pyrimidines

As part of a general study to evaluate the reaction of Grignard reagents with various heterocyclic N-oxides, it was found that various ring openings occur.[49] Thus addition of phenyl magnesium bromide to pyrimidine N-oxides **133** in cold tetrahydrofuran produced, after quenching, a residue that was presumed to be the adduct **134**. Acidic hydrolysis led to a good yield of benzaldehyde which was reported to arise from the ring opened intermediate **135**. An analog of the latter has been isolated by Kellogg[50] in another study involving pyridine N-oxides and Grignard reagents. These are discussed in a later chapter dealing with nitriles (Chapter 10, Section IV).

E. From 1,3-Dioxalanes and 1,3,5-Trioxanes

A one-carbon homologation of olefins has been accomplished[51] by photo-alkylation of 1,3-dioxalanes or 1,3,5-trioxanes to their respective acetals **136** and **137**. Since mild acid treatment of acetals leads to the corresponding alde-hydes **138** this alkylation reaction constitutes a new route for the synthesis of aldehydes from olefins.

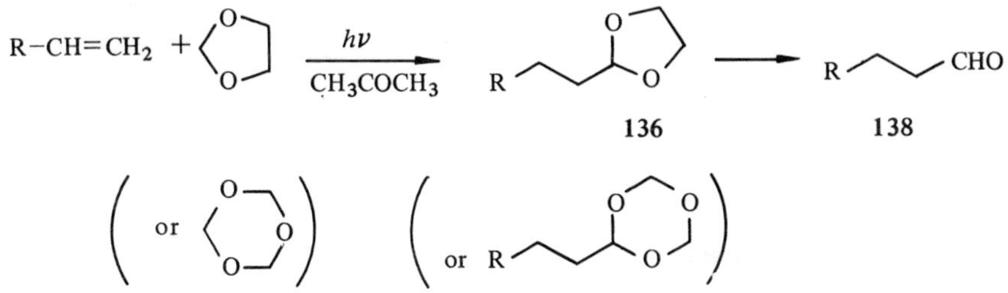

Olefin	1:1 Addition Product	%
	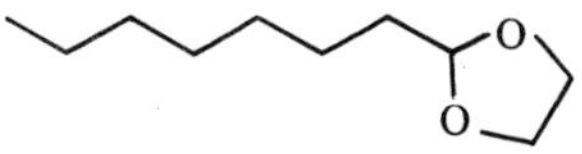	28-35
	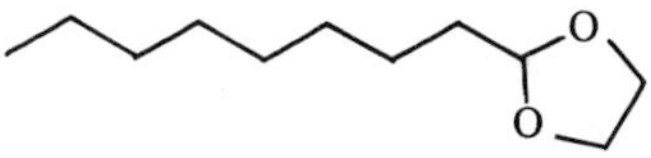	25
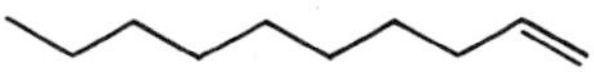	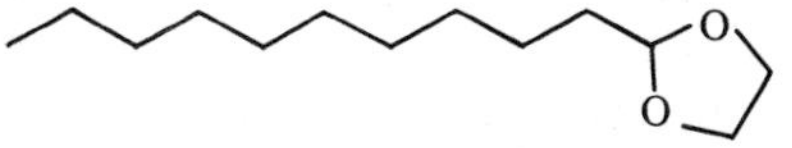	30-50
 MeO$_2$C	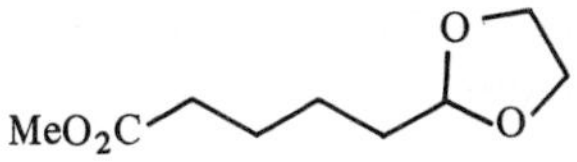 MeO$_2$C	18
 MeO$_2$C CO$_2$Me	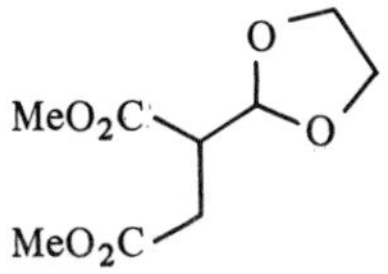 MeO$_2$C / MeO$_2$C	90

The alkylation involves the use of acetone, benzophenone, or acetophenone as an initiator of the photoalkylation producing 1:1 adducts (**136** or **137**) as the major product. Since the lowest energy transition in acetals involves a $n \to \sigma^*$ excitation, light of high energy (<2000 Å) would be required. However, in the presence of a ketone, whose lowest energy transition is of the $n \to \pi^*$ type, light of lower energy (>2900 Å) may be employed. This reaction has been classified as a free radical chain process and was formulated as follows:

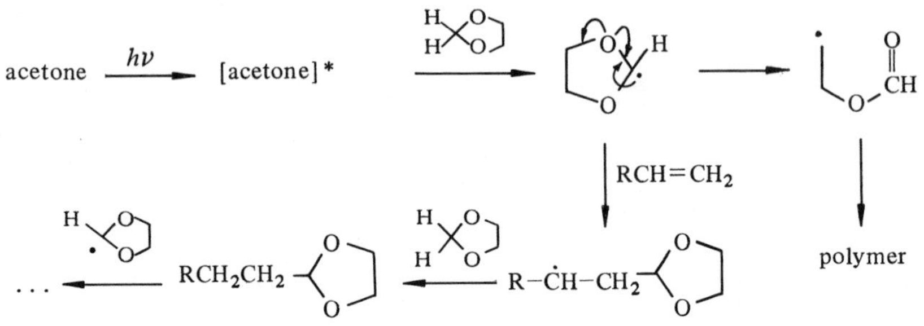

The dioxalane radical, if not efficiently trapped by the olefin, undergoes ring cleavage to formate esters and more complex derivatives therefrom. Another interesting product obtained during this reaction was the *bis*-dioxalane **139** and the 4-alkyl-1,3-dioxolane **140**. The latter arises from abstraction of the

139   140

somewhat lesser activated C–H bond in the dioxolane. The highest yield of alkylated acetal obtained occurred in the case of the maleate ester which gave the homologated acetal in 90% yield. This product has an excellent functionality array for synthesis. Although the method gives only moderate yields of acetals in most instances, its directness and simplicity should make it useful in synthetic programs.

F. From 1,3-Dithianes

In previous chapters, the synthetic utility of 1,3-dithianes toward the preparation of alkanes (Chapter 2, Section II), alkenes (Chapter 3, Section IX), and mercaptans (Chapter 7, Section IV.B) was discussed. The most important chemistry of dithianes, however, lies in their ability to act as a masked carbonyl group (A) capable of reacting as a nucleophilic acyl group (B) with an electrophile (E) leading, ultimately, to various carbonyl derivatives (C). The net

$$A \qquad\qquad B \qquad\qquad C$$

result of this scheme is a one-carbon homologation of electrophiles to carbonyl compounds. This impressive synthetic tool was first reported in 1965 by Corey and Seebach[52] and has since been reviewed in detail.[53] Aldehydes **141** or ketones **142** may be readily formed via the dithiane method by alkylation of lithio dithiane **143** or alkylation of lithio monoalkyldithiane **144**, respectively. Cleavage of the alkyl (**145**) or dialkyl (**146**) dithiane using a variety of techniques[53,54,55] releases the appropriate carbonyl products.

A number of recent variations has further demonstrated the utility of the 1,3-dithiane system.[56] The synthesis of optically active aldehydes and ketones[57,58] is also possible via this technique using the readily available fermentation alcohol, 2-methyl-1-butanol (**147**). Conversion of the latter to its iodide derivative (**148**) followed by reaction with lithiodithiane affords the elaborated dithiane (**149**). Hydrolysis under the usual neutral conditions ($HgO-HgCl_2$ or $CdCO_3$) releases the S-aldehyde or ketone in high optical yield. Furthermore, carbonyl compounds whose chiral center is alpha to the carbonyl group (**154**) may also be generated from the dithiane **153** with only 20% loss of activity. Thus 2-methylbutyraldehyde **151** prepared by oxidation

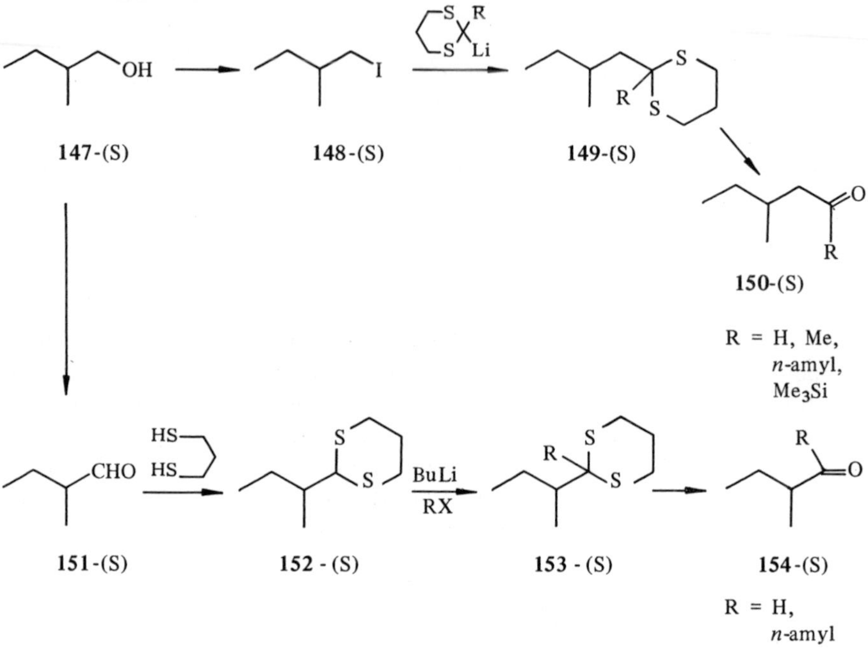

147-(S)   148-(S)   149-(S)

150-(S)

R = H, Me,
n-amyl,
Me₃Si

$R = H, Me, n\text{-amyl}, Me_3Si$

151-(S)   152 - (S)   153 - (S)   154-(S)

$R = H, n\text{-amyl}$

R = H,
n-amyl

of the alcohol is transformed into the dithiane **152** by treatment with 1,3-propanedithiol. Metallation and alkylation produces **153** which is then hydrolyzed under neutral conditions to the chiral aldehyde or ketone **(154)**. This approach to carbonyl compounds also allows the synthesis of either optical antipodes (R and S) by utilizing the corresponding alkyl halides (Scheme 2). The chiral ketones are available by two routes (path *A* or path *B*). The optical yields are about 10% higher using path *B* and this has been attributed to the higher degree of inversion in the alkylation of the halides with the 2-methyl lithiodithiane.

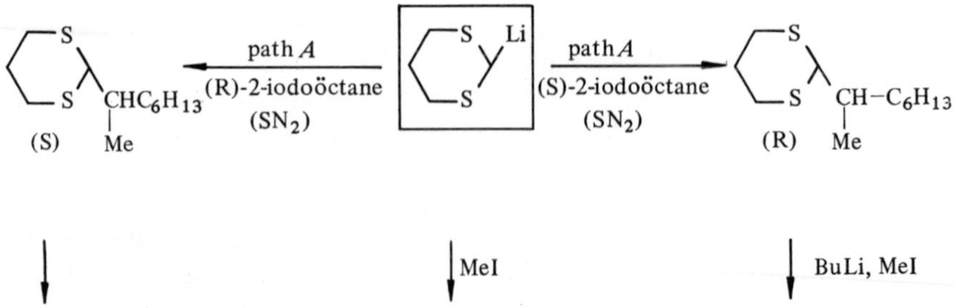

(S)   Me   path *A*   (S)   Li   path *A*   (R)   Me
$\xleftarrow{\text{(R)-2-iodoöctane}}$ $\xrightarrow{\text{(S)-2-iodoöctane}}$
(SN_2) (SN_2)

CHC₆H₁₃   CH—C₆H₁₃

MeI   BuLi, MeI

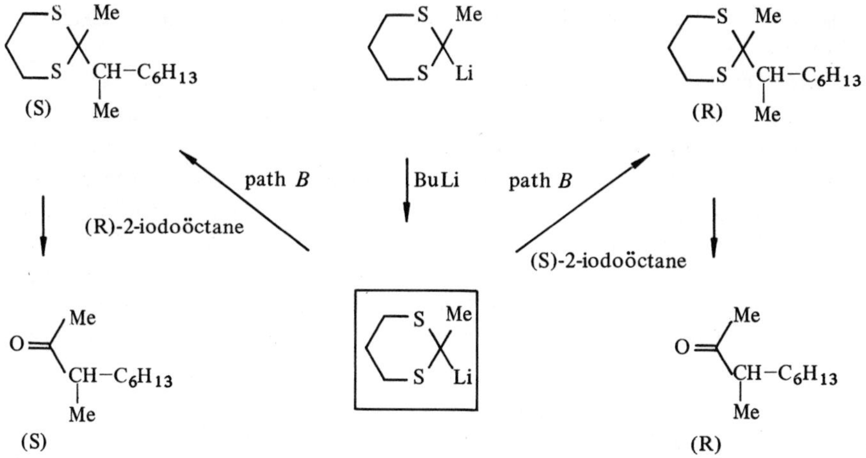

Various polyfunctional carbonyl compounds have also been reported[59,60] by passing through the dithiane heterocycle. Thus addition to cyclohexenone afforded the 1,2-addition product **155** which was highly acid sensitive. However, treatment with mercuric chloride buffered by calcium carbonate generated the carbonyl product **156**. In the absence of an acid scavenger (small amounts of hydrogen chloride are liberated during the hydrolysis) rearrangement to the isomeric dithiane **157** occurs. Cleavage then affords the γ-hydroxy ketone **158**.

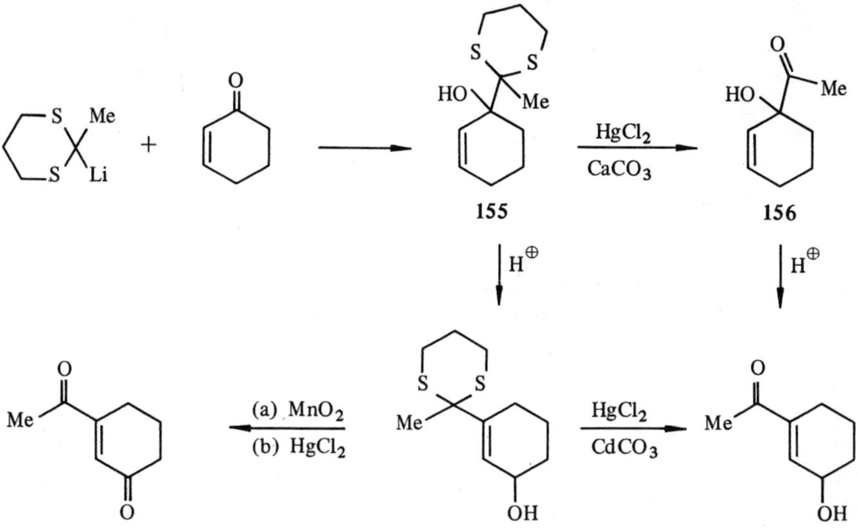

This method also allows the preparation of the diketone **159** by oxidation of **157** (or **158**).

Further extension of dithiane chemistry has opened routes into the prostaglandin series.[61] Alkylation of homosubstituted dithiane anions with bromoacetaldehyde diethyl acetal provides the dithiane acetal **160**. Since the sulfur heterocycle is stable to aqueous acid, only the acetal moiety is cleaved under these conditions generating the dithiane aldehyde **161**. Reaction of the latter with another lithio dithiane reagent leads to the *bis*-dithiane carbinol **162** which upon hydrolytic cleavage gives the 1,4-dicarbonyl compound (**163**). Cyclization to the cyclopentenone **164** occurs readily with dilute alkali. These last two sequences emphasize another valuable use of the dithiane molecule,

namely, it is now possible to mask two carbonyl functions in a molecule and regenerate *either one* by appropriate choice of conditions.

A recent innovation in dithiane chemistry was reported by Corey[62] which involved the 1,3-dithienium salt, **165**. This readily available crystalline material was found to undergo a smooth cycloaddition to conjugated dienes (butadiene, isoprene, etc.) affording the bicyclic adducts **166** in 85-95% yield. Treatment with butyllithium resulted in rearrangement to the vinylcyclopropane derivative **167** also in good yields. When heated (200°) in benzene solution the vinylcyclopropane underwent the expected rearrangement to the spirodithiane **168**

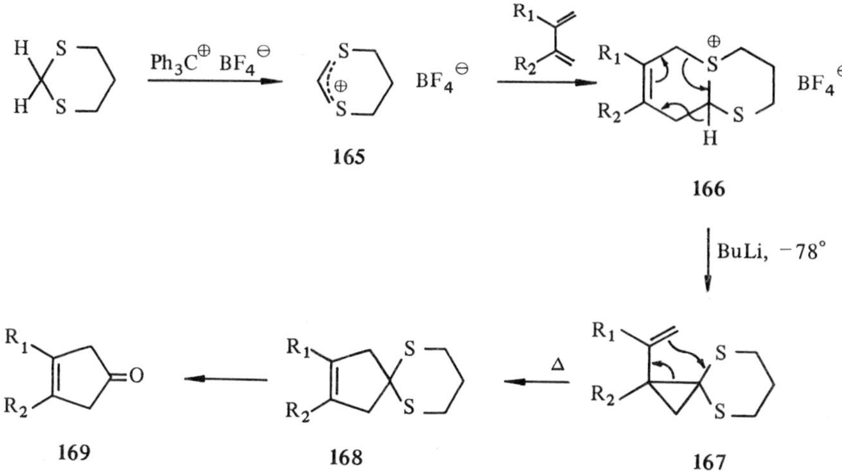

$R_1 = R_2 = H$
$R_1 = H, R_2 = Me$
$R_1 = R_2 = Me$

and the latter was ultimately cleaved (mercuric chloride-calcium carbonate) to the Δ^{-3}-cyclopentenone **169**. This technique possesses two significant features: *(a)* the use of the dithenium salt **165** as a masked substitute for carbon monoxide and *(b)* the neutral release of the ketone which avoided the production of the conjugated unsaturated isomer.

G. From 4-Boro-Δ^{-2}-oxazolines

A unique synthetic transformation has been reported[63] that, in effect, couples 2 moles of olefin with the synthetic equivalent of carbon monoxide and thus leads to ketones.

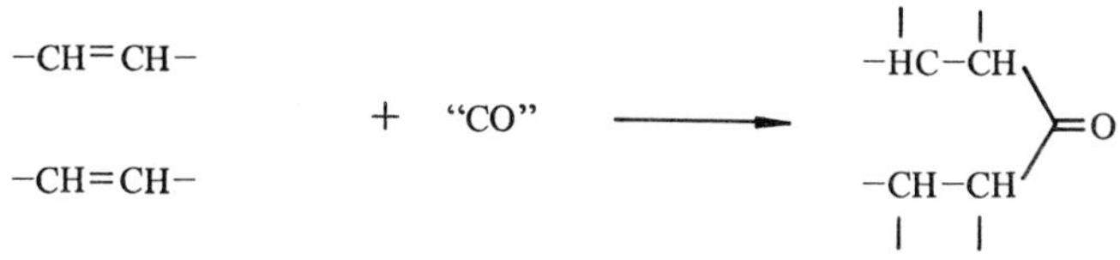

The method relies on the reaction of trialkylboranes (readily prepared from olefins and diborane) with sodium cyanide (the potential carbonyl group) to produce the cyanoborate **170**. Upon treatment with trifluoroacetic anhydride

170 is suggested to pass through the interesting heterocycle, 171, termed a

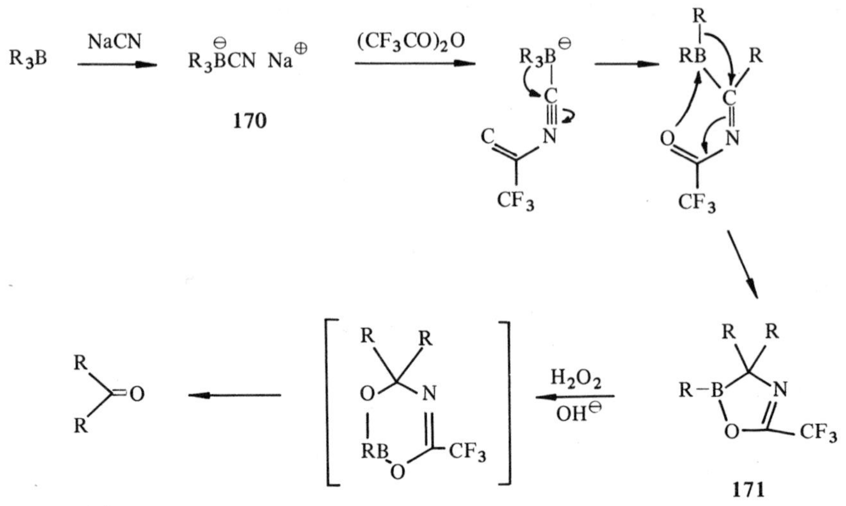

R	%
n-Octyl	95
n-Butyl	94
Cyclopentyl	84
Cyclohexyl	100
Norbornyl	99

boraoxazoline. Oxidation with alkaline hydrogen peroxide generates the ketones in very high yield. The entire process is carried out without isolation of any intermediates.

H. From Pyridine N-Oxides

A number of oxidative decarboxylations involving the use of pyridine N-oxide has been studied by Cohen.[64-66] By heating a solution of the α-bromo acids 172a-c in benzene, toluene, or xylene in the presence of pyridine N-oxide, the corresponding carbonyl compounds were obtained in fair yields. This one-carbon

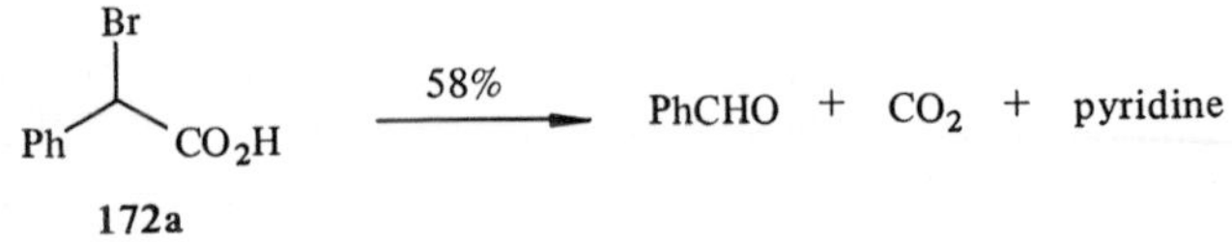

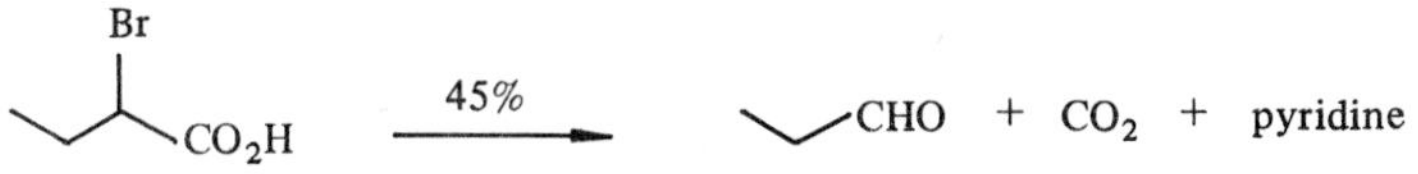

172b

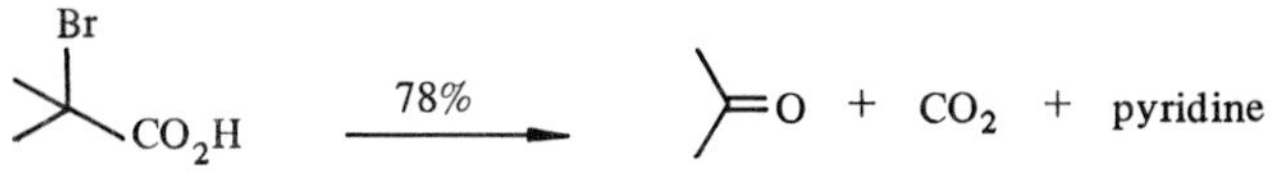

45% ⟶ CHO + CO$_2$ + pyridine

172c

78% ⟶ C=O + CO$_2$ + pyridine

degradative method appears to proceed via the intermediate **173**. Carboxylic

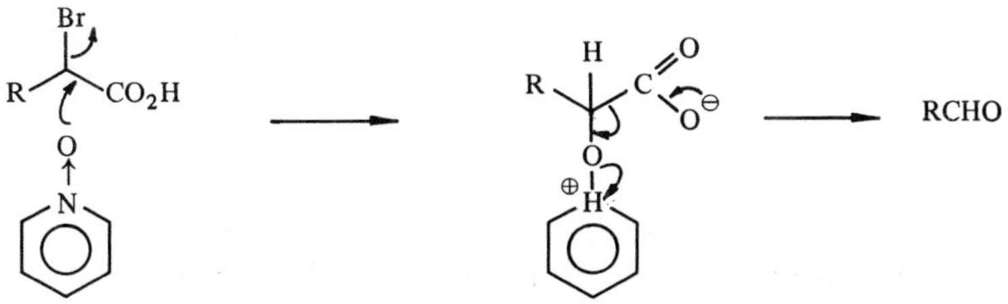

173

acids which do not contain an α-halo substituent have also been similarly degraded to their carbonyl derivatives using pyridine N-oxide. This sequence is somewhat less obvious and the mechanism seems to be more complex. Thus acids **174** have undergone a four election decarboxylative oxidation to the

Ph⌒CO$_2$H   68% ⟶   PhCHO + CO$_2$ +   pyridine

174a

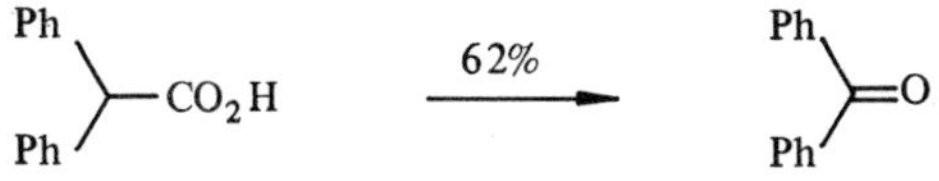

174b   62% ⟶

corresponding aldehydes or ketones. The reaction is performed in the presence of acetic anhydride and is postulated to proceed as follows:

R$-$CH$_2$CO$_2$H + O$\leftarrow$N⬡ $\longrightarrow$ [R$-$CH$_2$C(O)O–N⬡$^\oplus$ $\rightleftharpoons$ R$-$CH=C(OH)O–N⬡$^\oplus$] $\overset{\ominus}{\text{OH}}$ or O$\overset{\ominus}{\text{Ac}}$

(or mixed anhydride)

$\downarrow$ C$_5$H$_5$NO

C$_5$H$_5$N + CO$_2$ + RCHO $\longleftarrow$ R$-$CH$-$CO$_2$H + C$_5$H$_5$N

with O–N$^\oplus$⬡ substituent

175

The key step is the formation of intermediate **175** from the starting carboxylic acid since it is identical to the intermediate in the degradation of α-halo acids discussed above. The route by which **175** forms is still somewhat unclear but considerable support for the indicated mechanism has been accumulated. Anhydrides have also been oxidatively decarboxylated to aldehydes using pryidine N-oxide. Both Ruchardt[67] and Cohen[68] described simultaneously the reaction of anhydrides **176** with pyridine N-oxides affording the O-acyl

RCH$_2$$-$C(O)O$^\ominus$$-N^\oplus$⬡ ... RCH$_2$$-$C(O)$-$ $\longrightarrow$ RCH$_2$C(O)$-$O$-$N$^\oplus$⬡ $\longrightarrow$ RCH$_2$$^\oplus$

176

O$^\ominus$$-N^\oplus$⬡   **177**   **179**

⬡N$\rightarrow$O

RCHO $\xrightarrow[-\text{C}_5\text{H}_5\text{NH}^\oplus]{-\text{C}_5\text{H}_5\text{N}}$ RCH$_2$O$-$N$^\oplus$⬡

180   **178**

pyridine *N*-oxide **177**. The major feature of this intermediate is that it allows *nucleophilic* attack on the α-position of a carboxylic acid in contrast to the *electrophilic* substitution normally observed in acid derivatives. Thus pyridine *N*-oxide reacts nucleophilically at the α-position of **177** affording the *O*-alkyl pyridine *N*-oxide **178** with concomitant loss of carbon dioxide and pyridine. In certain instances, the carbonium ion **179** may be a discreet intermediate (when R = phenyl or when **179** is of the benzhydryl cation type) which is formed by spontaneous decarboxylation of **177**. The fragmentation of **178** to carbonyl compounds is not unexpected in view of previous work already mentioned in Section I.A of this chapter. Cohen found that the reaction may be made more convenient by using carboxylic acids and acetic anhydride forming *in situ* mixed anhydrides rather than the less accessible anhydrides of the carboxylic acid.

III. CARBONYL COMPOUNDS WITH TWO-CARBON CHANGE

$$(R-C) + (C-C=O) \longrightarrow R-C-C-C=O$$

This section deals with the synthesis of carbonyl compounds obtained via a two-carbon homologation of electrophiles or nucleophiles utilizing a heterocycle as a vehicle, precursor, or intermediate.

A. From Dihydro-1,3-Oxazines

A synthesis of carbonyl compounds (both aldehydes and ketones) was developed by the author and his co-workers based on the dihydro-1,3-oxazine system **181**.[69] The readily formed oxazine carbanion **182** was found to be highly nucleophilic at −78° and reacted with a number of electrophiles (alkyl halides, carbonyl compounds, epoxides, etc.) producing the elaborated oxazine **183**

181 A = H, Ph   182   183   A

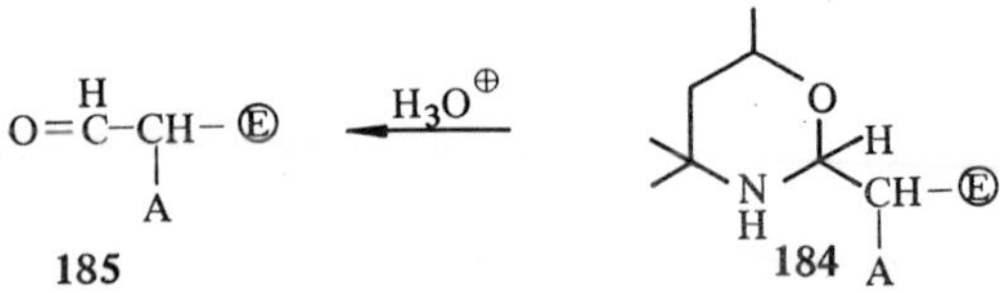

185 ⟵ H₃O⁺ 184

$$O=\overset{H}{\underset{A}{C}}-CH-\text{Ⓔ} \quad \xleftarrow{\ H_3O^{\oplus}\ } \quad 184$$

Ⓔ	Oxazine (181)	Aldehyde (185)
	H	
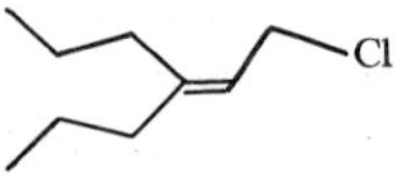	H	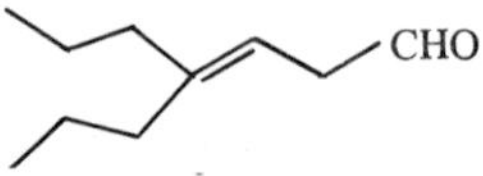
	Ph	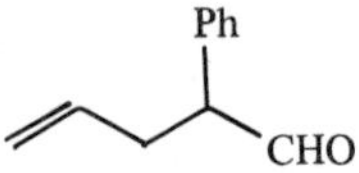
	Ph	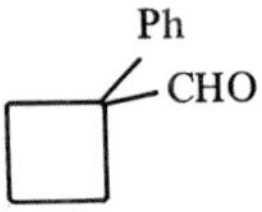
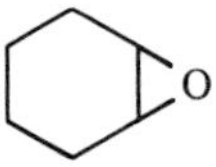	H	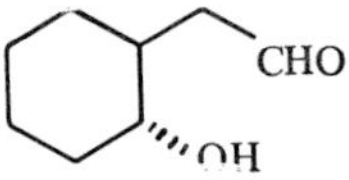
	Ph	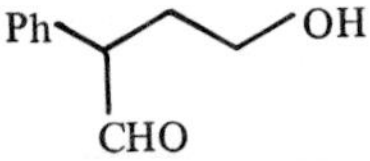
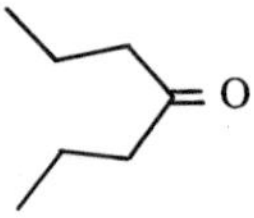	H	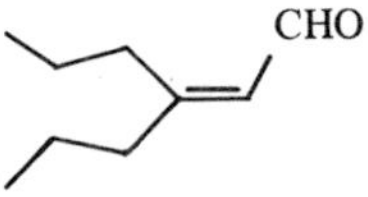
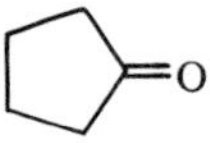	Ph	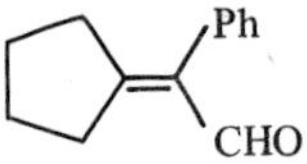
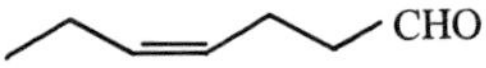	H	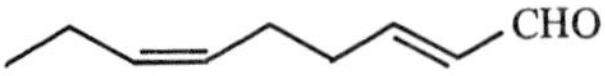
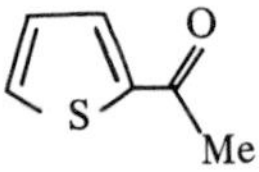	H	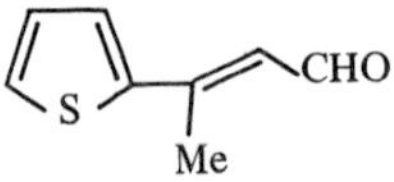
PhCHO	Ph	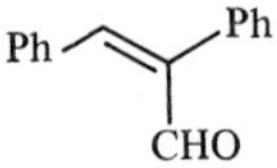

in high yield. The latter was reduced with aqueous sodium borohydride to the tetrahydro-1,3-oxazine **184** which is now recognizable merely as a protected aldehyde. Release of the aldehydic product **185** was accomplished by treatment with dilute aqueous acid, usually oxalic. In this fashion, a number of aldehydes were prepared including their C-1 deuterated derivatives when sodium boro-deuteride was employed in the reduction step. A representative list is given in tabulated form. This scheme is, in effect, a two-carbon homologation of electrophiles to aldehydes and may be considered as the aldehyde equivalent to the malonic ester, Knoevenagel, and Perkin-type reactions. Oxazines con-

186

187

R = alkyl, allyl, benzyl

taining the carboethoxy group **186** were also utilized as a source of α-formyl esters **187** by a similar sequence of manipulations. The value of this last scheme lies in the fact that it is tantamount to alkylation of protected α-formyl esters

188

189

190

188 with electrophiles. However, carbanions adjacent to the acetal linkage **189** are too fragile to be useful in synthesis because of acetal cleavage leading to structures of the type **190**.

Further applications of the oxazine-aldehyde synthesis allow the introduction of a wide range of functionality on the side chain. Since the oxazine is inert to the Grignard reagent, variations using Grignard chemistry are feasible. In this manner, the 2-bromoalkyl oxazine **191**, prepared by the addition of one

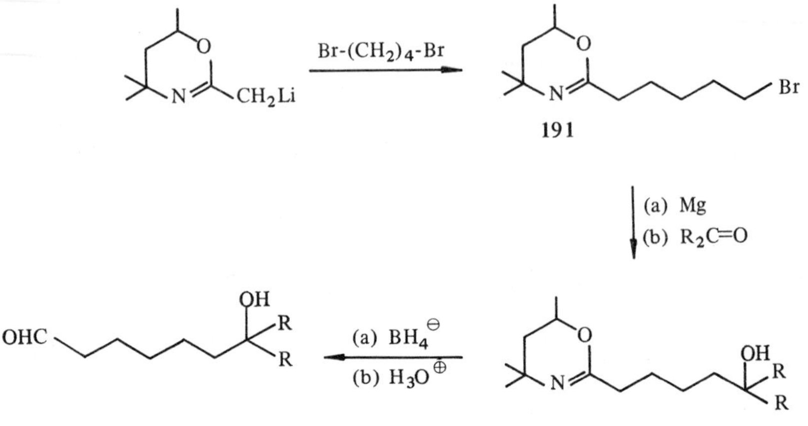

191

(a) Mg
(b) $R_2C=O$

193

(a) $BH_4^\ominus$
(b) $H_3O^\oplus$

192

equivalent of a dibromoalkane to the oxazine carbanion, may be transformed into its corresponding Grignard reagent and then into a tertiary (or secondary) alcohol **192** by introduction of a carbonyl compound. Reduction and hydrolysis generates the hydroxyaldehyde **193**.

More recent work on dihydro-1,3-oxazines has also revealed the synthetic utility of the 2-chloromethyl oxazine **194**.[70] The latter cleanly gave a carbanion

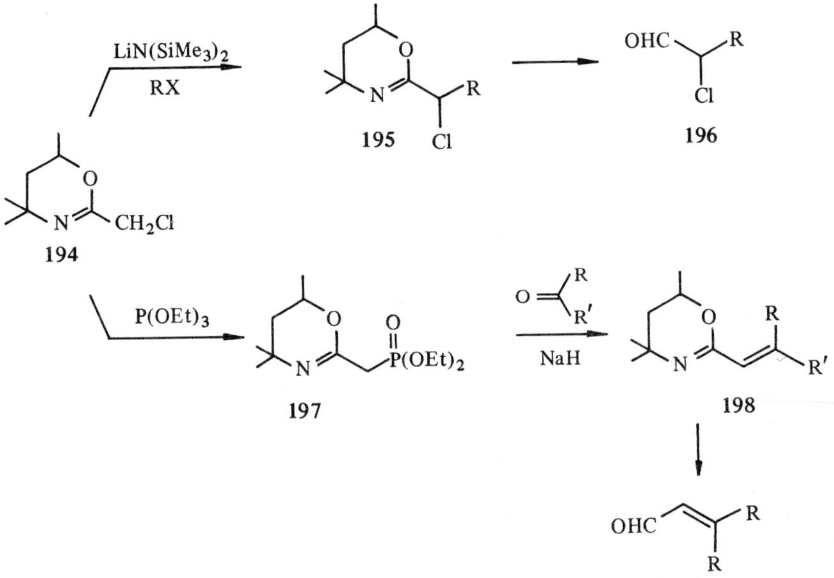

$LiN(SiMe_3)_2$
RX

195   Cl

196

194

$P(OEt)_3$

197

$$O=\overset{R}{\underset{R'}{\diagdown}}$$
NaH

198

199

when treated with lithio hexamethylsilazane and alkylated in high yield when added to alkyl halides. The resulting oxazine **195** was converted into the corresponding α-chloroaldehydes **196** when subjected to the usual borohydride reduction followed by hydrolysis. Furthermore, the 2-chloromethyl oxazine could be transformed into the phosphonate derivative **197** and, under the usual Wadsworth-Emmons technique,[71] coupled smoothly to the unsaturated oxazines **198**. Reduction and hydrolysis then afford the α,β-unsaturated aldehydes **199**.

A modification of the oxazine-aldehyde synthesis which eliminates the need for butyllithium and low temperatures has also been described.[72] The

(a) $BH_4^{\ominus}$/EtOH

(b) $H_3O^{\oplus}$

$OHC-CH_2R$

204

methiodide salt **200** of the 2-methyl oxazine was converted into its corresponding ketene-*N,O*-acetal **201** by treatment with sodium hydride in THF or DMF. This reagent is a highly nucleophilic enamine which may be alkylated with various alkyl iodides at room temperature affording the oxazinium salt **202**. The latter is effectively neutralized to its corresponding ketene *N,O*-acetal **203** by a second equivalent of sodium hydride introduced at the start of the sequence. Addition of ethanolic sodium borohydride reduces **203** to the tetrahydro-1,3-oxazine which is then hydrolyzed to the final aldehydic product **204**. The entire sequence, except for the hydrolysis step, is carried out in a single reaction vessel and close to ambient temperatures. The synthetic utility of the oxazine-ketene-*N,O*-acetal **201** toward the synthesis of acetic esters will be described in the next chapter which deals with carboxylic acids.

Ketones are also available from the dihydro-1,3-oxazine route by invoking some simple structural modifications. For example, the alkylation of the oxazine carbanion **205** by alkyl halides leading to **206** has already been de-

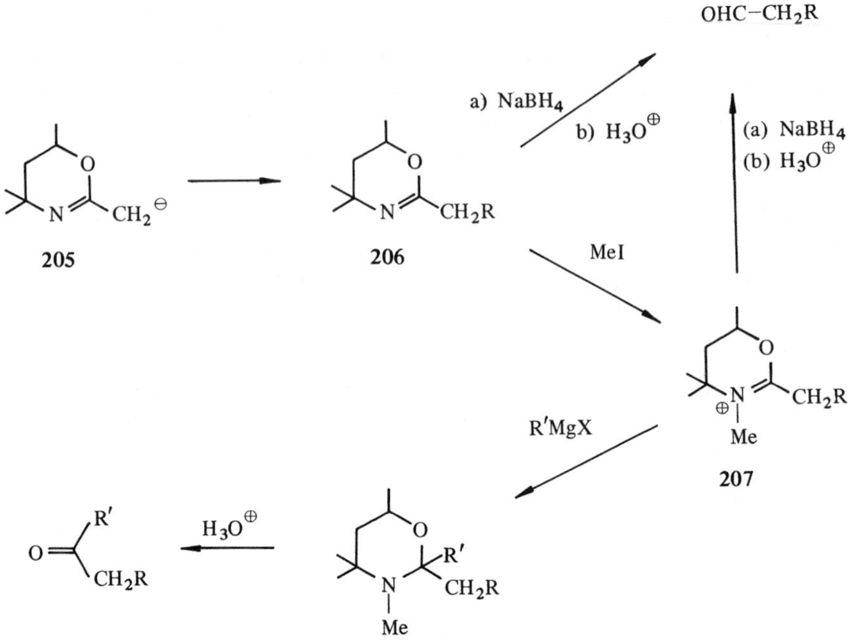

OHC–CH$_2$R

scribed. Introduction of hydride ion into the salt **207** produces, after hydrolysis, the homologated acetaldehydes, whereas introduction of a carbanionic species should produce ketones **209**. This route was, indeed, reported[73] from the methiodide salt **207** and Grignard (or organolithium) reagents which added to the iminium bond giving the ketone precursor **208**. The methiodide salt could be prepared *in situ* after formation of the elaborated oxazine **206**. Hydrolysis of **208** liberates the ketones **209** in good yield. As already mentioned, the C=N link of the oxazines (**206**) is inert to Grignard reagents and only by transforming this group into the more reactive iminium species (C=N⁺–Me) can addition be made to take place. An efficient synthesis of cyclopentenones **210** has been devised[74] using this oxazine ketone synthesis (Scheme 3).

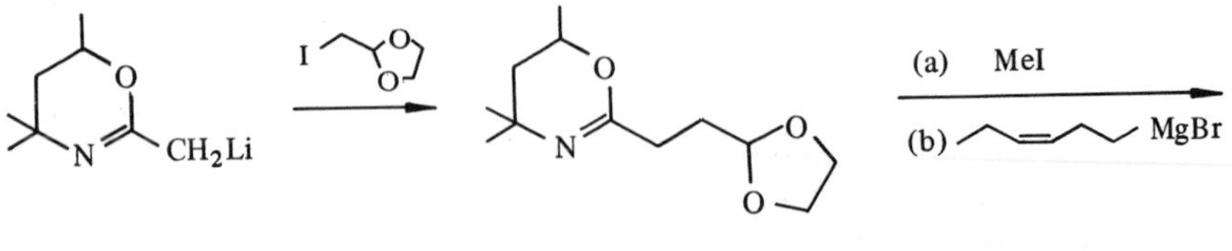

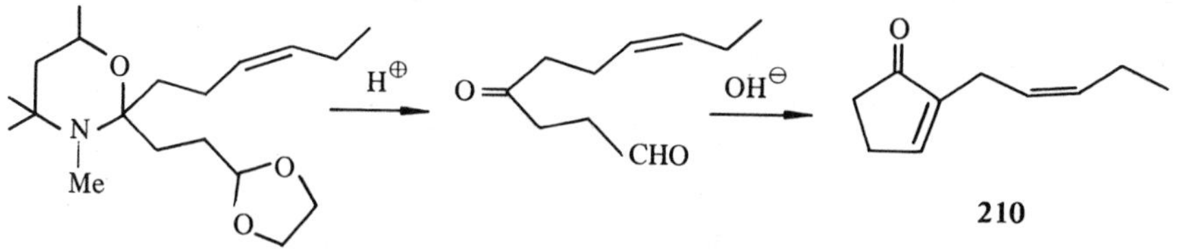

210

Scheme 3

Certain other structural features on the oxazines permit another ketone synthesis[75,76] which has considerable potential. Since the α-carbanion of 2-(dialkyl)oxazines **211** was not stable at the temperature formed (∿0-10°) a rapid rearrangement to the ketenimines **212** takes place. If two equivalents of

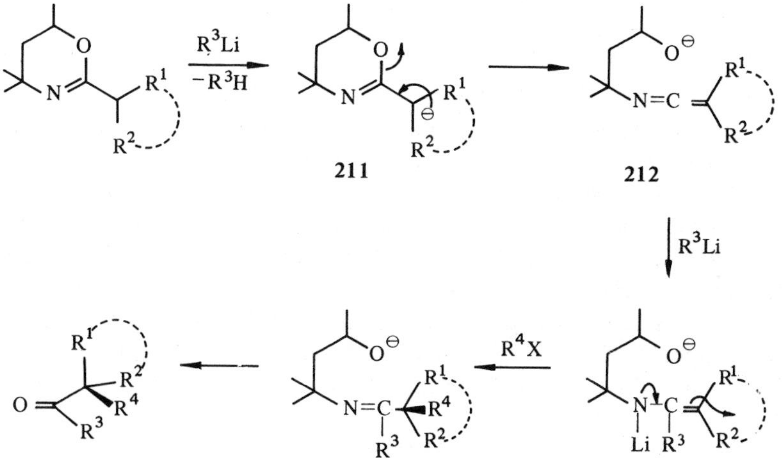

211 212 213 214 215

Oxazine				
R^1	R^2	R^3Li	R^4X	Ketone
Me	Me	Ph	Et	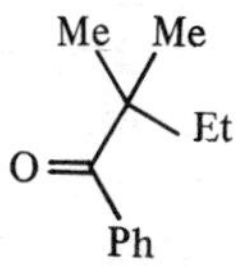

Me	Me	n-Bu	Me	
		n-Bu	Me	
Ph	n-Amyl	Et	Me	

the organolithium base are present, the remaining base adds quickly to the ketenimine producing the adduct **213** which in essence represents a metallated enamine. The well-known property of metallated enamines[77] was now exhibited by smooth alkylation of **213** to the alkyl imine **214**. Hydrolysis leads to the α-(quaternary carbon) ketones **215**. It is worthy to mention that, once again, the entire process, save the hydrolysis, is performed in a sequential manner in a single reaction vessel. Preparation of the starting oxazines for use in synthesis of carbonyl compounds is accomplished as mentioned in Section I.F. In a subsequent section of this chapter (IV.A) a three-carbon homologation to aldehydes and ketones will be discussed.

B.   From 1,2-Isoxazines

The cycloaddition of α-chloronitrones **216** to simple olefins under the influence of silver salts leads to high yields of isoxazinium salts **217**. Eschenmoser[78,79] has demonstrated the synthetic prowess of this novel class of heterocycles by the development of an aldehyde synthesis. The neutralization of **217** with carbonate base results in the isoxazine **218** which undergoes smooth fragmentation to the imine **219**. Hydrolysis produces the unsaturated aldehyde **220** in good yield. The net result of this sequence is an oxidative cleavage of the olefinic linkage with a two-carbon homologation of one of the carbonyl groups. Of equal significance is the fact that the product **220** is a trisubstituted olefin—a class of compounds that represents an important precursor for complex synthetic programs.

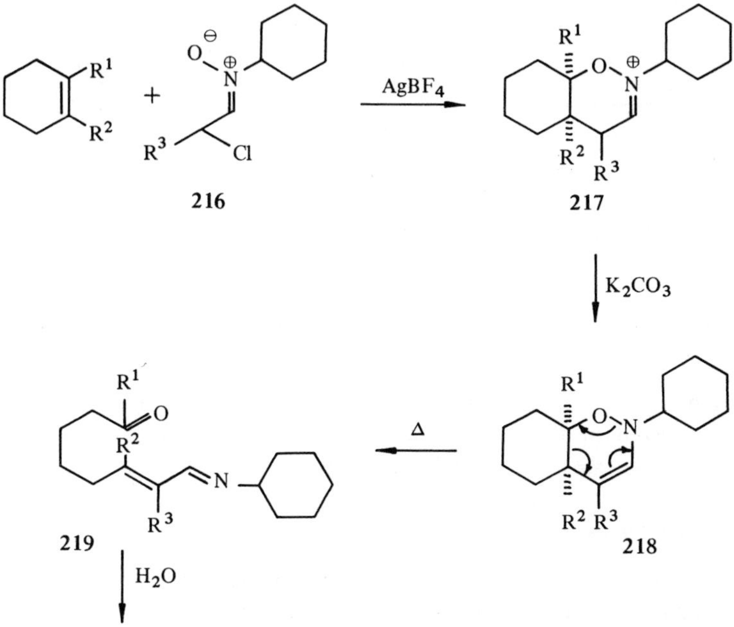

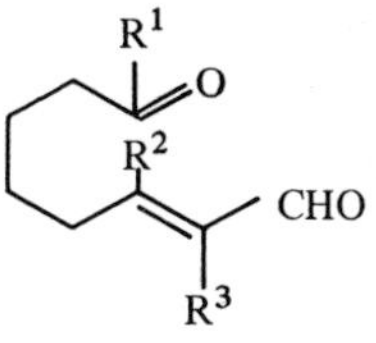

R^1	R^2	R^3
Me	H	H
H	H	Me
Me	Me	H
Me	Me	Me

C. From Thiazoles

An aldehyde synthesis based on the thiazole system[80] has been described which is conceptually very similar to the oxazine-aldehyde synthesis (Section III.A). Metallation of the 2-methyl thiazole **221** using *n*-butyllithium generates the lithiomethyl thiazole which was alkylated with benzyl bromide affording the 2-phenethyl derivative **222**. The thiazole was converted to its *N*-methyl quaternary salt **223** so that the C=N and C=C bonds could be reduced (sodium borohydride) to the thiazolidine **224**. The aldehyde **225** was released from the latter, by employing mercuric salts in an aqueous solution. The key feature of this method lies in the neutral release of the aldehyde thus avoiding acidic conditions utilized in the oxazine-aldehyde synthesis. This aspect would be important in those cases where aldehydes were acid sensitive.

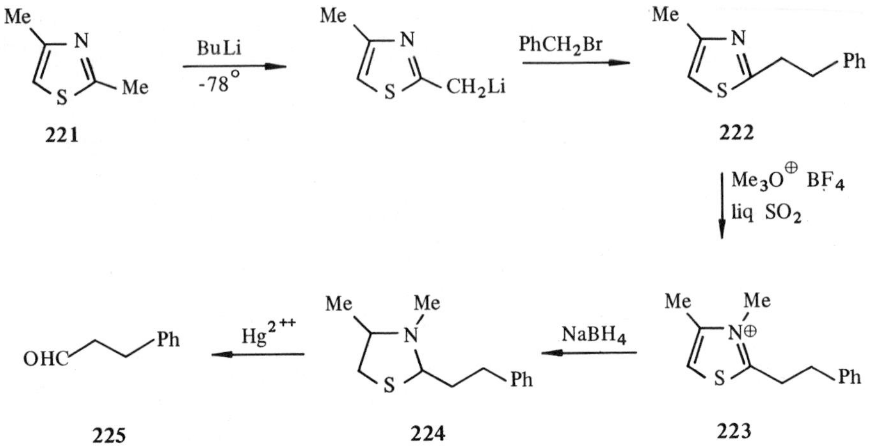

D. From Thiazolines

Aldehydes have also been obtained[81] by employing the simple 2-thiazoline derivative **226**. Treatment with butyllithium to produce the lithio thiazoline was followed by introduction of various alkyl bromides or iodides generating

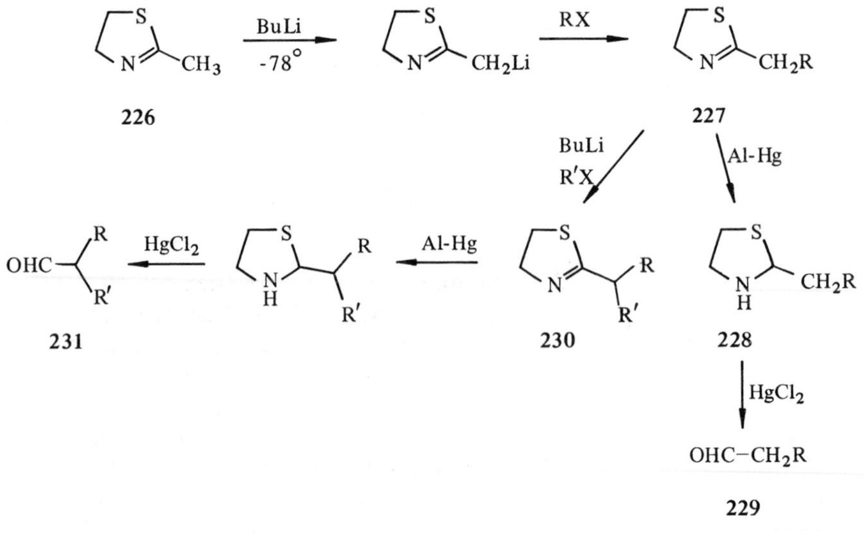

R	R'	Aldehyde
PhCH$_2$	–	 Ph⌒⌒CHO
PhCH$_2$	Et	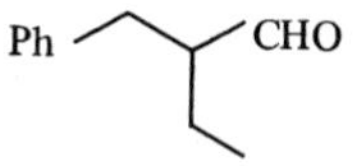 Ph⌒⌒CHO (branched)
n-Bu	Et	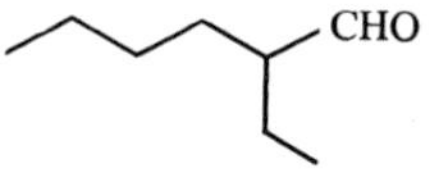 ⌒⌒⌒CHO (branched)

the alkylated thiazoline **227**. The C=N link was cleanly reduced using aluminum-amalgam in moist ether providing the thiazolidine **228**. Treatment with mercuric chloride in aqueous acetonitrile gave the aldehyde **229** in 40–50% overall yield.

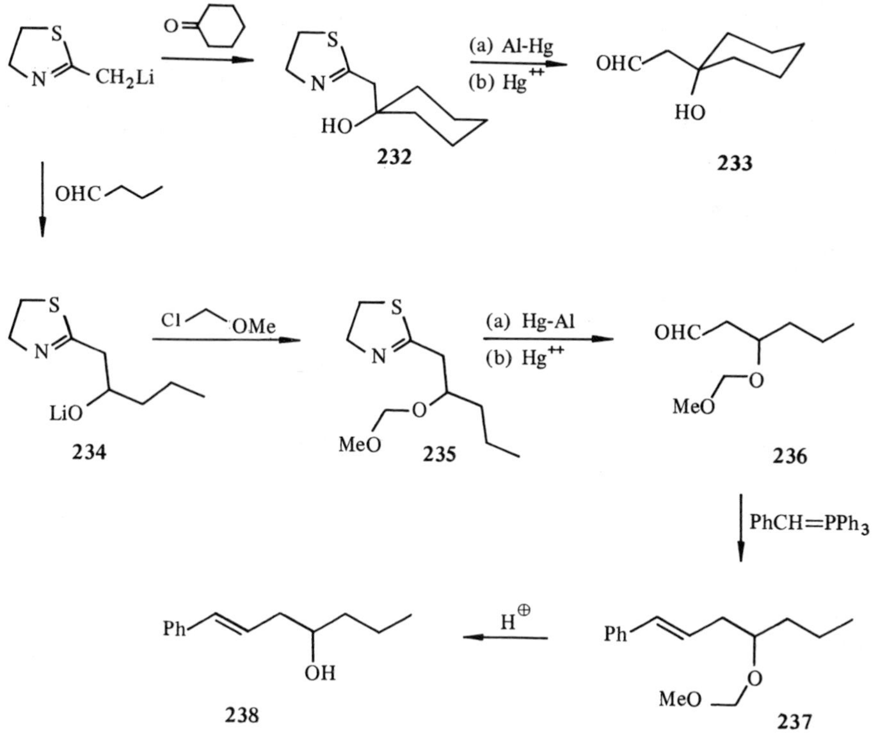

232

233

234

235

236

237

238

This method, like that described in the preceding section, possesses the advantage of releasing the aldehydic product under neutral conditions. Furthermore, the initially alkylated thiazoline may be treated with a second equivalent of base and the lithio salt alkylated again producing the dialkylthiazoline 230 which eventually gives the 2,2-dialkylacetaldehydes 231.

An interesting aspect of the thiazoline-aldehyde synthesis is seen by the products (232) obtained when the lithio thiazoline is treated with a carbonyl compound. The fact that the aldehydic product is released under neutral conditions now allows the preparation of β-hydroxy aldehydes 233—a heretofore elusive class of compounds. The tendency of β-hydroxyaldehydes to either lose water and pass on to their α,β-unsaturated derivatives or revert back to acetaldehyde and the carbonyl component is a well-known fact. However, in many cases observed in this sequence, the β-hydroxyaldehyde could be isolated. A synthesis of homoallylic alcohols[82] has been developed by taking advantage of this fact. The technique may be exemplified by the reaction of n-butyraldehyde with the lithio thiazoline which gives the adduct 234 as the *in situ* initial product. Introduction of chloromethyl ether in the same vessel transforms the O-lithio salt into the ether 235, which is reduced with aluminum amalgam and cleaved with mercuric chloride, furnishing the ether-aldehyde 236. Coupling of the latter with a typical phosphorane leads to the olefin 237 and after hydrolysis to remove the labile ether masking group, affords the homoallylic alcohol 238. There are many useful reactions that can be performed with the protected β-hydroxyaldehyde 236 and thiazoline 235 and only continued study will determine its full potential.

E.   From Sulfolenes

The well-known extrusion of sulfur dioxide from sulfolenes leading to dienes has already been described (Chapter 4, Section I). This behavior provides the basis for a new synthesis of 1,4-diaryl-2,3-butanediones 242 from primary alkyl

chlorides.[83] The latter are transformed into dialkyl sulfones 239, via oxidation of their corresponding sulfides, and are then condensed with ethyl oxalate producing the diketosulfolanes 240 in equilibrium with their enol tautomers. Reduction using zinc and acetic acid produces the tautomers 241. In the presence of sodium acetate, an acetic acid solution of **241** rapidly evolved sulfur dioxide upon heating and gave the diketone **242** in quantitative yield. The method was attempted successfully with four benzylic chloride derivatives and may be considered as a two-carbon insertive coupling reaction (dotted arrow) leading to a-diketones. Once again, the sulfolene has proven to be an extremely versatile vehicle for synthesis of functionalized molecules. It should also be noted that there is considerable potential for further elaboration in this process since the sulfolene may be alkylated[84] at the carbon adjacent to both the carbonyl and sulfone groups (**240 → 243**). This modification would, when the carried out to completion, lead to dialkylated diketones **244**.

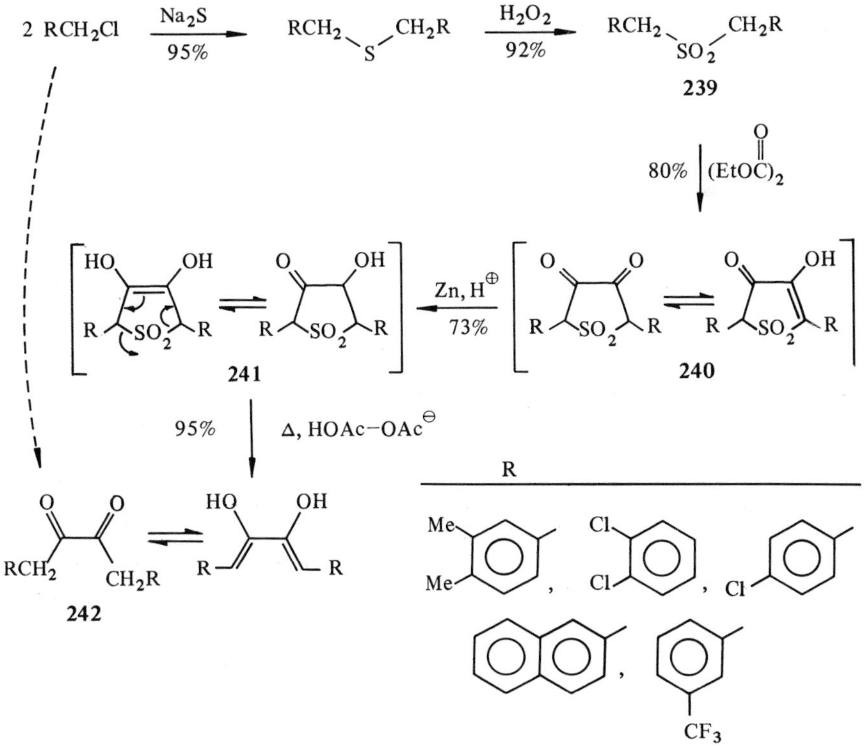

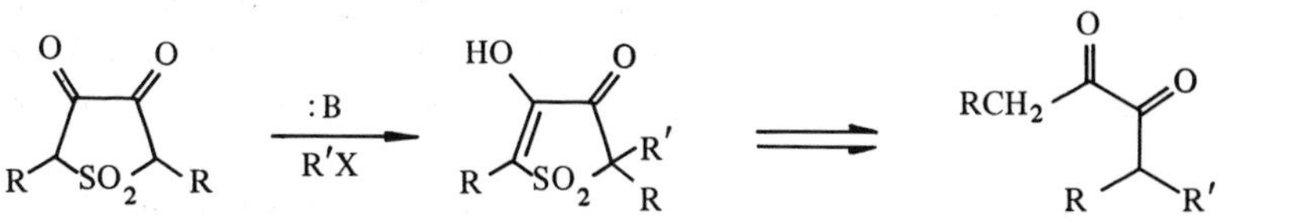

F. From Thiiranes

The now famous "sulfide contraction" (Scheme 4), which arose from the total synthetic effort of Vitamin B-12 by the Harvard-ETH groups (Chapter 1, Section I.D), has been used as a general tool in the synthesis of 1,3-dicarbonyl compounds 245 and enaminoketones 246.[85]

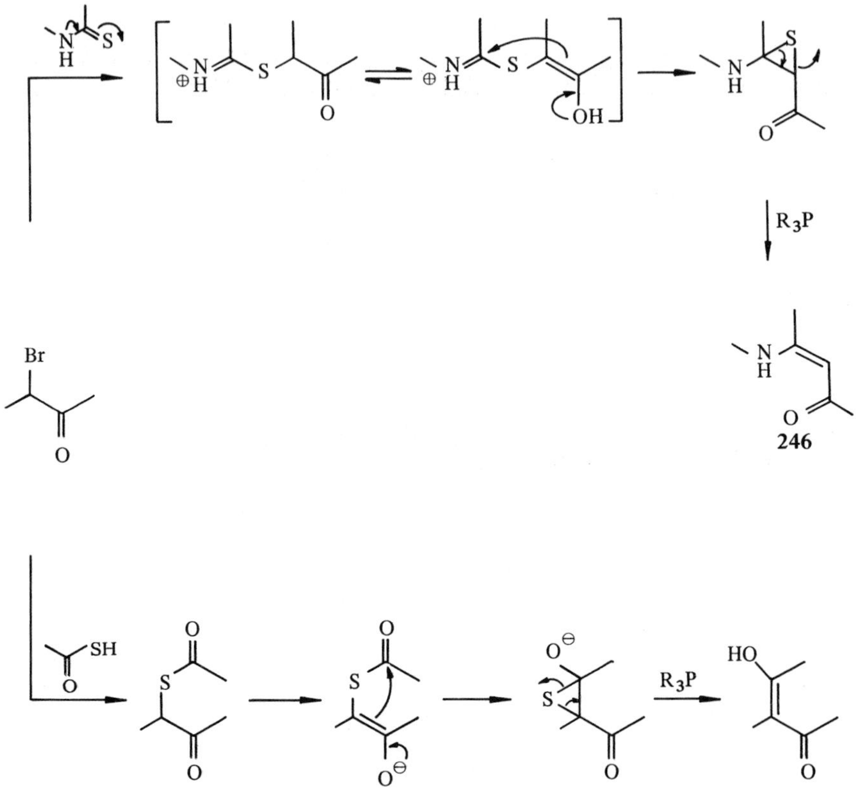

Scheme 4

The utility of the heterocycle, thiirane, in this transformation is quite evident from the scheme even though its presence is merely transient. The important property of thiiranes, namely its ability to lose sulfur under the influence of tervalent phosphorous compounds, has already been described in Chapter 3, Section I.

The route to 1,3-dicarbonyl compounds requires the initial formation of a thioester 247 which is then contracted according to Scheme 4. A variety of

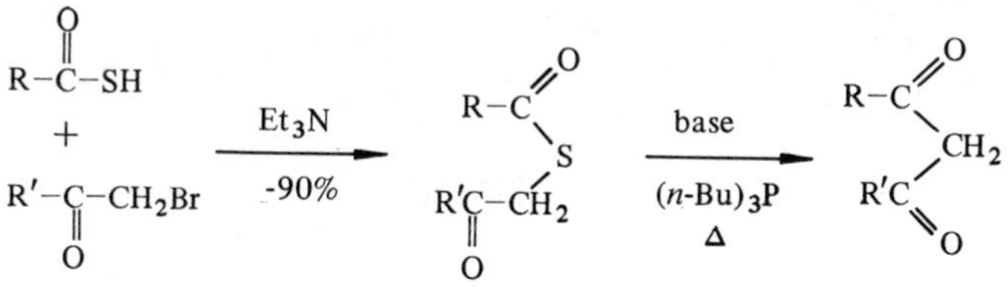

247

thioacids and α-haloketones or aldehydes may be used as starting materials. Alkylated diketones **248** akin to those formed in the classical enolate alkylations are also formed in good yield by using secondary bromoketones.

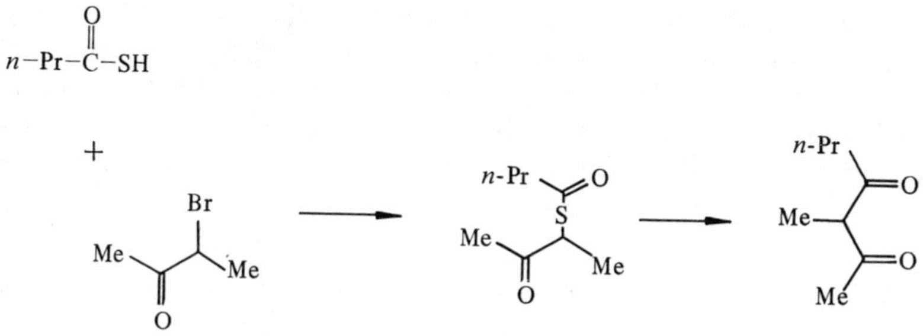

248

If thioamides **249** are employed with halo carbonyl compounds the reaction leads to vinylogous amides (enaminoketones) **250**.

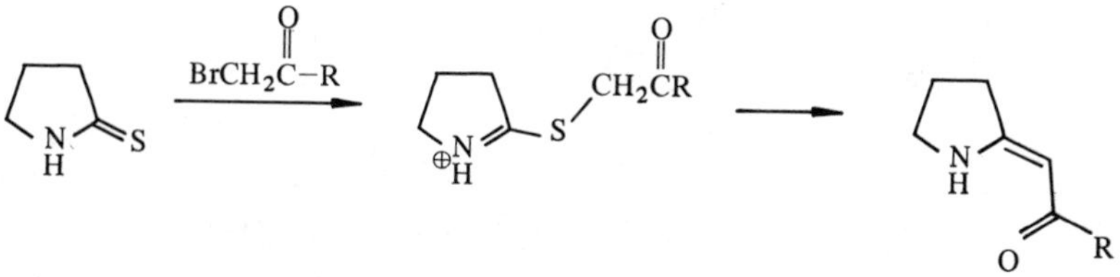

249

250

G. From Oxiranes

The temporary construction of an oxirane moiety **253** in vinyl silanes allows a novel approach to aldehydes[86] **254** under mild conditions. Thus 1-octyne **(251,** R = *n*-Hex) is transformed into the vinyl silane **252** by addition of

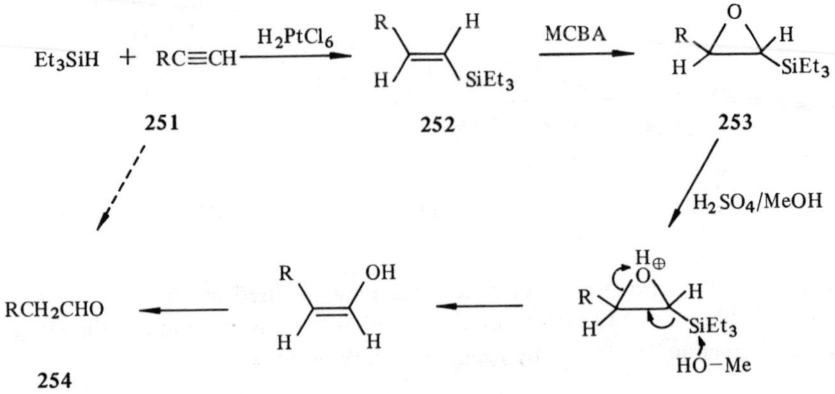

$$Et_3SiH \ + \ RC{\equiv}CH \xrightarrow{H_2PtCl_6}$$

251   252   253

RCH_2CHO

254

triethylsilane in the presence of chloroplatinic acid. Epoxidation using *m*-chloroperbenzoic acid leads to the epoxysilane **253** which rearranges smoothly in the presence of methanolic sulfuric acid to the aldehyde **254**. The transformation of terminal acetylenes to aldehydes, as depicted above, represents an alternative to the hydroboration technique which also converts acetylenes to aldehydes.[87]

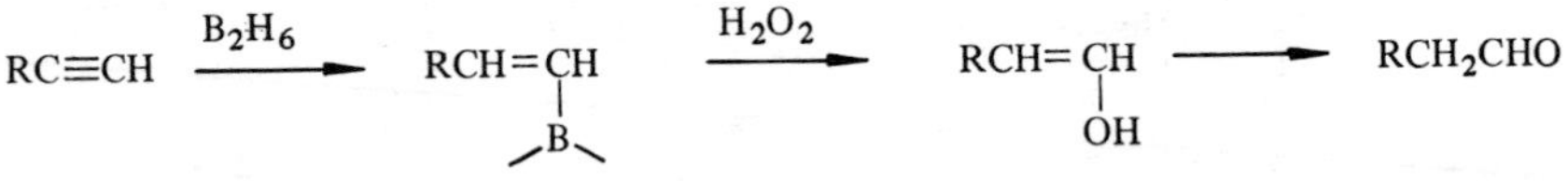

$$RC{\equiv}CH \xrightarrow{B_2H_6} RCH{=}\underset{B}{\underset{|}{CH}} \xrightarrow{H_2O_2} RCH{=}\underset{OH}{\underset{|}{CH}} \longrightarrow RCH_2CHO$$

A two- or more-carbon homologation via the epoxy silane is also possible by starting with trimethylsilyl acetylene **255**. This system may be readily elabo-

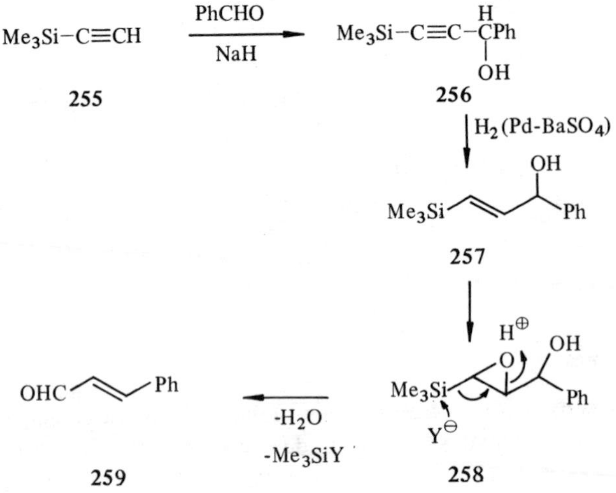

$$Me_3Si{-}C{\equiv}CH \xrightarrow[NaH]{PhCHO} Me_3Si{-}C{\equiv}C{-}\underset{OH}{\underset{|}{\overset{H}{C}Ph}}$$

255   256

$$\downarrow H_2 \ (Pd{-}BaSO_4)$$

257

OHC=CH—Ph $\xleftarrow[-Me_3SiY]{-H_2O}$ 258

259

rated to the acetylenic alcohol **256** using its sodium or other metal salts. Partial reduction to the olefinic alcohol **257** is followed by epoxidation to **258**. Acidic rearrangement then produces the unsaturated aldehyde **259**.

IV. CARBONYL COMPOUNDS WITH THREE-CARBON CHANGE

$$(R-C) + (C-C-C=O) \longrightarrow R-C-C-C-C=O$$

Only a few cases have been reported where a heterocycle has assisted in three-carbon homologations leading to carbonyl compounds.

A. From Dihydro-1,3-oxazines

Carbonyl compound syntheses from oxazines involving zero-, one-, and two-carbon change in chain length have been described in earlier sections of this chapter. The versatile oxazine system is also capable of effecting a three-carbon chain extension leading to aldehydes and ketones. Addition of organometallic reagents to the 2-vinyl oxazine **260** invariably leads to polymeric products presumably via rapid reaction of the magnesio-salt **261** with the vinyl oxazine. However, if an alkyl iodide is placed in solution with the vinyl oxazine

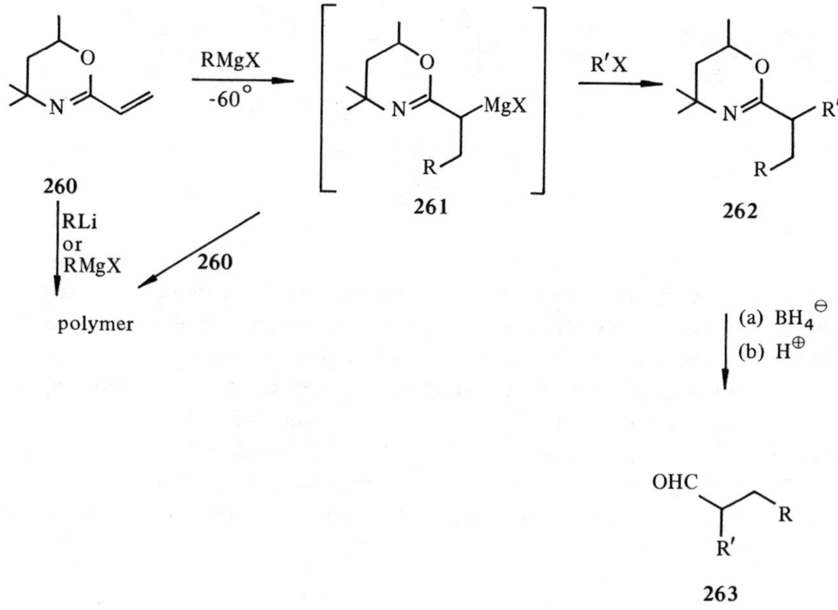

RMgX	R'X
Ph	Me
Ph	PhCH$_2$
Ph	Allyl
Ph	Et ⎫
Me	PhCH$_2$ ⎬ same product
Et	PhCH$_2$
Allyl	PhCH$_2$

and treated with a Grignard reagent, the initially formed magnesio-salt **261** is efficiently intercepted affording the α,β-dialkyl oxazine **262** in high yield.[69] Reduction with sodium borohydride and hydrolytic cleavage releases the aldehyde **263**. The net result of this scheme may be envisioned as a 1,4-addition of Grignard reagents to acrolein followed by alkylation of the intermediate enolate (Scheme 5). Since acrolein is well known to produce 1,2-addition

Scheme 5

products[88] with organometallics, this behavior of the vinyl oxazine takes on added significance. A number of α,β-dialkylpropionaldehydes were prepared using this technique and these are presented in tabular form.

The method may be modified to allow the preparation[89] of various ketones **267** and **269** as well as aldehydes **271**. By starting with the 2-isopropylidene **264** (R^1 = Me) or the 2-(α-styryl) **264**, (R^1 = Ph) oxazines, addition of organometallics (RMgX or RLi) occurs without polymerization affording the keten-imine **265**. This behavior is in stark contrast to the addition of organometallics

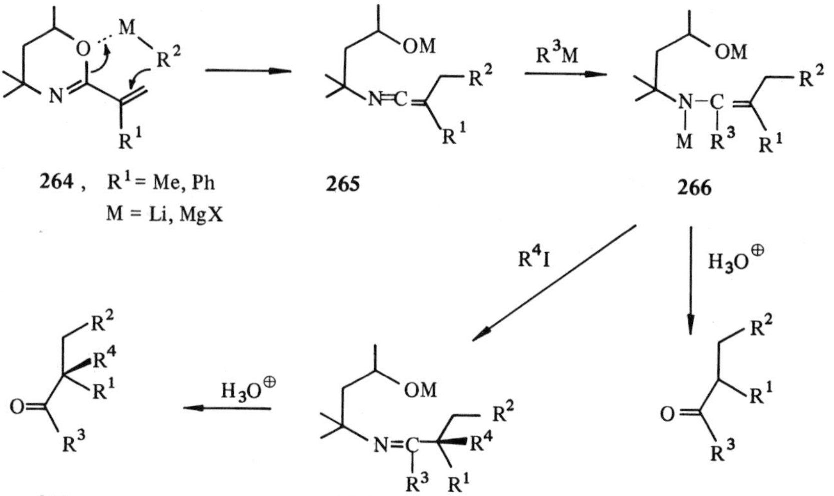

264 ,   R^1 = Me, Ph
M = Li, MgX

265

266

269

268

267

to the 2-vinyl oxazine **260** which polymerizes readily. Addition of a second equivalent of organometallic (either the same or different from the first) produces the metallo enamine **266** which may be *(a)* hydrolyzed to the ketone **267** or *(b)* treated with an alkyl iodide generating the α-quaternary imine **268**. Oxalic acid treatment of the latter produces the highly substituted ketone **269** in good overall yield. On the other hand, if the initially formed ketenimine **265** is quenched, recyclization to the elaborated oxazine **270** occurs and reduction followed by hydrolysis provides the α-methyl or α-phenyl aldehydes **271**. The versatility of this process is demonstrated by a few representative examples given below. The synthetic equivalent of **264** is shown at the head ot the table so that the reader may more clearly appreciate the three-carbon unit, "O=C–C=C," which is actually involved in the construction of the carbonyl compounds.

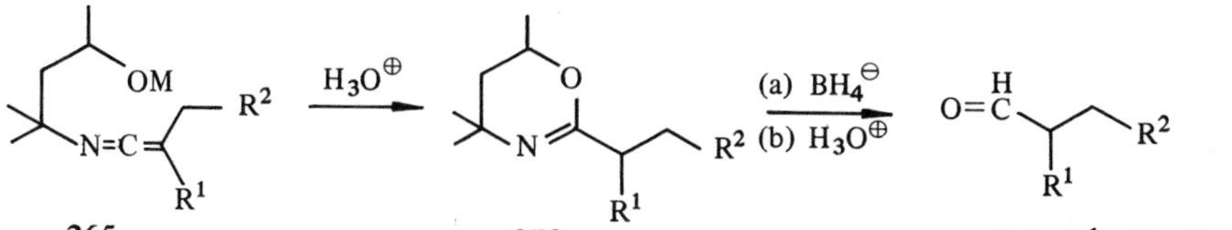

265

270

271   R^1 = Me, Ph

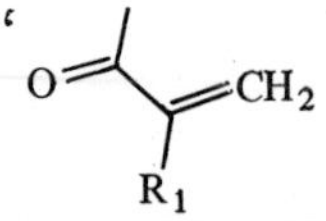

"O=C(R₁)−CH₂" structure header

R_1	R^2M	R^3M	R^4I	Product	
Me	t-BuLi	NaBH₄	–	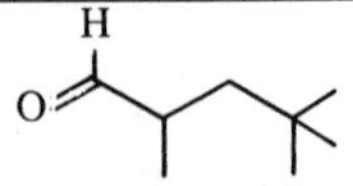	(271)
Me	CyclohexMgBr	NaBH₄	–	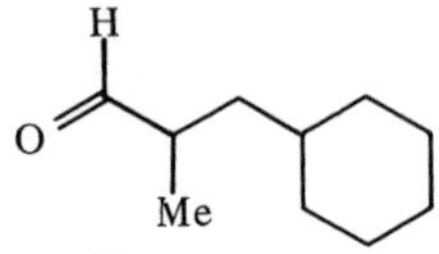	(271)
Ph	CyclohexMgBr	NaBH₄	–	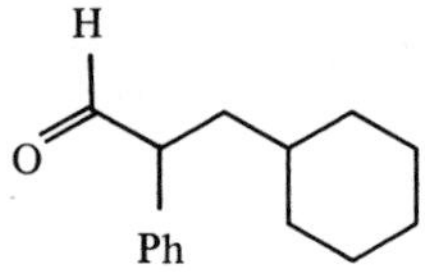	(271)
Me	CyclohexMgBr	MeLi	–	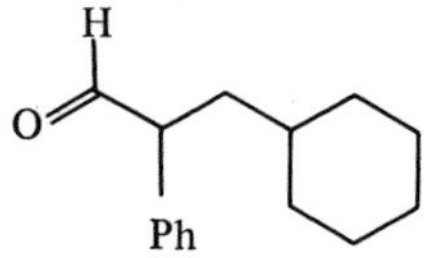	(267)
Ph	t-BuLi	MeLi	–	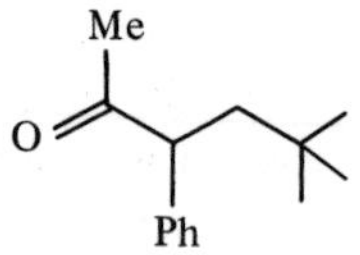	(267)
Ph	n-BuMgBr	n-BuMgBr	–	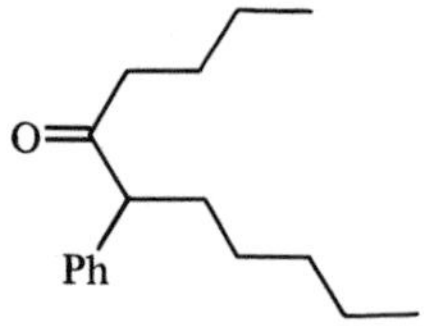	(267)
Me	CyclohexMgBr	MeLi	MeI	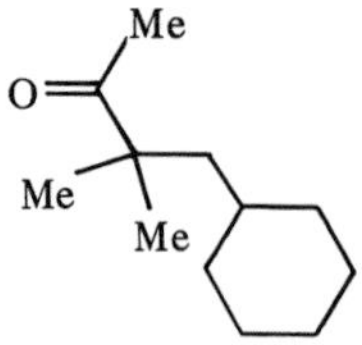	(269)

B.　From 2-Oxa-3-borenes

A recent method[90] for preparing a,β-unsaturated ketones **273** starting from 1-alkynes **275** involves the oxaborene **272**. The latter is prepared in 65% yield by treating the trialkyl(1-alkynyl)borate **274** with acetyl chloride resulting in alkyl B to C shifts. Jones oxidation of the heterocycle **272** generates the unsaturated ketone in approximately 30-40% overall yield. This is an interesting

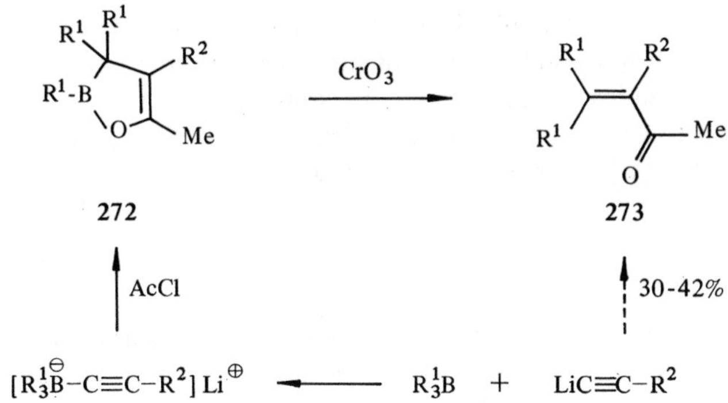

272 → 273

$[R_3^1\overset{\ominus}{B}-C\equiv C-R^2]\,Li^{\oplus}$ ⟵ R_3^1B + $LiC\equiv C-R^2$

274   275

R^1	R^2
n-Amyl	i-Pr
n-Amyl	n-Bu
Ph	i-Pr
Ph	n-Bu

method which, once again, calls for the temporary construction of a useful heterocycle to achieve the synthetic goal. The net transformation (dotted arrow) involves gem-alkylation at C-1 and acylation at C-2 of the starting acetylene.

V.   CARBONYL COMPOUNDS WITH FOUR-CARBON CHANGE

$$(R-) + (C-C-C-C=O) \longrightarrow R-C-C-C-C=O$$

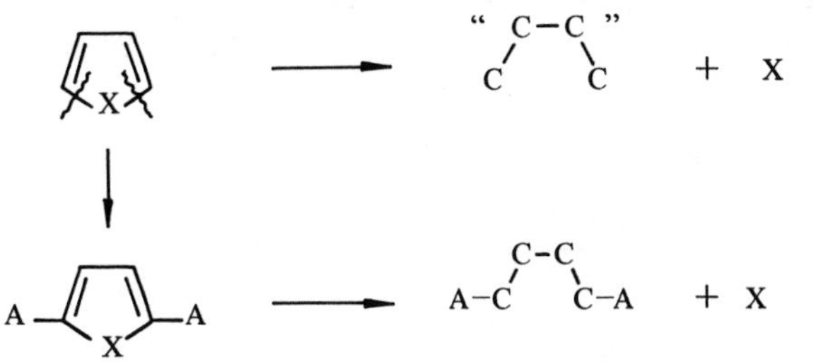

Certain five-membered heterocyclic systems (e.g., furan, thiophene) contain structural features that allow the ring to be elaborated with substituents, A, and then fragmented to a four-carbon unit by elimination of the heteroatom, X, (as S, H_2S, or H_2O). In this fashion a number of carbonyl compounds have been obtained by using thiophenes and furans as a template upon which were introduced the appropriate structural moieties. Destruction of the heterocycle via the scheme above then generated the desired products.

This concept of synthesis has been termed "latent functionality" by Lednicer in a recent excellent review[91] where a discussion of many functionalized molecules is presented.

A. From Furans

Ring opening reactions of furans has been reviewed in an extensive discussion on furan chemistry by Bossard and Eugster.[92] These authors, as well as Lednicer,[91] have emphasized the poor preparative value of direct hydrolysis of furans to 1,4-diketones **276** because of extensive polymerization. They suggest

the initial conversion to 2,5-dimethoxy dihydrofurans **277** and then to the tetra-hydrofuran **278** prior to hydrolysis. The transformation of furans to **277** may be done chemically (bromine, methanol) or electrochemically (ammonium bromide-methanol) and its synthetic value has been reviewed by Elming.[93] Nevertheless, in recent years, there have been several successful procedures to convert furans to their ring opened carbonyl derivatives which now render the molecule an even more valuable precursor in synthesis.

Johnson[94,95] has utilized the furan system **279** as a convenient starting point in a biogenetic synthesis of (±)-16, 17 dehydroprogesterone. The key inter-mediate required for this approach was the triene-dione **283** and was constructed by coupling two fragments, 281 and 282. The synthesis of the former is of direct concern to the subject matter at hand and demonstrates an improved method for obtaining dicarbonyl compounds. Metallation of 2-methylfuran followed by alkylation gave the necessary carbon skeleton in the form of the bromopropyl furan 280. Since direct aqueous hydrolysis of furans was known to be precarious, the cleavage was performed in ethylene glycol-benzene containing a small quantity of p-toluene sulfonic acid. In this manner, yields of the bis-ketal 281 were obtained which were consistently in the 70-90% range. Coupling with 282, followed by mild hydrolysis, produced the diketone 283 in good yields. These experiments indicate that by trapping the dicarbonyl product arising from the direct acid cleavage of furans, polymerization and undesired side products may be effectively minimized.

The more popular approach to 1,4-dicarbonyl compounds from furans, namely conversion to the 2,5-dialkoxy dihydrofurans 277 prior to cleavage has been employed in reaching the prostaglandin precursor 289.[96] Friedel-Crafts acylation of furan-2-acetic ester 284 introduced the proper carbon content onto the furan ring system 285 which was subjected to Wolff-Kishner reduction and alkoxyl interconversion to the dimethyl ester 286. Addition of bromine in methanol transformed the furan into its dihydro derivative 287. Reduction and hydrolysis provided the diketone 288 which was readily cyclized to the target molecule, 289.

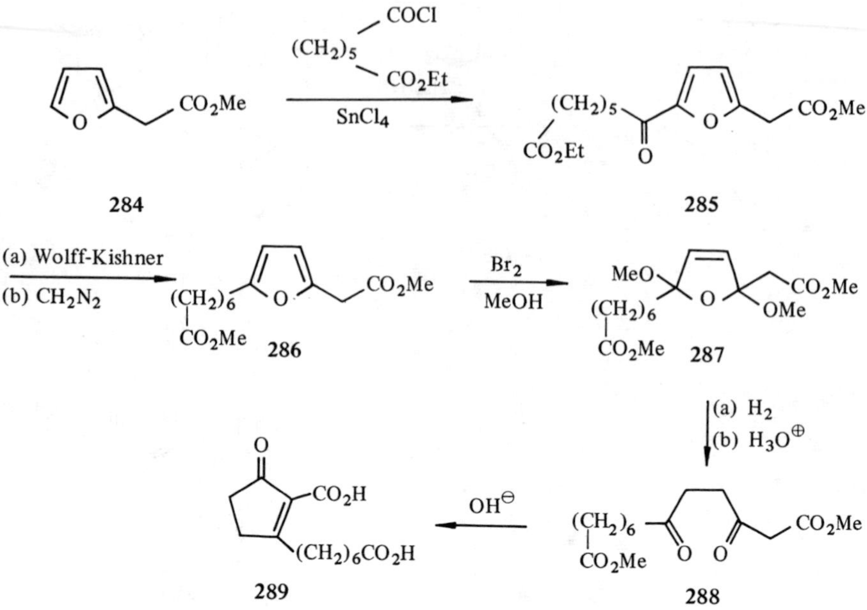
COCl
$(CH_2)_5$
CO_2Et
CO_2Me
$SnCl_4$
284
285
(a) Wolff-Kishner
(b) CH_2N_2
286
Br_2
MeOH
287
(a) H_2
(b) $H_3O^{\oplus}$
288
$OH^{\ominus}$
CO_2H
$(CH_2)_6CO_2H$
289

Bicyclic enediones **294** are obtained in a two-step reaction starting from furans and α,β-unsaturated ketones, **290**.[97] The acid catalyzed condensation of unsaturated ketones to furans, a well-known reaction,[98] provided the substituted furan **291** with the necessary carbon content. Refluxing the ketofuran **291** in acetic acid containing a trace of sulfuric acid produced the triketone

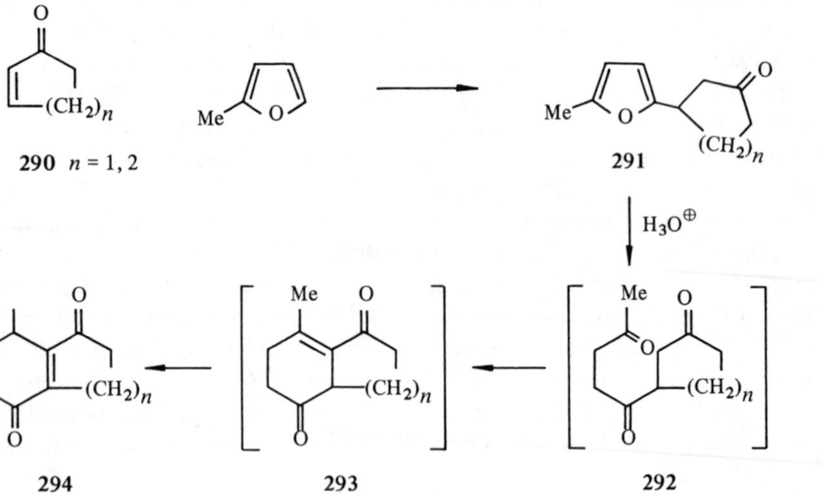
$(CH_2)_n$
290 $n = 1, 2$
Me
291
$H_3O^{\oplus}$
294
293
292

292 which spontaneously cyclized to **293** and ultimately to the product **294**. The ring opening of the furan and subsequent cyclization to the bicyclic enedione took place in yields of 30-45%. In view of the superior technique mentioned earlier leading to the *bis*-ketal **281**, it would be of interest to repeat this sequence by cleavage of **291** in ethylene glycol-benzene.

Buchi[99] reported the synthesis of Jasmone **298** in 40-45% overall yields starting from 2-methylfuran **295**. After metallation of the latter and alkylation

to the hexylfuran **296**, the carbonyl groups were released by treatment with a trace of sulfuric acid in acetic acid. The diketone thus produced (∿50% yield) was cyclized to *cis*-Jasmone **298** under the influence of ethanolic sodium hydroxide. A similar approach to *cis*-Jasmone was also reported by Crombie.[100]

The synthesis of a bicyclo(5,3,0)decanone **301** was accomplished in 17% overall yield from 1-methylcycloheptanone and furfural.[101] The condensation product **299** was opened up to the diketo acid **300** in ethanolic hydrogen chloride and then cyclized to the bicyclic ketone. The ring opening step (**299** → **300**) proceeded in only 28% yield but involved an unusual intramolecular oxidation reduction. It is seen that the double bond in **299** is reduced while the terminal aldehyde resulting from furan cleavage is oxidized affording the carboxyl group. No comment by the authors was made regarding this transformation. An explanation for this reaction, however, came a few years later in the analogous study that follows.

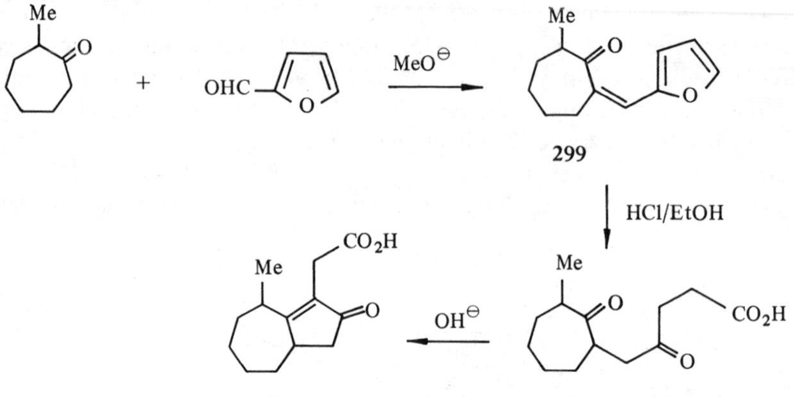

299

HCl/EtOH

301 ← OH⊖ ← **300**

A convenient synthesis of substituted levulinic esters **305**[102,103] is available from α-furfuryl alcohols **302** readily prepared by the Grignard reaction on furfural. When **302** is treated with alcoholic hydrogen chloride, ring opening takes place to **303** followed by rearrangement to the precursor of levulinic ester **304**. Here again, an intramolecular oxidation-reduction has taken place which may be used to comprehend the process (**299** → **300**).

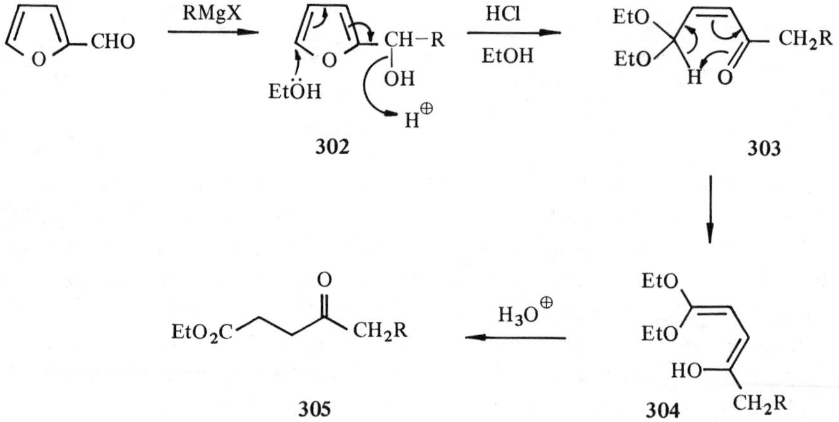

302   **303**

305 ← H₃O⊕ ← **304**

β-Carboxyacrolein acetals **311**, a highly functionalized and useful system, can be produced by nucleophilic attack of methoxide ion on 2-nitrofuran **306**. The presumed initial adduct **307** undergoes a Nef-type reaction to the unsaturated lactone **308**.[104]

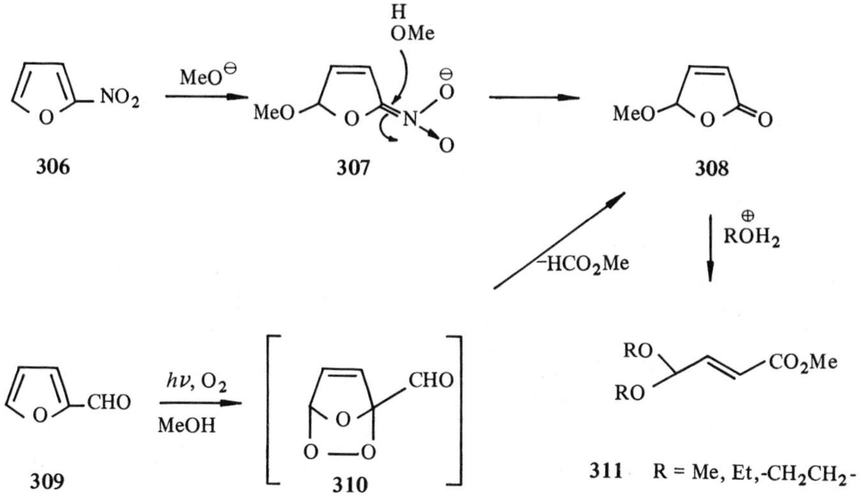

The same lactone has been prepared more efficiently by the photooxidation of furfural **309** in alcoholic solvents producing first the dioxa-adduct **310** which rearranges to the lactone.[105] The latter is converted smoothly to the unsaturated acetal **311** by acidic catalysis in alcoholic medium.[106]

The light induced rupture of furan derivatives is actually a rather general method for obtaining dicarbonyl compounds. Thus 2-methylfuran gives levulinaldehyde,[107] and 2,5-dimethylfuran gives acetonyl acetone.[107] A new and interesting application of the synthetic utility of furans was recently described by Schechter[108] who studied the thermal decomposition of α-diazofurans **313** prepared easily from the tosyl hydrazones **312** of the 2-acyl furans. A

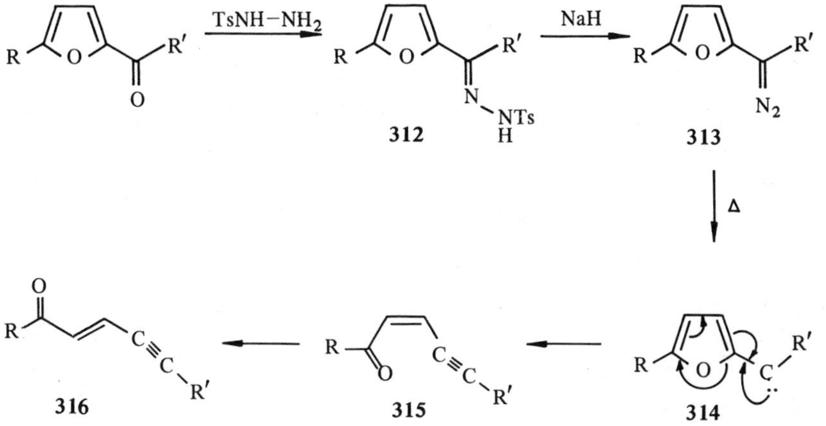

R	R'	%
H	H	66 [81:19 (cis-trans)]
Me	H	43 [87:13 (cis-trans)]
H	Ph	43 [53:47 (cis-trans)]
H	Me	36 [73:27 (cis-trans)]
H	Et	47 [68:32 (cis-trans)]

remarkable reorganization of the diazofuran took place giving moderate yields of the acetylenic-α,β-unsaturated ketones 315 and 316. The process probably proceeds via the carbenoid intermediate 314 in an electrocyclic process shown by the arrows in the structure. It is likely that the decomposition of 314 is stereospecific giving initially the *cis* ketone 315 but because of the extreme lability of these systems, isomerization results in product mixtures. The intermediate 314 was successfully trapped using cyclooctane and styrene affording

317   318

the substituted furans 317 and 318, respectively. Further studies on diazofurans and other diazo heterocycles should provide interesting new areas for synthesis when one recalls that the furan ring is also a "masked 1,4-dicarbonyl" compound.

B. From Thiophenes

The manifold of reactions available to the thiophene nucleus, coupled with the fact that the sulfur may later be extruded, provides the chemist with a highly versatile tool for synthesis. Some of these reactions and their uses in hydrocarbon syntheses have already been discussed (Chapter 2, Section I).

With respect to carbonyl compound preparations, the thiophene system allows for considerable variations in structure. Thus Friedel-Crafts acylation of thiophene leads to good yields of the 2-acyl derivatives 319 which, upon Raney nickel desulfurization, generates the four-carbon-homologated ketone 320.[111] In some instances, depending on the specific nature of the desulfurization catalyst, the carbonyl groups may be reduced[109] and it is, therefore, best to

319 **320**

321 **322**

mask the carbonyl (e.g., ethylene ketal) in the thiophene **319** prior to Raney nickel treatment. Since Friedel-Crafts alkylation on the deactivated 2-acyl thiophene proceeds exclusively (or nearly so)[110] at the 4-position **321**, this leads, after sulfur extrusion, to the branched ketones **322**.

Polyketone derivatives have also been prepared using thiophenes as vehicles. Wynberg[111] reported the acylation of bithienyl **323** which led to diacyl derivatives **324** in good yield. Upon Raney nickel desulfurization, the long chain diketone **325** was produced. Similarly, terthienyl **326** led to the diketone

323 **324** **325**

326 **327**

327. It is noteworthy that the successive addition of Raney nickel and chromium trioxide circumvented the problem of mixtures of ketones and alcohols by *in situ* oxidation of any alcoholic side products. This technique, therefore, may be used as an alternative to masking the carbonyl groups prior to desulfurization. Of course, any substituents that are sensitive to oxidizing agents (olefins, amines) would preclude the use of this method. Mention should also be made of the fact that the diacyl polythienyls **324** are very useful precursors to alcohols **329** by simply adding Grignard reagent to form the hydroxy derivatives **328** prior to desulfurization.[111]

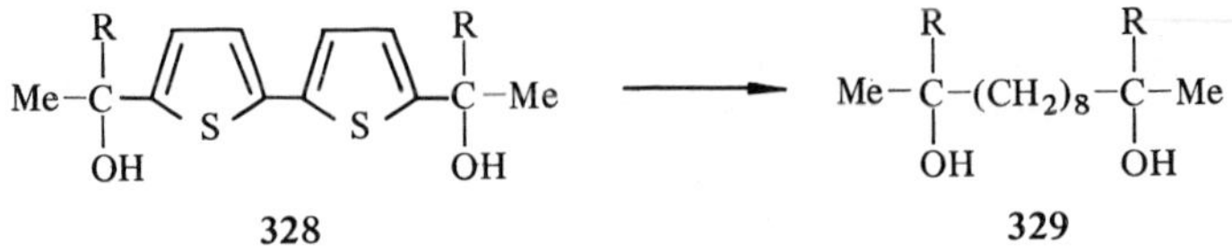

328

329

Cagniant and co-workers[112] have extensively investigated the cyclization of ω-thienylcarboxylic acids to thieno[b]cycloalkeniones and found that this reaction proceeds smoothly in good yield. Desulfurization leads to 2-alkyl, 2,3-dialkyl, and 3-alkyl cycloalkanones. The value of such transformations is evident when one is faced with alkylation of cyclic ketones. Although this route to 2-alkylcycloalkanones has been virtually overlooked, the difficulties en-

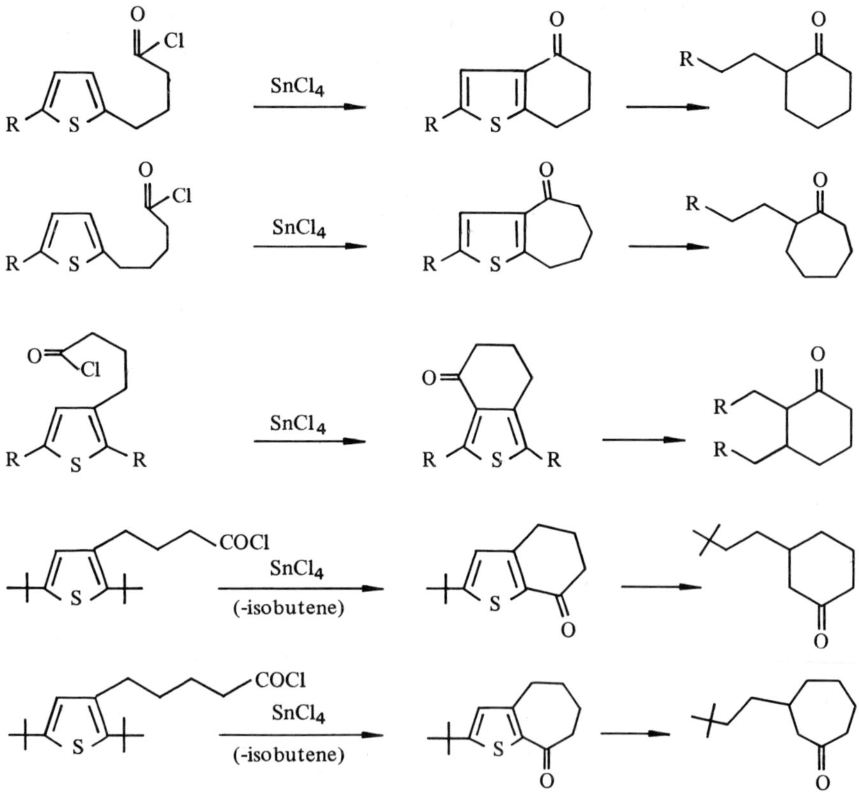

countered at times with enamine alkylations may encourage chemists to take a serious look at this alternative. As for the 3-alkylcycloalkanones, this method may have some merit, when compared to the more elegant and modern techniques using organometallic reagents[113] (i.e., alkylaluminums and lithio alkylcuprates).

C. From Dihydrothiophenes

A new approach to polycyclic ketones with angular substituents **335** and **337** was described by Stork[114] which relied on the dihydrothiophene system **331**. The latter may be readily obtained from the muconic ester **330**[115] or by condensing β-mercaptopropionic ester with dimethyl maleate to the thio keto-ester **332**. Carbonyl reduction, followed by base catalyzed elimination, affords the requisite dihydrothiophene. Diels-Alder addition of butadiene (R = H) or 2-ethoxybutadiene (R = OEt) gave the cycloadduct **333** which was transformed into the dicarboxylic acid derivatives **334** and **336**, respectively, using W-5 Raney nickel. These dicarboxylic acids are obvious precursors to cyclic ketones (**335** and **337**) via the base catalyzed condensation, and thus a route to *trans*-CD-steroid ring system is opened. Also worthy of mention is the fact that

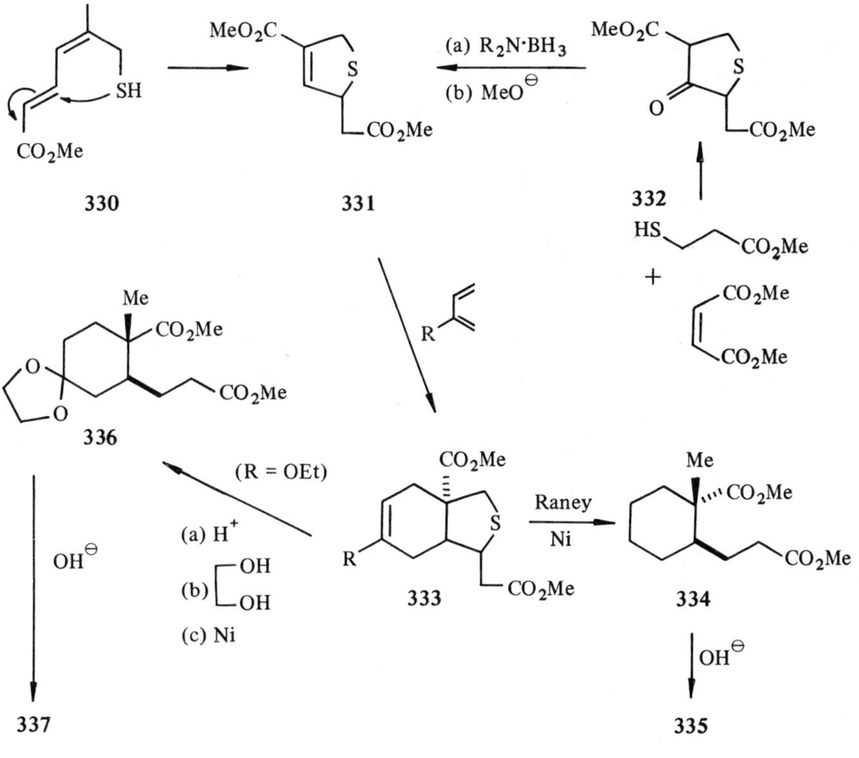

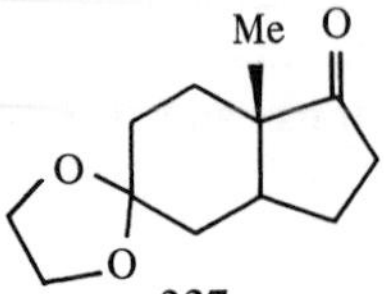

337

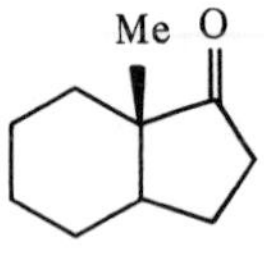

335

the Diels-Alder adducts **333** represent, after desulfurization, *cis*-1,2-dialkyl cyclohexane carboxylic acids of general structure **339** if the scheme is initiated with dihydrothiophenes of general structure **338**.

338

339

D.   From Chromans

A novel annelation reaction via temporarily constructed chromans **340** was reported by Dolby.[116] The method is based upon the cycloaddition[117] of 2-methoxy-butadiene to various phenolic systems at elevated temperatures. In this manner, good yields of the chromans are obtained which serve as precursors to various polycyclic ketones. Thus Birch reduction of **340** gives the enol ether **341** whose lability to acid produces the diketone **342**, which was cyclized to the bicyclic unsaturated ketone **343**. The method appears to have some generality as evidenced by the formation of polycyclic ketones **344-346** from the corresponding phenolic systems after initially constructing the chroman intermediates. This method is considered as a four-carbon homologation because of the four-carbon butadiene moiety which is used to elaborate the phenolic derivative.

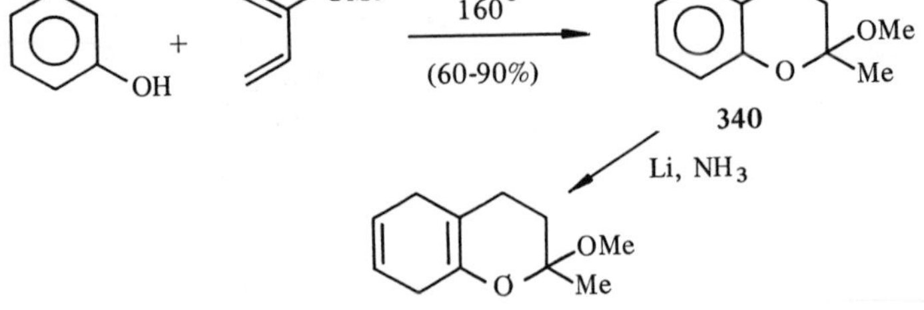

340

341

342 343

344

345

346

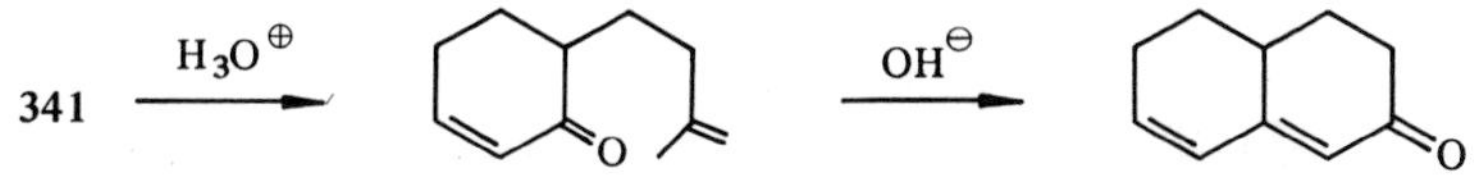

E. From Isoxazoles

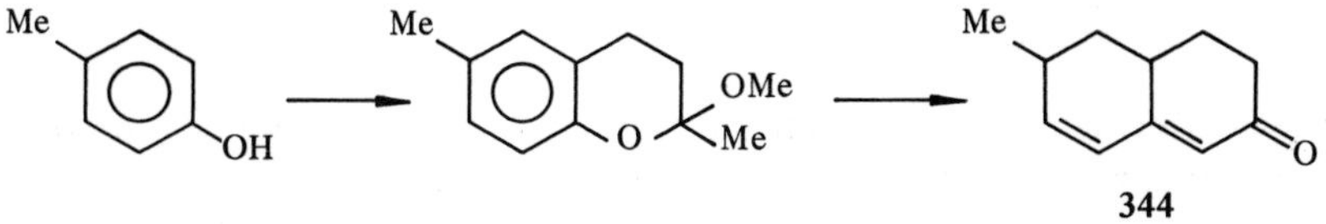

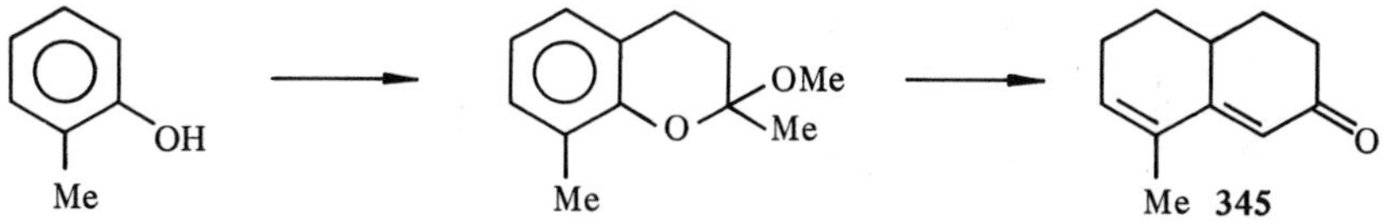

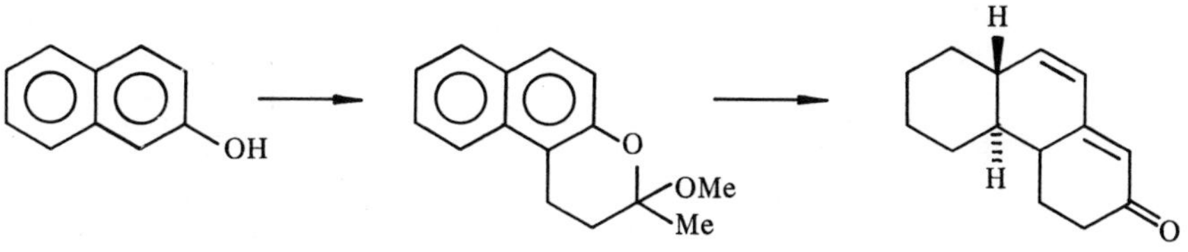

The isoxazole system has been discussed earlier (Chapter 5, Section I.K) with regard to its utility as a precursor to aromatic derivatives. This highly versatile heterocycle, however, has achieved its greatest prominence in the synthesis of polycyclic carbonyl compounds. The requisite isoxazole **349** is prepared by chloromethylation of 3,5-dimethylisoxazole **348** which, in turn, is obtained by

347 348 349

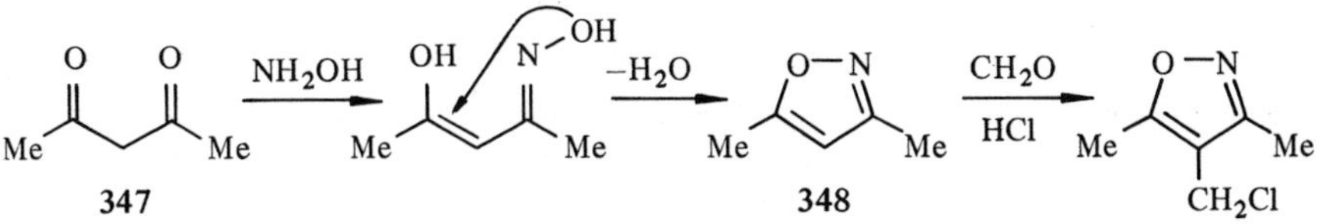

the well-known route of condensing the diketones **347** with hydroxylamine.[118] The basic four-carbon unit **350** (asterisks) excludes the 5-carbon[119] in the

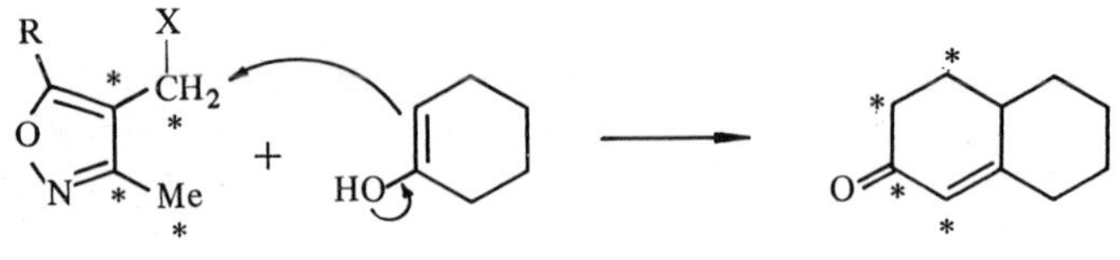

350   **351**

isoxazole ring and the following mechanism (Scheme 6), leading to the 2-octalone **351**, will assist the reader in confirming this fact.

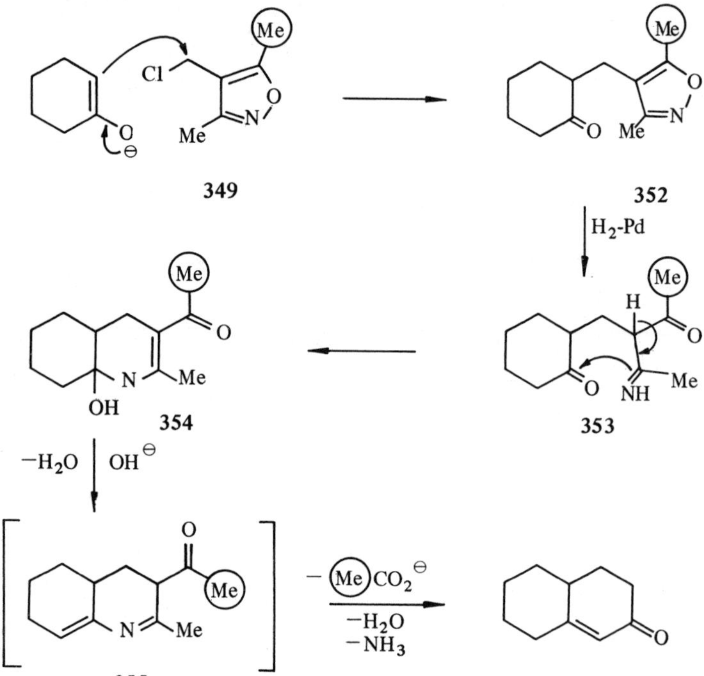

349

352

354

353

$-H_2O$ | $OH^\ominus$

355

$- \text{(Me)} CO_2^\ominus$

$-H_2O$
$-NH_3$

351

Scheme 6

The process is initiated by treating **349** with the enolate of cyclohexanone affording the elaborated isoxazole **352**. Catalytic hydrogenation ruptures the highly labile N–O linkage producing *in situ* the imino diketone **353** which, after tautomerization, cyclizes rapidly to the carbinol amine **354**. The latter is hydrolyzed in alkali generating the masked triketone **355** and cyclization ensues to the product **351**. The loss of methyl acetate (originally the 5-carbon and its methyl substituent shown encircled throughout the scheme) may occur prior to final cyclization or after **351** is formed. This remarkable example of a heterocycle displaying its synthetic prowess has become a valuable tool in the construction of complex molecules. For example, a total synthesis of *dl*-homotestosterone **360** was achieved by constructing the AB rings via the isoxazole route.[120] The enolate ion of the bicyclic ketone was alkylated with the chloromethyl isoxazole **356** prepared by a novel general method[121] to produce the key intermediate **357**. The usual sequence (reduction, hydrolysis) led to the transient diketone **358** which completed the annelation to **359**. Standard experiments released the carbonyl group after reduction of the olefinic linkage and the steroid **360** was in hand.

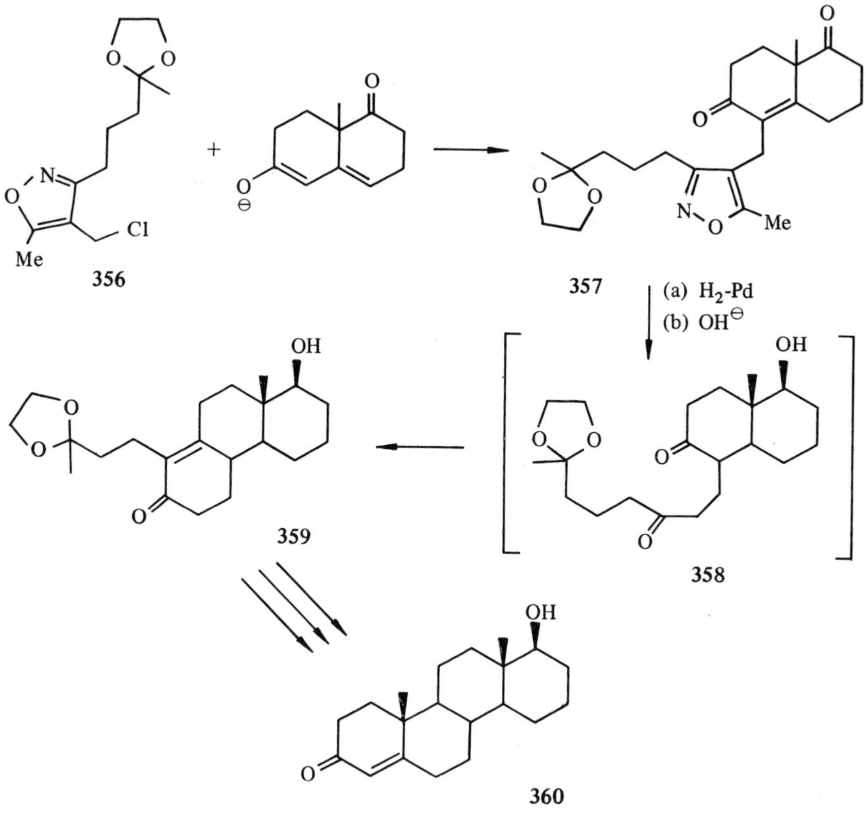

In another interesting example depicting the use of isoxazoles in synthesis, Saucy[122] employed this system in combination with his own earlier work on the preparation of the tricyclic ketone **361**. The replacement of a suitable ketone containing moiety in place of the *n*-propyl group in **362** would provide the necessary component to annelate **361** into the corresponding steroid.

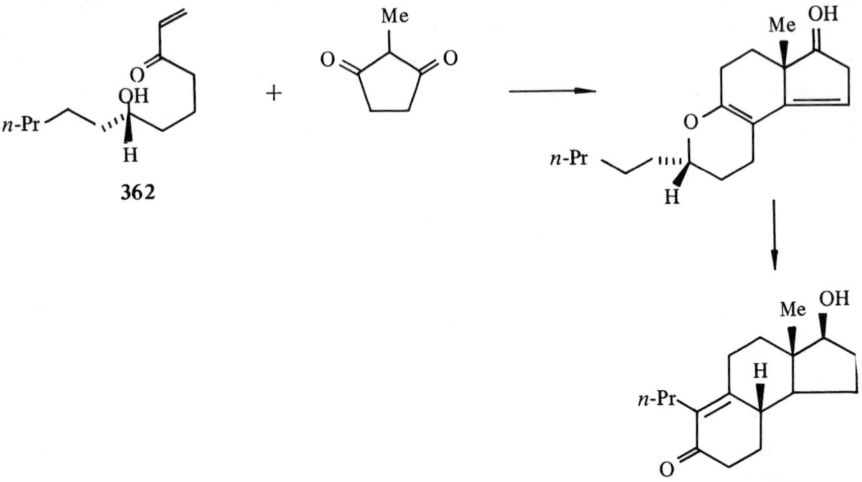

This ketonic moiety may be introduced via the isoxazole derivative **363**. The preparation of the latter was accomplished beginning with the usual

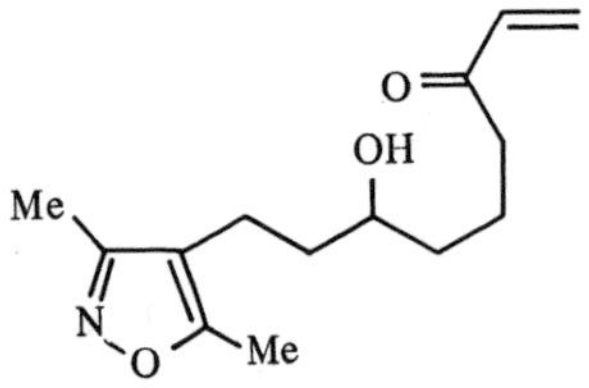

dimethylchloromethyl isoxazole **(349)** and subjecting it to a Wittig reaction via the phosphonium salt **364** and the 2-formyl dihydropyran. The resulting olefin **365** was hydrated to **366** and then oxidized to the lactone **367**. Treatment with vinyl magnesium chloride produced the desired isoxazole **363** in possession of all the requisite functionality. Repeating the reaction mentioned earlier the isoxazole **368** was formed which underwent the rearrangement to the tricyclic system **369**. The isoxazole was reductively cleaved and transformed into the steroid system **370** after hydrolysis of the carbinol amine intermediate. It should be emphasized that the wide variety of transformations that were carried out in this study proceeded without any detrimental effects on the isoxazole ring and only at the proper step in the sequence did the isoxazole release its treasured carbonyl moiety.

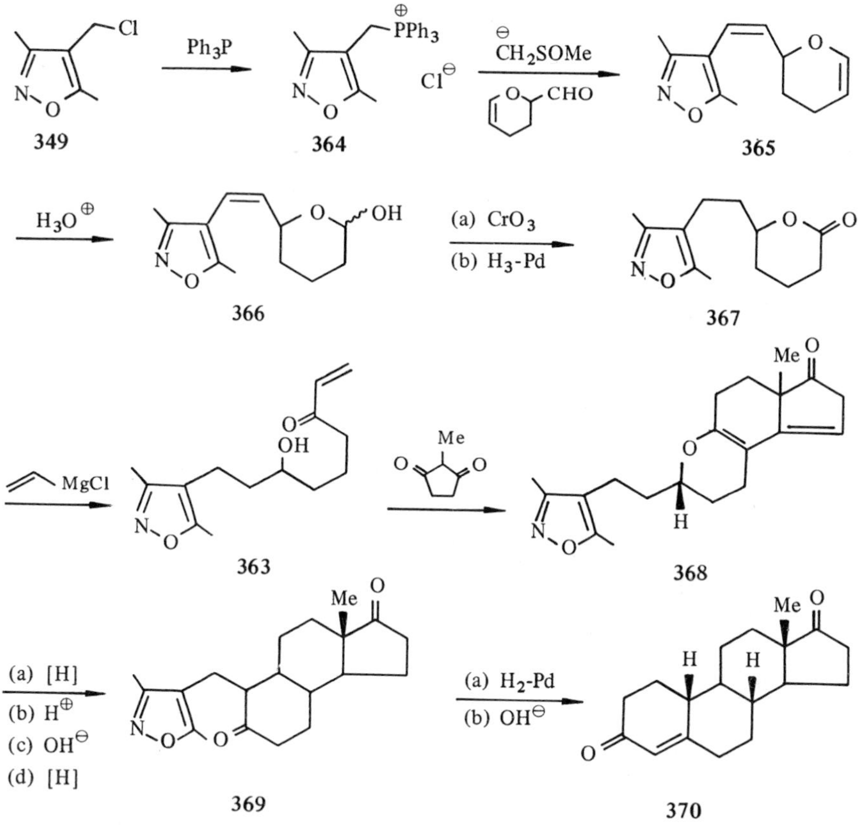

Isoxazoles may be considered to be synthetic equivalents of β-dicarbonyl compounds and used as such (**371** → **372**). If proper elaboration of the isoxazole

371   372

ring is performed, it is possible to obtain polyketo compounds **375**. A recent study[123] designed to obtain synthetic units for biogenetic pathways, which may

373   374

375

involve the β-penta-acetate **375**, focused on the preparation of the *bis*-isoxazole **374**. This compound may be regarded as the synthetic equivalent to **375** since hydrogenation and hydrolysis would be expected to generate it. By metallating (butyllithium, $-78°$) the 5-methyl group in the isoxazole, the lithio salt **373** was treated with the isoxazole ester and provided the suitably elaborated precursor to polyketones.

REFERENCES

1. F. Kröhnke and E. Borner, *Berichte,* **69**, 2006 (1936).
2. For a review on this subject see F. Kröhnke, *Angew. Chem. Int. Ed.,* **2**, 380 (1963), and references cited therein.
3. L. A. Paquette, *Tetrahedron,* **22**, 25 (1966); V. Boekelheide and W. E. Feely, *J. Org. Chem.,* **22**, 1135 (1957).
4. A. R. Katritzky and E. Lunt, *Tetrahedron,* **25**, 4291 (1969); V. Boekelheide and W. J. Linn, *J. Am. Chem. Soc.,* **76**, 1286 (1954); J. F. Vozza, *J. Org. Chem.,* **27**, 3856 (1962).
5. W. N. Marmer and D. Swern, *Tetrahedron Lett.,* 531 (1969).
6. S. Danishefsky and R. Cavanaugh, *J. Am. Chem. Soc.,* **90**, 520 (1968).
7. S. Danishefsky and A. Nagel, *Chem. Commun.,* 373 (1972).

8. S. Danishefsky, A. Nagel, and D. Peterson, *ibid.*, 374 (1972).

9. K. E. Wilzbach and D. J. Rausch, *J. Am. Chem. Soc.*, **92**, 2178 (1970).

10. R. A. Olofson and D. M. Zimmerman, *ibid.*, **89**, 5057 (1967).

11. V. Calo, L. Lopez, and P. E. Todesco, *J. Chem. Soc., Perkin I*, 1652 (1972).

12. E. J. Corey and K. Achiwa, *J. Am. Chem. Soc.*, **91**, 1429 (1969).

13. H. Ruschig, W. Fritsch, J. Schmidt-Thome, and W. Haede, *Chem. Ber.*, **88**, 883, 1955; W. E. Bachmann, M. P. Cava, and A. S. Dreiding, *J. Am. Chem. Soc.*, **76**, 5554, 1954; S. S. Rawalay and H. Schechter, *J. Org. Chem.*, **32**, 3129 (1967); R. G. R. Bacon and D. Stewart, *J. Chem. Soc. (C)*, 1384, 1966; R. G. R. Bacon, W. J. W. Hanna, and D. Stewart, *ibid.*, p. 1388; R. G. R. Bacon and W. J. W. Hanna, *J. Chem. Soc.*, 1965, 1962.

14. I. C. Nordin, *J. Heterocycl. Chem.*, **3**, 531 (1966).

15. D. Haidukewych and A. I. Meyers, *Tetrahedron Lett.*, 3031 (1972).

16. H. C. Brown and A. Tsukamoto, *J. Am. Chem. Soc.*, **83**, 4549 (1961).

17. For an excellent review of this subject see H. A. Staab, *Angew. Chem. Int. Ed.*, **1**, 351 (1962).

18. J. J. Ritter and E. J. Tillmanns, *J. Org. Chem.*, **22**, 839 (1957).

19. For modification on this preparation see ref. 69 below.

20. A. I. Meyers and A. Nabeya, *Chem. Commun.*, 1163 (1967).

21. R. V. Stevens, J. M. Fitzpatrick, M. Kaplan, and R. L. Zimmerman, *ibid.*, 857 (1971).

22. R. C. Cookson, I. D. R. Stevens, and C. T. Watts, *ibid.*, 744 (1966).

23. J. Thiele and O. Stange, *Annals*, **283** (1894); J. Stolle, *Berichte*, **45**, 273 (1912); R. C. Cookson, S. S. H. Gilani, and I. D. R. Stevens, *Tetrahedron Lett.*, 615 (1962).

24. C. W. Rees and R. C. Storr, *Chem. Commun.*, 1305 (1968).

25. F. D. Marsh and M. E. Hermes, *J. Am. Chem. Soc.*, **86**, 4506 (1964).

26. J. E. McMurry, *ibid.*, **91**, 3676 (1969); J. E. McMurry and A. P. Coppolino, *J. Org. Chem.*, **38**, 2821 (1973).

27. W. E. McEwen and R. L. Cobb, *Chem. Rev.*, **55**, 511 (1955).

28. F. D. Popp, *Adv. In Heterocycl. Chem.*, **9**, A. R. Katritsky, ed., 1 (1968), Academic Press, New York.

29. R. L. Cobb and W. E. McEwen, *J. Am. Chem. Soc.*, 77, 5042 (1955).

30. H. Shirai and N. Oda, *Chem. Pharm. Bull. (Tokyo)*, **8**, 744 (1960).

31. E. Cuingnet and M. Adalberon, *Compt. Rend.*, **258**, 3053 (1964).

32. P. T. Izzo and A. S. Kende, *Tetrahedron Lett.*, 5731 (1966).

33. G. Stork and E. Colvin, *J. Am. Chem. Soc.*, **93**, 2080 (1971).

34. A. Hassner and F. W. Fowler, *ibid.*, **90**, 2869 (1968); F. W. Fowler, "Synthesis and Reaction of 1-Azirines," in *Adv. Heterocycl. Chem.*, **13**, A. R. Katritzky, ed., Academic Press, New York, 1971, p. 45.

35. A. Hassner and F. W. Fowler, *J. Org. Chem.*, **33**, 2686 (1968).

36. P. W. Neber and A. Burgard, *Annals*, **493**, 381 (1932); C. O'Brien, *Chem. Rev.*, **64**, 81 (1964).

37. D. Felix, R. K. Muller, U. Horn, R. Joos, J. Schreiber, and A. Eschenmoser, *Helv. Chim. Acta*, **55**, 1276 (1972).

38. R. K. Muller, D. Felix, J. Schreiber, and A. Eschenmoser, *ibid.*, **53**, 1479 (1970).

39. G. Buchi and J. C. Vederas, *J. Am. Chem. Soc.*, **94**, 9128 (1972).

40. J. A. Marshall and H. Roebke, *Synthesis*, 444 (1971); *J. Org. Chem.*, **34**, 4188 (1969).

41. (a) N. A. LeBel, N. D. Ojha, J. R. Menke, and R. J. Newland, *J. Org. Chem.*, **37**, 2896 (1972); (b) I. Ikeda, G. Takemoto, and S. Komori, *Kogyo Kagaku Zasshi*, **74**, 419 (1971).

42. T. Mukaiyama, M. Araki, and H. Takei, *J. Am. Chem. Soc.*, **94**, 4763 (1973).

42a. W. S. Trahanovsky and M. Park, *J. Am. Chem. Soc.*, **95**, 5412 (1973).

43. A. I. Meyers and E. W. Collington, *ibid.*, **92**, 6676 (1970); A. I. Meyers, R. Brinkmeyer, and E. W. Collington, *Org. Syn.*, in press.

44. H. M. Fales, *J. Am. Chem. Soc.*, **77**, 5118 (1955).

45. E. L. Eliel and F. W. Nader, *ibid.*, **92**, 584 (1970).

46. H. W. Post, *The Chemistry of Aliphatic Ortho Esters*, Rheinhold, New York, 1943.

47. M. Kharasch and O. Reinmuth, *Grignard Reactions of Nonmetallic Substances*, Prentice-Hall, Englewood Cliffs, N. J., 1954.

48. J. Carnduff, *Quart. Rev.*, **20**, 169 (1966).

49. T. Kato, H. Yamanaka, T. Adachi, and H. Hiranuma, *J. Org. Chem.*, **32**, 3788 (1967).

50. T. J. van Bergen and R. M. Kellogg, *ibid.*, **36**, 1705 (1971).

51. I. Rosenthal and D. Elad, *ibid.*, **33**, 805 (1968).

52. E. J. Corey and D. Seebach, *Angew. Chem. Int. Ed.*, **4**, 1075, 1077 (1965).

53. D. Seebach, *Synthesis*, **1**, 17 (1969).

54. E. Vedejs and P. L. Fuchs, *J. Org. Chem.*, **36**, 366 (1971).

55. E. J. Corey and D. Crouse, *ibid.*, **33**, 298 (1968); S. J. Daum and R. L. Clarke, *Tetrahedron Lett.*, 165 (1967); K. Ogura and G. Tsuchihashi, *ibid.*, 3151 (1971); T. Oishi, K. Kamemoto, and Y. Ban, *ibid.*, 1085 (1972); K. Narasaka, T. Sakashita, and T. Mukaiyama, *Bull. Chem. Soc., Japan*, **45**, 3724 (1972).

56. An additional review on recent dithiane chemistry will appear shortly (Professor D. Seebach, personal communication, May 1973).

57. D. Seebach and D. Steinmuller, *Angew. Chem.*, **80**, 617 (1968).

58. D. Seebach, D. Steinmuller, and F. Demuth, *ibid.*, **80**, 618 (1968).

59. E. J. Corey and B. W. Erickson, *J. Org. Chem.*, **36**, 3553 (1971).

60. E. J. Corey and D. Crouse, *ibid.*, **33**, 298 (1968).

61. W. D. Woessner and R. A. Ellison, *Tetrahedron Lett.*, 3735 (1972).

62. E. J. Corey and S. W. Walinsky, *J. Am. Chem. Soc.*, **94**, 8932 (1972).

63. A. Pelter, M. G. Hutchings, and K. Smith, *Chem. Commun.*, 1529 (1970).

64. T. Cohen, I. H. Song, J. H. Fager, and G. L. Deets, *J. Am. Chem. Soc.*, **89**, 4968 (1967).

65. T. Cohen and I. H. Song, *J. Org. Chem.*, **31**, 3058 (1966).

66. T. Cohen and I. H. Song, *J. Am. Chem. Soc.*, **87**, 3780 (1965).

67. C. Ruchardt, S. Eichler, and O. Kratz, *Tetrahedron Lett.*, 233 (1965).

68. T. Cohen, I. H. Song, and J. H. Fager, *ibid.*, 237 (1965).

69. A. I. Meyers, A. Nabeya, H. W. Adickes, I. R. Politzer, G. R. Malone, A. C. Kovelesky, R. L. Nolen, and R. C. Portnoy, *J. Org. Chem.*, **38**, 36 (1973).

70. G. R. Malone, Ph.D. thesis, Wayne State University (1972); G. R. Malone and A. I.

Meyers, *J. Org. Chem.*, March 1974; R. Brinkmeyer, research in progress.

71. W. S. Wadsworth and W. D. Emmons, *J. Am. Chem. Soc.*, **83**, 1735 (1961).

72. A. I. Meyers and N. Nazarenko, *ibid.*, **94**, 3243 (1972).

73. A. I. Meyers and E. M. Smith, *J. Org. Chem.*, **37**, 4289 (1972).

74. A. I. Meyers and N. Nazarenko, *ibid.*, **38**, 175 (1973).

75. A. I. Meyers, E. M. Smith, and M. S. Ao, *ibid.*, **38**, 2129 (1973).

76. C. Lion and J. E. Dubois, *Tetrahedron*, **29**, 3417 (1973)

77. G. Stork and S. Dowd, *J. Am. Chem. Soc.*, **85**, 2178 (1963).

78. U. M. Kempe, T. K. Das Gupta, K. Blatt, P. Gygax, D. Felix, and A. Eschenmoser, *Helv. Chim. Acta.*, **55**, 2187 (1972).

79. P. Gygax, T. K. Das Gupta, and A. Eschenmoser, *ibid.*, **55**, 2205 (1972).

80. L. J. Altman and S. L. Richheimer, *Tetrahedron Lett.*, 4709 (1971).

81. A. I. Meyers, R. A. Munavu, and J. L. Durandetta, *ibid.*, 3929 (1972).

82. A. I. Meyers and J. L. Durandetta, unpublished results.

83. M. Chaykovsky, M. H. Lin, and A. Rosowsky, *J. Org. Chem.*, **37**, 2018 (1972).

84. C. G. Overberger and J. M. Hoyt, *J. Am. Chem. Soc.*, **73**, 3305, 3957 (1951).

85. M. Roth, P. Dubs, E. Gotschi, and A. Eschenmoser, *Helv. Chim. Acta.*, **54**, 710 (1971).

86. G. Stork and E. Colvin, *J. Am. Chem. Soc.*, **93**, 2080 (1971).

87. H. C. Brown, *Hydroboration*, Benjamin, New York, 1962, p. 233.

88. M. S. Kharasch and O. Reinmuth, *Grignard Reactions of Nonmetallic Substances*, Prentice-Hall, Englewood Cliffs, N. J., 1954.

89. A. I. Meyers, A. C. Kovelesky, and A. F. Jurjevich, *J. Org. Chem.*, **38**, 2136 (1973).

90. M. Naruse, T. Tomita, K. Utimoto, and H. Nozaki, *Tetrahedron Lett.*, 795 (1973).

91. D. Lednicer, "Latent Functionality in Organic Synthesis," in *Advances in Organic Chemistry*, Vol. 8, E. C. Taylor, ed., Wiley-Interscience, New York, 1972, p. 179.

92. P. Bosshard and C. H. Eugster, "Development of the Chemistry of Furans, 1952-1963," in *Advances in Heterocyclic Chemistry*, Vol. 7, A. R. Katritzky ed., Academic Press, New York, 1966, p. 377.

93. N. Elming, "Dialkyoxy and Diacyloxydihydrofurans," in *Advances in Organic Chemistry*, Vol. 2, R. A. Raphael, E. C. Taylor, and H. Wynberg, ed., Interscience, New York, 1960, p. 67.

94. W. S. Johnson, T. Li, C. A. Harbert, W. R. Bartlett, T. R. Herrin, B. Staskun, and D. H. Rich, *J. Am. Chem. Soc.*, **92**, 4461 (1970).

95. W. S. Johnson, M. B. Gravestock, and B. E. McCarry, *ibid.*, **93**, 4332 (1971).

96. N. Finch, J. J. Fitt, and I. H. C. Hsu, *J. Org. Chem.*, **36**, 3191 (1971).

97. M. A. Tobias, *ibid.*, **35**, 267 (1970).

98. K. Alder and S. Schmidt, *Chem. Ber.*, **76**, 183 (1943).

99. G. Buchi and H. Wuest, *J. Org. Chem.*, **31**, 977 (1966).

100. L. Crombie, P. Hemesly, and G. Pattenden, *J. Chem. Soc.*, *(C)*, 1024 (1969).

101. A. M. Islam and M. T. Zemaity, *J. Am. Chem. Soc.*, **79**, 6023 (1957).

102. R. Lukes and J. Srogl, *Coll. Czech. Chem. Commun.*, **26**, 2238 (1961).

103. K. G. Lewis, *J. Chem. Soc.*, 4690 (1961).

104. T. Irie, E. Kurosawa, and T. Hamada, *J. Fac. Sci. Hokkaido Univ.*, **5**, 6; *CA.*, **52**, 16328 (1958).

105. F. Farina, M. Lora-Tomayo, and M. V. Martin, *Ann. Real. Soc. Espana, Fiz Quim (Madrid)*, **B60**, 715 (1964); *CA*, **63**, 4213c (1965); J. I. DeGraw, *Tetrahedron*, **28**, 967 (1972).

106. A. I. Meyers, E. W. Collington, T. A. Narwid, R. L. Nolen, and R. C. Strickland, *J. Org. Chem.*, **38**, 1974 (1973).

107. G. O. Schenk and K.Gollnick, *1,4-Cycloaddition Reactions*, J. Hamer, ed., Academic Press, New York, 1967, p. 255; C. S. Foote, M. T. Wuesthoff, S. Wexler, I. G. Burstain, R. Denny, G. O. Schenck, and K.-H. Schulte-Elte, *Tetrahedron*, **23**, 2583 (1967); C. S. Foote and G. Uhde, in *Organic Photochemical Syntheses*, Vol. 1, Wiley-Interscience, New York, 1971, p. 70.

108. R. V. Hoffman and H. Shechter, *J. Am. Chem. Soc.*, **93**, 5940 (1971).

109. S. Gronowitz, "Chemistry of Thiophenes," in *Advances in Heterocyclic Chemistry*, Vol. 1, A. R. Katritzky, ed., Academic Press, New York, 1963, p. 111.

110. E. C. Spaeth and C. B. Germain, *J. Am. Chem. Soc.*, **77**, 4066 (1955).

111. H. Wynberg and A. Logothetis, *ibid.*, **78**, 1958 (1956).

112. G. Murad, D. Cagniant, and P. Cagniant, *Bull. Soc. Chim. (France)*, part 2, 343 (1973); P. Cagniant and D. Cagniant, *ibid.*, **62** (1953); *ibid.*, 680 (1955).

113. G. H. Posner, "Conjugate Addition Reaction of Organo Copper Reagents," *Organic Reactions*, Vol. 19, Wiley, New York, 1972.

114. G. Stork and P. L. Stotter, *J. Am. Chem. Soc.*, **91**, 7780 (1969).

115. P. L. Stotter, S. A. Roman, and C. L. Edwards, *Tetrahedron Lett.*, 4071 (1972).

116. L. J. Dolby and E. Adler, *ibid.*, 3803 (1971).

117. L. J. Dolby, C. A. Elliger, S. Esfandiari, and K. S. Marshall, *J. Org. Chem.*, **33**, 4508 (1968).

118. N. K. Kochetkov and S. D. Sokolov, "Recent Developments in Isoxazole Chemistry," *Advances in Heterocyclic Chemistry*, Vol. 2, A. R. Katritzky, ed., Academic Press, New York, 1963, p. 365.

119. G. Stork and J. E. McMurry, *J. Am. Chem. Soc.*, **89**, 5463 (1967).

120. G. Stork and J. E. McMurry, *ibid.*, **89**, 5464 (1967).

121. G. Stork and J. E. McMurry, *ibid.*, **89**, 5461 (1967).

122. J. W. Scott and G. Saucy, *J. Org. Chem.*, **37**, 1652 (1972); J. W. Scott, R. Borer, and G. Saucy, *ibid.*, **37**, 1659 (1972).

123. T. Tanaka, M. Miyazaki, and I. Iijima, *Chem. Commun.*, 233 (1973).

10 CARBOXYLIC ACIDS AND THEIR DERIVATIVES —AMINO ACIDS, AMIDES, PEPTIDES, AND NITRILES

The employment of heterocycles as useful sources in synthesis has also been very evident in the preparation of the titled class of compounds. Although many excellent methods, both traditional and new, are available for the formation of these derivatives, it behooves the practicing chemist as well as the student to become familiar with approaches based solely on heterocyclic systems. The reader will immediately recognize that some of the following methods are, indeed, those of choice and are widely used, whereas others are less frequently considered as viable routes to carboxyl containing molecules.

This chapter deals with carboxylic acid derivatives subdivided into four sections:

1. Carboxylic acids.
2. Amino acids.
3. Amides and peptides.
4. Nitriles.

I. CARBOXYLIC ACIDS

A. From Thiophenes

Utilizing the same basic approach that led to carbonyl compounds (Chapter 9, Section V.B), thiophenes may also be manipulated to produce carboxylic acids. Thus the route described by Wynberg[1] exemplifies this technique which, in effect, homologates even-numbered fatty acids by five carbons to their corresponding odd-numbered derivatives. The scheme initially involves acylation

243

$$\text{RCO}_2\text{H} \xrightarrow[\text{P}_2\text{O}_5]{\overset{\displaystyle\langle\!\!\langle\text{S}\rangle\!\!\rangle}{}} \underset{\textbf{1}}{\text{RC}\!-\!\langle\text{S}\rangle} \xrightarrow{\text{Zn(Hg), H}^{\oplus}} \underset{\textbf{2}}{\text{RCH}_2\!-\!\langle\text{S}\rangle}$$

(+ 5 carbons)

$\text{Ac}_2\text{O, H}_3\text{PO}_4$

$$\underset{\textbf{5}}{\text{RCH}_2(\text{CH}_2)_4\text{CO}_2\text{H}} \xleftarrow{\text{Ni}} \underset{\textbf{4}}{\text{RCH}_2\!-\!\langle\text{S}\rangle\!-\!\text{CO}_2\text{H}} \xleftarrow{\text{NaOCl}} \underset{\textbf{3}}{\text{RCH}_2\!-\!\langle\text{S}\rangle\!-\!\overset{\text{O}}{\text{C}}\!-\!\text{Me}}$$

of thiophene using a carboxylic acid to afford the acyl thiophene **1**. Clemmensen reduction to the 2-alkyl thiophene **2** follows and reacylation with acetic anhydride then generates the 2-acetyl-5-alkyl thiophene **3**. By performing the haloform reaction on **3**, the carboxylic acid is formed which leads, after desulfurization, to the homologated acid **5**. In this manner, a variety of extended carboxylic acids were prepared and the overall yields for the five-step process were 25-30%. Homologation with other[2] acyl derivatives furnishes even longer chain carboxylic acids. Acylation of **2** with succinic anhydride leads to the keto acid **6** which is subjected first to reduction (**7**) and then to nickel desulfurization producing the eight-carbon homolog **8**.

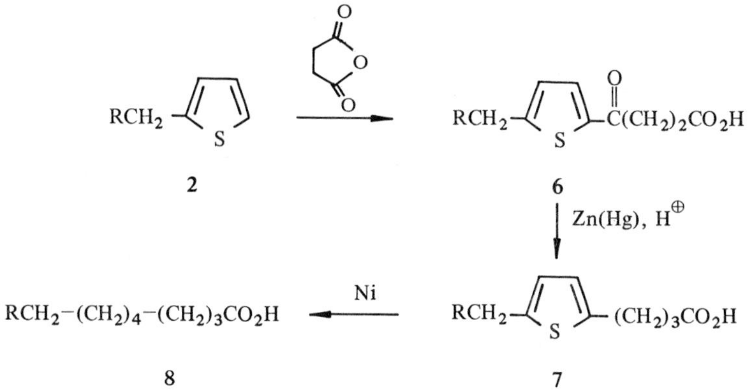

It is not a difficult matter to modify this acid synthesis to include the homologation of dicarboxylic acids (9). These have been prepared by the sequential acylation and Wolff-Kishner reduction of the thiophene with half-ester chlorides

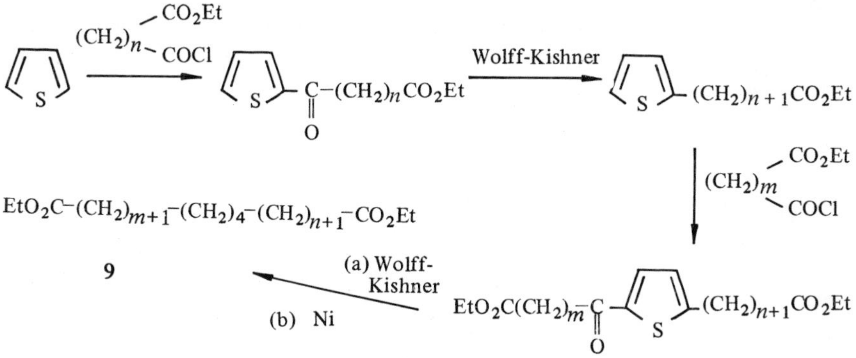

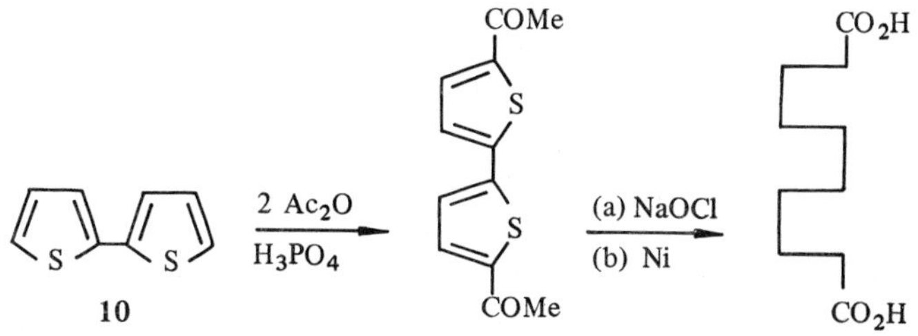

of dicarboxylic acids. By employing bithienyls 10 or di-2-thienyl methanes 11, one may introduce eight[1] or nine carbons,[3] respectively. Further examples describing the use of thiophenes in the synthesis of carboxylic acids may be found in the excellent review by Gronowitz.[4]

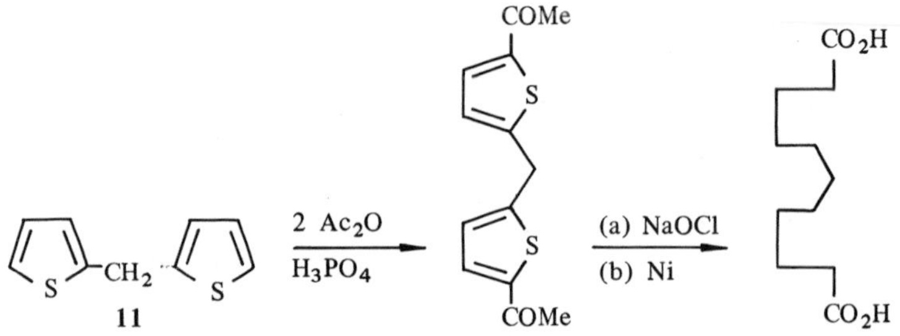

B. From Indoxyls

A novel degradation of cyclic ketones **12** was reported[5] by interim formation of an indoxyl derivative **13** followed by chromic acid oxidation to the dicarboxylic acid **14**. This process represents a two-step cleavage of cyclic ketones with loss of one carbon atom. Although the reaction was demonstrated only with steroidal derivatives, it should be applicable to most other cyclic carbonyl compounds. The indoxyl **13** is generated by treatment of the ketone with o-nitrobenzaldehyde in alkaline medium and its formation has been postulated

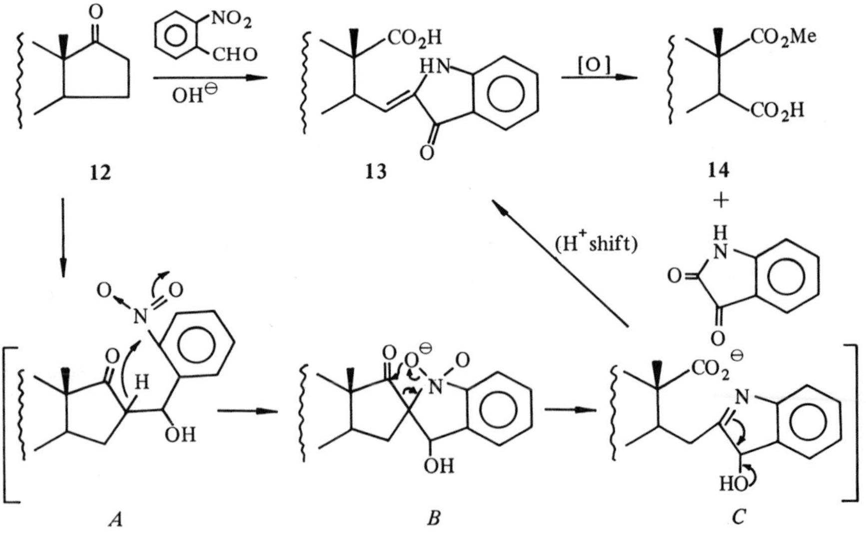

to proceed through the intermediates *A-C*. The latter, after dehydration, will lead to the indoxyl **13**. A major limitation of this degradative procedure could arise if the cyclic ketone possesses enolizable protons on the opposite side of the carbonyl group. This would then interfere with the formation of the spiro system *B*.

C. From Oxazolines

The reaction of carboxylic acids and amino alcohols producing 2-oxazolines 15 and the heterocycle's ready reversal by hydrolysis into the starting materials is a well-known sequence.[6] However, the oxazolines, which are in a sense masked carboxyl groups, possess properties that are very different and, in many ways, highly advantageous with respect to carboxylic acids. Wehrmeister[7]

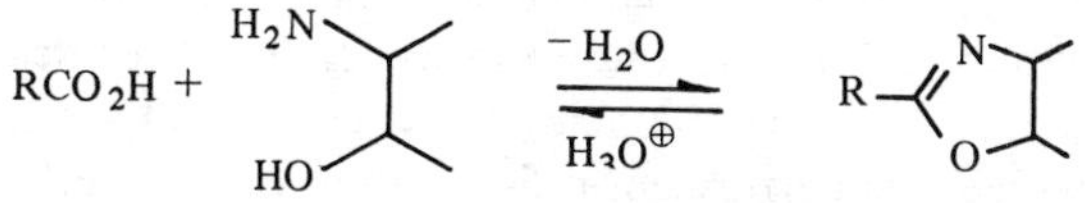

showed in 1962 that the 2-methyl-2-oxazoline 16 condenses cleanly with aromatic aldehydes in the presence of acidic catalysts affording the adducts 17 in good yield. Acidic hydrolysis subsequently produced the cinnamic acids 18. The method was equally applicable to α-methyl cinnamic acids 20 by starting the sequence with the 2-ethyl-2-oxazoline 19. This use of a simple, readily

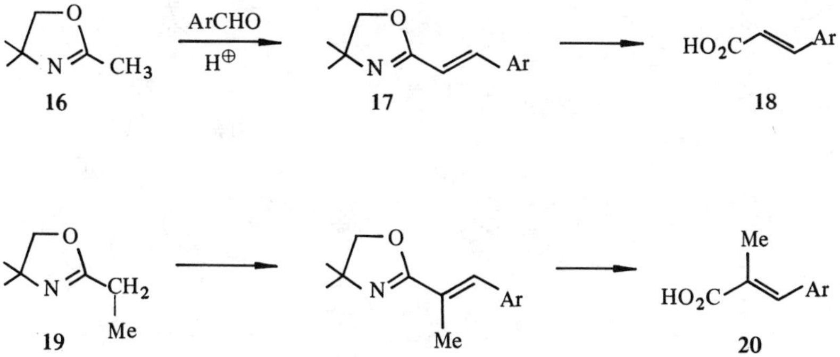

available oxazoline 16 or 19 may serve as an alternative to the classical approaches to cinnamic acids via the Perkin, Doebner, or Wittig methods. Examples of unsaturated acids prepared from 2-oxazolines are listed below in tabluated form.

Oxazoline	Acid, 18	Acid, 20	% Yield
16	Ar = Ph		97
16	Ar = p-Cl-Ph		100
19		Ar = Ph	92
19		Ar = p-Cl-Ph	92
19		Ar = o-Cl-Ph	99
19		Ar = p-(Me₂N)-Ph	80

Another synthetically useful process based on the 2-methyl-2-oxazoline **16** was reported[8] recently. Treatment with one equivalent of butyllithium at low temperature generated the lithio-oxazoline **21** which could be efficiently mono-alkylated with alkyl bromides and iodides affording **22**. The amount of dialkylated oxazoline **25** obtained never exceeded 7%. The oxazoline **22** readily ruptured in aqueous hydrochloric acid to the elaborated acetic acid **23** or the esters **24** if cleavage was performed in 5-8% sulfuric acid in ethanol. The overall yields of acids **23** or esters **24** ranged from 70 to 90%. Alternatively, cleavage in other primary or secondary alcohols furnishes the corresponding esters in equally good yield.[8a] a-Alkyl acetic acids **26** or their corresponding esters **27** were also prepared by subjecting the monoalkylated oxazine **22** to further metallation and alkylation producing the 2-(dialkyl)-2-oxazoline **25**. The homologation of acetic acids via the oxazoline offers some advantages over the direct alkylation of metallated acetic acids or esters.[9,10]

Specifically, the oxazoline route uses *n*-butyllithium as the base, whereas the other methods require lithium dialkylamides, prepared from the appropriate amine and butyllithium. Furthermore, the oxazoline is an excellent masking group for carboxylic acids against both lithium aluminum hydride and Grignard reagents. This aspect may be appreciated by the conversion of *p*-bromobenzoic acid to its oxazoline derivative **28** and ultimately to the "Grignard reagent of *p*-bromobenzoic acid" **29**.[11] Many typical Grignard reactions were performed

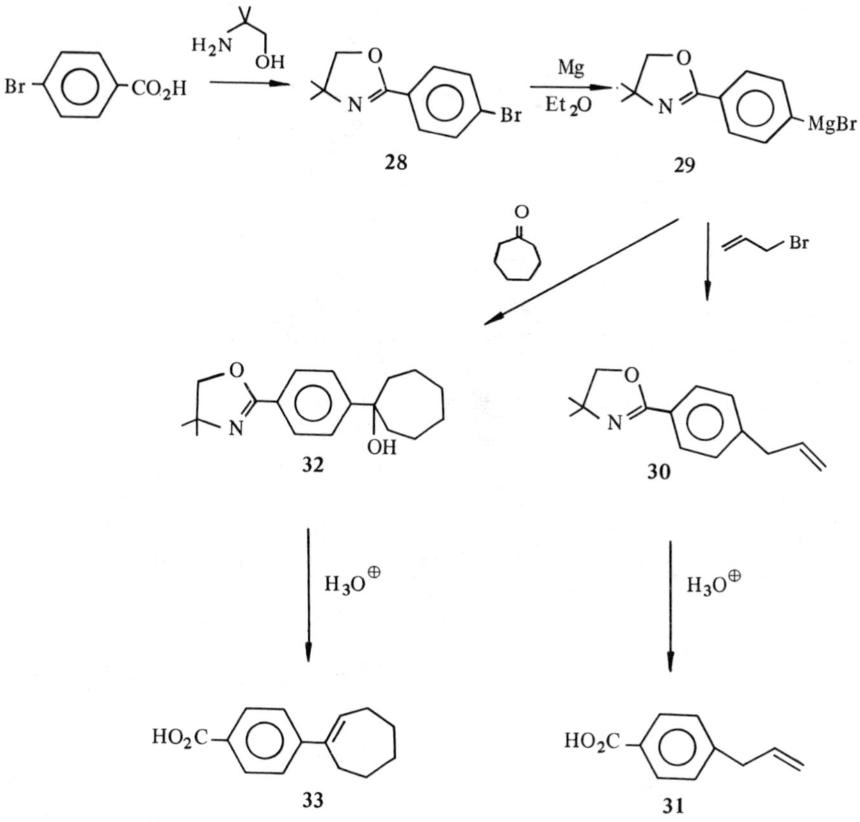

and two are shown: *(a)* the synthesis of *p*-allyl benzoic acid **31** via its oxazoline precursor **30** and *(b)* *p*-cycloheptenyl benzoic acid **33** via its oxazoline precursor **32**. Similar results were obtained using *o*-bromobenzoic acid and its oxazoline derivative. The versatility of the 2-methyl oxazoline **16** was further demonstrated[8] by its ability to serve as a convenient reagent in the formation of Reformatsky esters **35**. High yields of the latter were consistently obtained by treatment of the lithio salt of **16** with various aldehydes or ketones producing

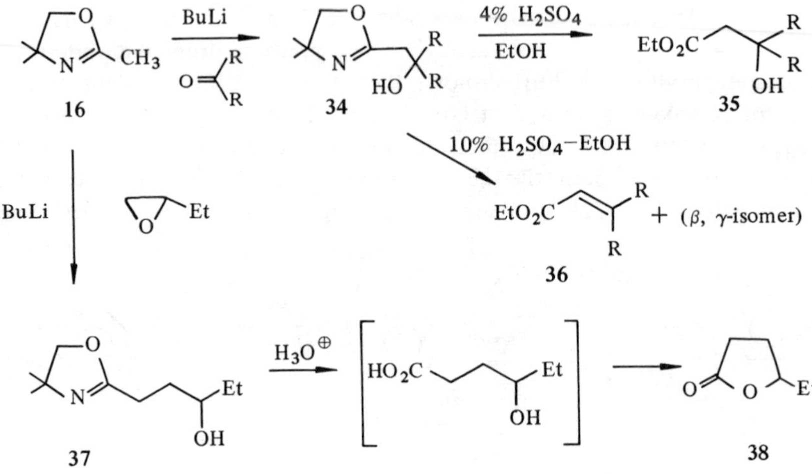

34 which was hydrolyzed in 4% sulfuric acid-ethanol to the β-hydroxy esters. Hydrolysis of **34** in 8-10% sulfuric acid-ethanol resulted in dehydration to the unsaturated esters **36** as a thermodynamic mixture of α,β- and β,γ-isomers. Epoxides were also smoothly alkylated by the oxazoline anion generating, in the case of butene oxide, the adduct **37**. Hydrolysis in aqueous acid led to good yields of the lactone **38**, via the initially formed γ-hydroxy acid. A major limitation of the alternate Reformatsky synthesis leading to β-hydroxy esters, is the inability to efficiently prepare α-alkyl-β-hydroxy esters **39**. Thus the anion of 2-(*n*-propyl)-2-oxazoline **39**, whose metallation and addition to cyclo-

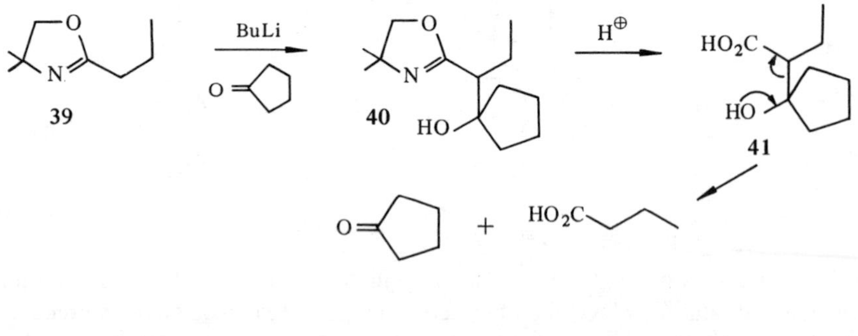

pentanone proceeded well (to **40**), gave, upon hydrolysis, poor yields (15-20%) of the β-hydroxy acid **41** (or corresponding ethyl ester). The major products were the starting ketone and *n*-butyric acid (or ethyl ester). This behavior has been previously observed in attempts to prepare such highly substituted β-hydroxy acid derivatives by either the classical Reformatsky reaction[12] or using metallated carboxylic acids.[13]

Since the 2-oxazolines are normally prepared by distillation from a mixture of carboxylic acid and 2-methyl-2-aminopropanol,[14] the method suffers in those cases where the required oxazoline cannot be distilled because of high molecular weight and low volatility. Aromatic carboxylic acids, on the other hand, are efficiently transformed into their oxazoline derivatives by use of thionyl chloride induced amide formation followed by cyclization.[15] The need for a mild and general conversion of carboxylic acids to oxazolines was recently answered by forming N-acyl derivatives 43 of 2,2-dimethyl aziridine 42.[16] These systems are readily prepared at room temperature from carboxylic acids and the dimethyl aziridine 42 using dicyclohexylcarbodiimide (DCC). Alternatively, esters are transformed into the acylaziridines by reaction with the magnesium

RCO₂H

DCC

$Me \overbrace{}^{NH}$ Me

42

$R-\underset{O}{\overset{\|}{C}}-N \overset{Me}{\underset{Me}{\diagdown}}$

43

RCO₂Et

$Me \overbrace{}^{NMgX}$ Me

45

H⊕

Me
Me $\overbrace{}$ O
N — R

44

halide salt of dimethylaziridine 45. Conversion of 43 to the oxazoline derivative 44 was accomplished using a variety of mild acid catalysts (potassium bisulfite, pyridinium tosylate, or the sulfate salt of the 2-methyl-2-oxazoline 16) in ether or dichloromethane at room temperature. In this manner, a variety of functionalized aliphatic and aromatic acids were protected as their oxazoline derivatives. Although 44 is a 5,5-dimethyl oxazoline and the previous systems discussed were 4,4-derivatives, there was little difference in their synthetic usefulness.[17]

It is intriguing to contemplate the plethora of potentially useful synthetic methods that may arise from the 2-oxazoline system. For example, an optically active oxazoline **46** would provide an asymmetric environment for alkylation and this could lead to optically active acids or esters **47**. In addition, the stability of oxazolines to lithium aluminum hydride and Grignard reagents could lead to complexes of an asymmetric nature **48** and thus direct the addition of this

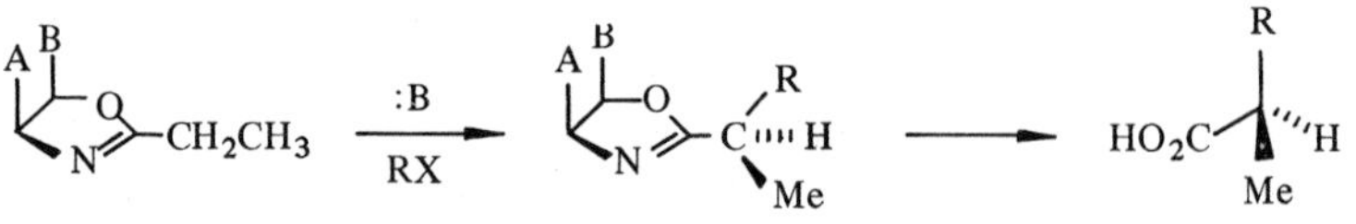

(+)- **46**   (+)- **47**

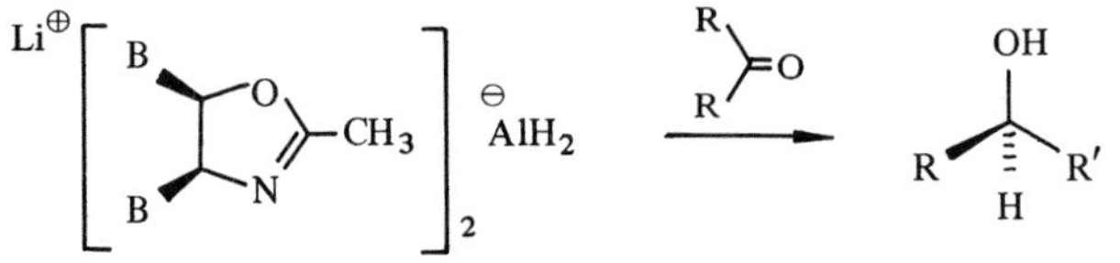

(+) - **48**   (+)- **49**

process in a biased manner. Only time will reveal whether these thoughts can be brought to fruition.

Another useful and novel reaction which involves only the transient inter-mediacy of 2-oxazolines has led to the transformation of ketones **50** to their homologated carboxylic acids **51**.

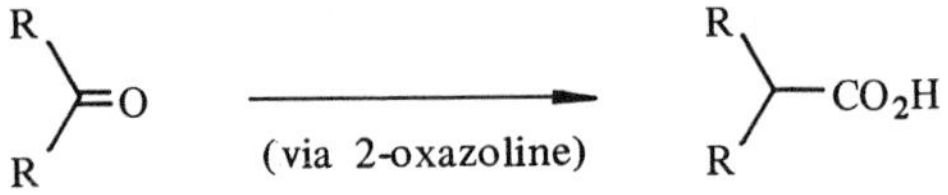

50   **51**

Schöllkopf has extensively studied the condensation of α-carbanions of various isonitriles and carbonyl compounds which leads, *in situ*, to 2-oxazolines

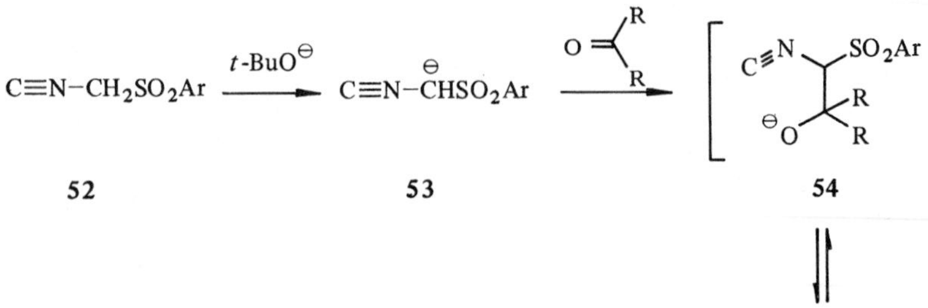

52   **53**   **54**

55

55 existing in ring chain equilibrium with their isonitrile counterparts 54. The utility of this sequence has already been mentioned with respect to olefin syntheses (Chapter 3, Section VII). In a more recent study,[18] the sulfonyl-methyl isonitrile 52, was converted to its α-carbanion 53 using potassium tert-butoxide in tert-butanol. Addition of a carbonyl compound produced the adduct 54 which may, as already stated, be in "equilibrium" with its valence tautomer 55. Proton abstraction from the solvent leads to the 2-H oxazoline 56, which loses the more acidic proton α- to the sulfonyl group. Alternately, the proton may be intramolecularly transferred in 54 since β-hydroxy iso-nitriles are also readily cyclized to oxazolines.[19] Rearrangement of 56 occurs leading to the α-sulfonyl enamide 57. Hydrolysis of the reaction mixture produces the carboxylic acid 58 in good yield. This particular transformation of ketones to carboxylic acids (50 → 51) is virtually without competition by other routes. The method of Ziegler,[20] wherein ketones are transformed into nitriles of one additional carbon by decomposition of dialkyl cyanodiazene-carboxylates, is noteworthy in this respect.

D. From Oxazoles

Certain 4,5-polymethylene oxazoles 59 are sensitive to photooxidation and were reported[21] to produce ω-cyano anhydrides 60 in good yield. This latter product may be easily hydrolyzed to the ω-cyano carboxylic acid 61 and thus the route affords a novel entry into this polyfunctional system. The reaction was shown to proceed initially to the azonide 62 which underwent double scission (via 63) to the ω-cyano anhydride 60. When the 2-substituent was aryl (59, R = phenyl) the reaction took a different course, producing (a) N-aroylisoimides 64 in

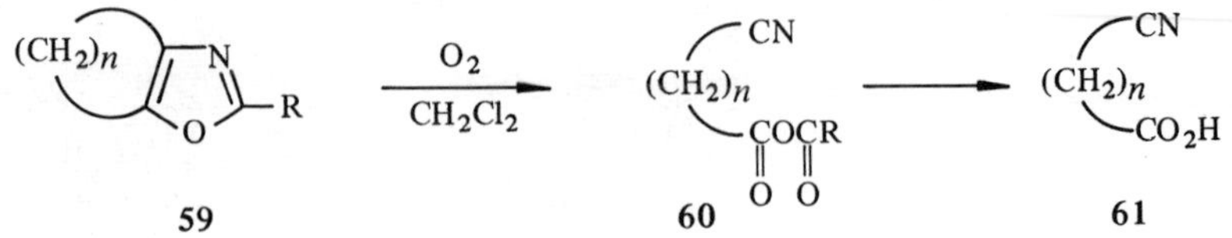

R	n	% Cyano Acid (61)
H	4	77
H	5	72
H	6	80
H	10	94
Me	4	61
Me	5	60
Me	6	65
Me	8	73
Me	10	76

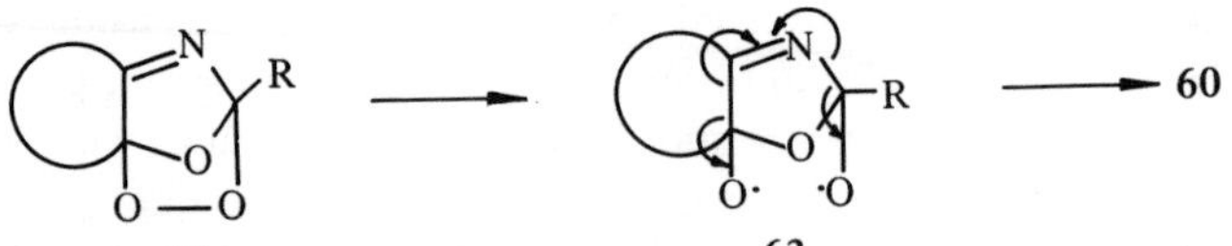

methylene chloride and (b) ω-imide esters 65 in methanol solvent. Of further interest in synthesis is the fact that the oxazole precursors 59 are conveniently prepared in high yield from the appropriate acyloins 66 and formamide in sulfuric acid.[22]

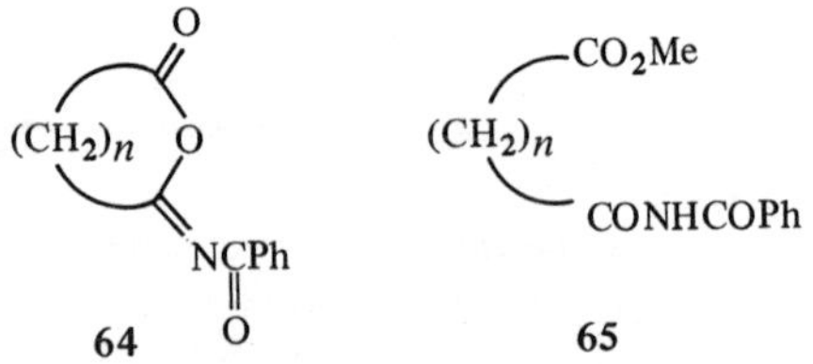

Since the cyclic acyloins are derived from long chain dicarboxylic esters, this sequence provides a method for transforming one of the carboxyl groups into a cyano group.

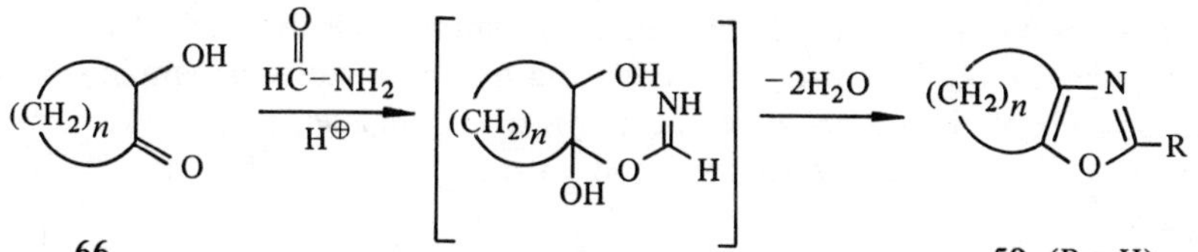

66 → **59, (R = H)**

E. From Oxazol-5-ones

Homologation of a carboxylic acid by three carbons has been accomplished using the 2-substituted oxazol-5-one **68** as a vehicle.[23] The heterocycle is constructed for subsequent use by condensing the appropriate carboxylic acid with *dl*-valine **67** in a standard cyclodehydrative sequence. The key feature of the oxazol-5-one is its ability to form a highly delocalized anion **69** when treated with triethylamine.

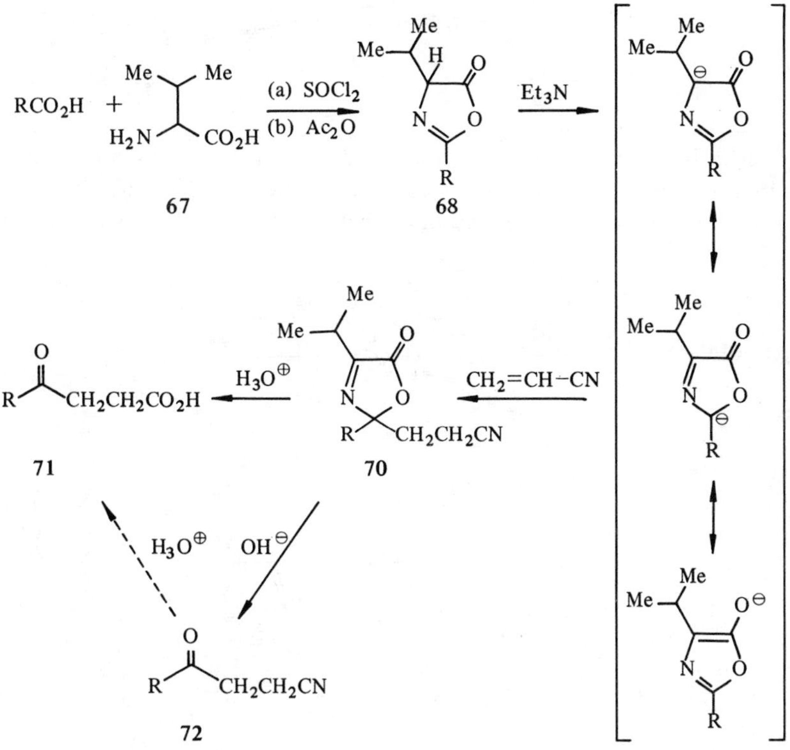

Addition of acrylonitrile results in exclusive alkylation at C-2 providing the elaborated oxazolone **70**. This last step is tantamount to a nucleophilic acylation by the carbonyl group of a carboxylic acid. Simple hydrolysis of **70** affords the γ-keto acid **71** or the nitrile **72** depending on the conditions employed. The overall yields of the homologated acid or nitrile from a wide variety of carboxylic acids ranges from 55 to 89% and introduces a valuable new synthetic technique.

F. From Pyrazolones

In 1958, Carpino described a new synthesis of acetylenic **(73)**[24] and olefinic **(74)**[25] carboxylic acids from β-keto esters. The method involved the intermediate formation of the well-known system 5-pyrazolone **75**. The pyrazolones

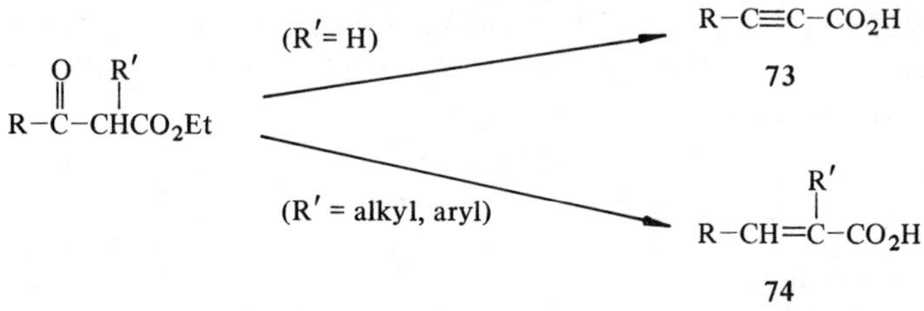

are routinely obtained by condensing hydrazine with β-keto esters and the required precursor to unsaturated acids, the a,a-dichloro **(76)** and the a-chloro **(77)** pyrazalones were formed by direct halogenation on the ring. The chloro-

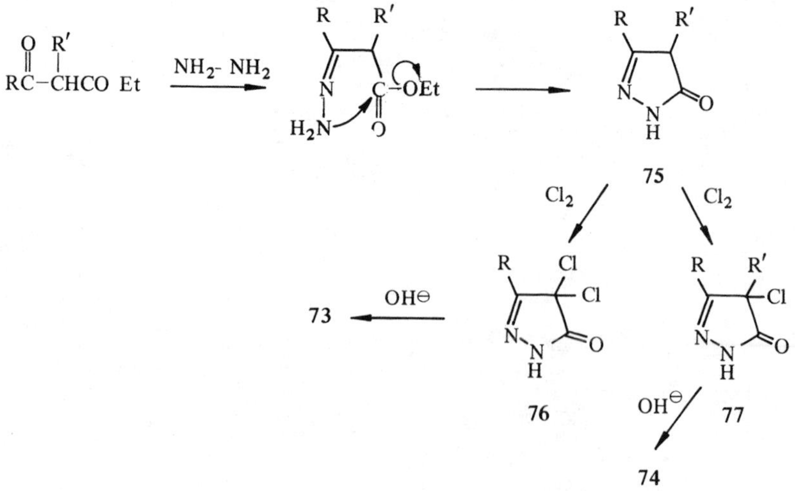

pyrazolones **76** and **77** smoothly decomposed in aqueous alkali to the acetylenic acids **73** and olefinic acids **74**, respectively. The pathway taken by this process has been shown[26] to proceed via the anion **78** which rearranges with concomitant expulsion of chloride ion to the diazacyclopentadienone 79. Hydroxide ion addition occurs with extrusion of nitrogen generating the acetylene carboxylic acid 73. In a similar fashion, the anion 80 derived from the 4-halo-5-pyrazolone rearranges to the diazacyclopentadienone 81 with elimination of chloride ion and the ensuing ring cleavage gives the vinyl carbanion 82 which protonates rapidly in the aqueous medium. Thus the cis-(83a) and trans-(83b) acids are formed in mixtures consisting predominantly of the former isomer (66-90%). If 80 is generated in $D_2O\text{-}OD^{\ominus}$, the acid products 83 (a and b) possess deuterium at the vinyl carbon which attests to the existence of a vinyl carbanion 82 during this reaction. Proof for the existence of the diazacyclopentadienones 79 and 81 as intermediates was also gathered, albeit indirectly, by trapping experiments with butadiene and cyclopentadiene. In both instances, the Diels-Alder adducts 84 and 85 were isolated and characterized.

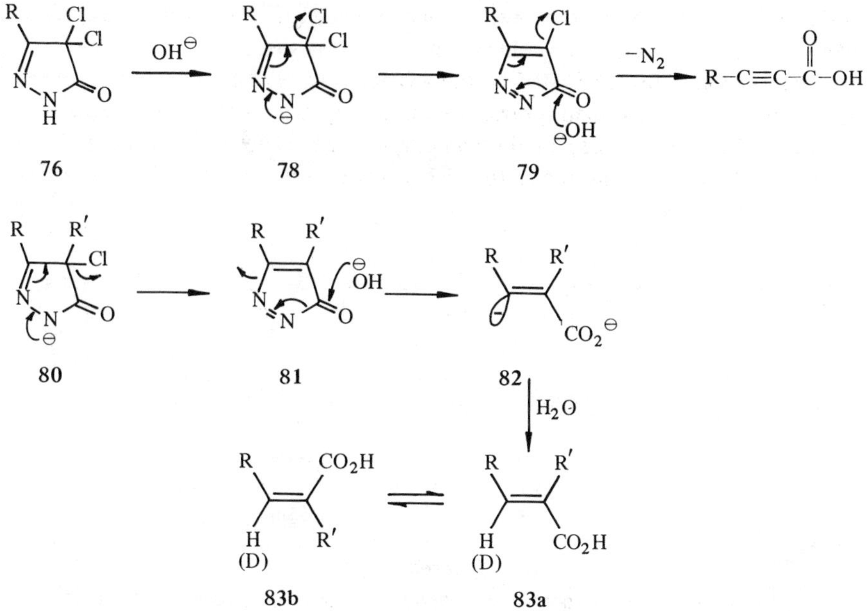

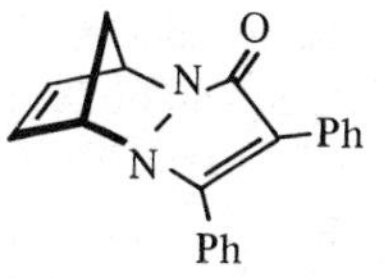

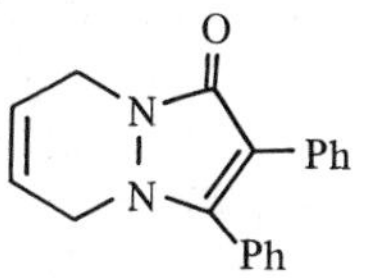

This method has recently[27] been extended to 3,4-polymethylene derivatives of

5-pyrazolones **87**, which were, expectedly, prepared from 2-carboethoxy cyclo-alkanones **86**. Halogenation (chlorine in dichloromethane or bromination using *N*-bromosuccinimide) of **87** led to the 4-halopyrazolone **88** whose ring system was ruptured as described previously affording the cycloalkenyl carboxylic acids **89**. Only in the case of the trimethylene pyrazolone (**87**, $n = 3$) did the reaction fail and this was due to the highly unstable nature of the corresponding halopyrazolone **88**.

$$86 \ (n = 3\text{-}6) \qquad 87 \qquad 88$$

$$89 \ (n \neq 3)$$

The above synthesis of acetylenic carboyxlic acids has recently been stream-lined considerably by the Taylor and McKillop groups[28] whose fortunes in thallium chemistry are rapidly becoming legend. Treatment of pyrazolones **75**, preformed in the usual manner or prepared *in situ*, with thallium(III) nitrate (TTN) in methanol produces excellent yields of the acetylenic esters **90**

$$75 \qquad\qquad 90$$

after 30 minutes at room temperature. Thus the halogenation step required in the Carpino synthesis is circumvented. Even more significant is the fact that the pyrazolone need only be formed as a transient species for the reaction to proceed to the desired product. This would be equivalent to a simple dehydration of the enol form of β-keto esters to acetylenic esters. That this highly useful synthetic transformation occurs as easily as shown is due to the remarkable oxidative properties of thallium salts. Reaction of the tautomer of **75** with TTN in the indicated fashion leads to the thallated pyrazolone salt **91** which now returns to the tautomer **92** via a prototropic shift. Since TTN is still present in the medium, it oxidizes the latter to the diazacyclopentadienone **93** which is

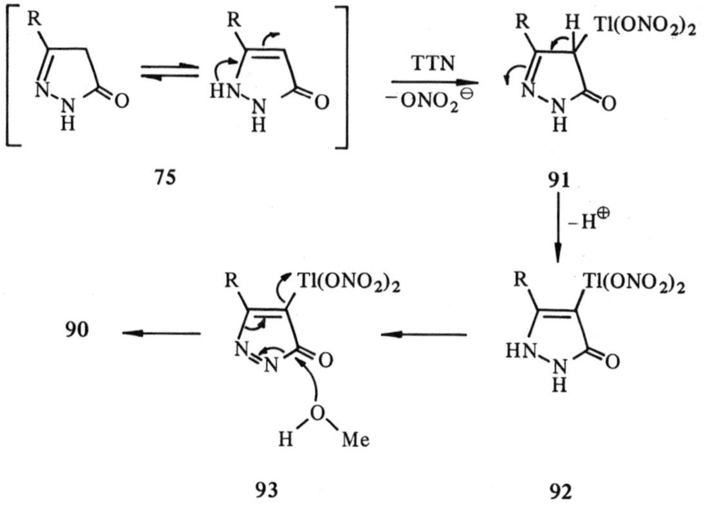

immediately recognized as the synthetic equivalent of the 4-halo diazacyclo-
pentadienane **79** in the Carpino synthesis. Methanolysis completes the reaction
by expulsion of both thallium(I) nitrate and nitrogen leaving the acetylenic
ester **90** in hand.

Under the same conditions, a-alkyl-β-keto esters **94** are converted to allenic
esters[29] **95** because of a structural rearrangement in the intermediate thallated

pyrazolone **96**. Since this species cannot rearrange to the tautomer corresponding to **92**, the *exocyclic* proton engages in the prototropic shift producing **97**.
Subsequent oxidation with TTN to **98** followed by methanolysis gives the
allenic ester in good yield. This technique also does not require the use of the
pyrazolone as a discreet starting compound and its *in situ* formation suffices to
produce the final product.

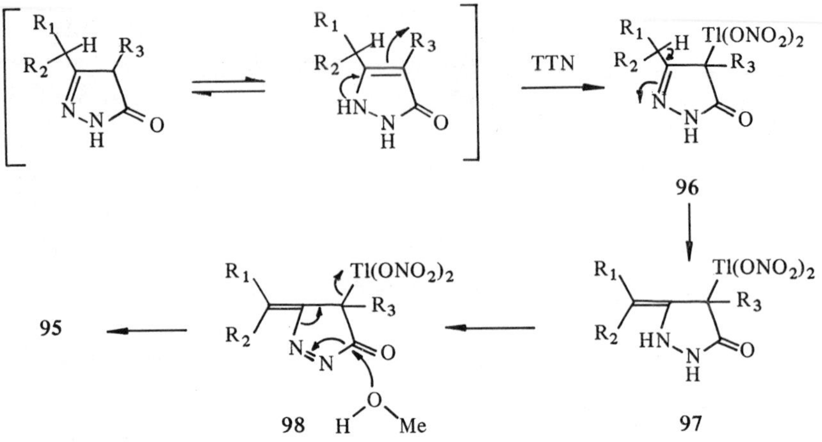

96

95 ←——— **98** H··Me   ←——   **97**

G.   From 1,2-Dioxalan-3,5-diones

Photolytic decomposition of the cyclic diacyl peroxide (1,2-dioxalan-3,5-dione) **99**, prepared in 78% yield by treatment of di-n-butylmalonic acid with 98% hydrogen peroxide in methane sulfonic acid, was shown by Adam[30] to give the α-methoxy carboxylic acid **101** in 88% yield. This result is considered as proof that the α-lactone is initially generated as a discreet intermediate, which is efficiently captured by the methanol solvent. The high yield of α-methoxy-α-butyl caproic acid makes this route rather attractive for preparation of α-alkoxy acids from readily accessible disubstituted malonic acids. Of additional interest

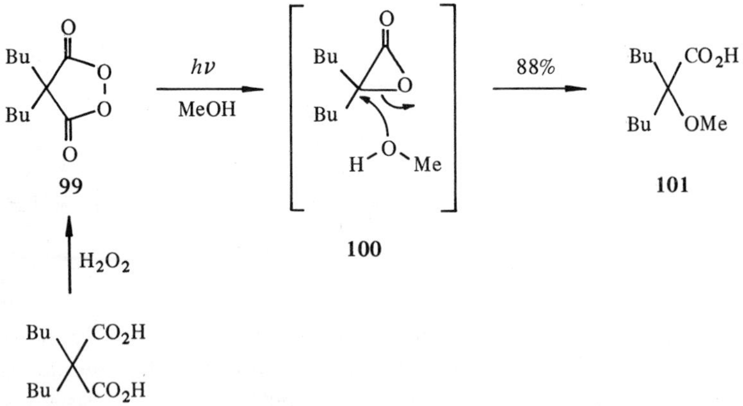

is the fact that this technique appears to allow further studies on the elusive α-lactone system, which in itself is an intriguing heterocycle. Future reports by the Puerto Rican group should open new vistas in α-lactone chemistry.

H. From Dithianes

Toward the synthesis of carboxylic acids, the dithiane system has also demonstrated its versatility. A one-carbon homologation of aldehydes **102** to carboxylic acids **103** was offered by Corey[31] which made use of the formation of a ketene thioacetal **106**. This latter heterocycle was readily obtained from the well-known[31] trithioncarbonate **104** and trimethyl phosphite which produced the dithiane phosphorane **105**. Introduction of an aldehyde to the phosphorane generated the ketene thioacetal **106** quantitatively via a usual Wittig coupling

104   **105**

102 —RCHO

(+ 1 carbon)

106

HgCl$_2$-H$_2$O

HO$_2$C–CH$_2$R

103

107   **108**

process. Since ketene thioacetals, or in general substituted dithianes, readily undergo cleavage in the presence of mercuric salts, the method provides a route to the homologated carboxylic acid **103**. The dithiane phosphorane **105** was shown to be virtually inert to ketones but this aspect of its chemistry may be considered more of an advantage than a shortcoming. It, therefore, should be feasible to homologate a keto-aldehyde of the type **107** to its carboxylic acid derivative **108** using this approach. This manipulation of polycarbonyl groups is not a trivial matter since there are not too many organic reagents that will perform selectively on aldehydes or ketones. Another technique which may be mentioned in this context is the conversion of keto-aldehydes **109** to their benzodithiepine derivative **110** leaving the keto function untouched.[32] Typical carbonyl reactions may then be carried out (RMgX, LiAlH$_4$, etc.) affording, in the case of Grignard addition, the alcohol **111**. Hydrolytic cleavage of the sulfur heterocycle regenerates the functionalized aldehyde **112**.

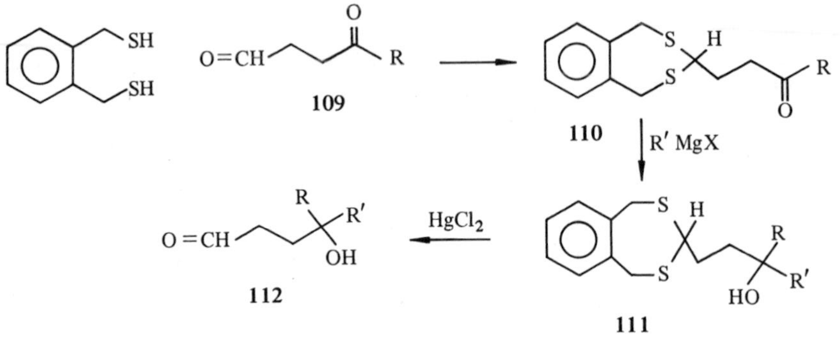

Another synthesis of carboxylic acids which invokes the dithiane system was reported by Eliel.[33] The key intermediate in this process is the carboethoxy 1,3-dithiane **113** easily prepared from 1,3-propanedithiol and diethoxy acetic ester.[34] Treatment with sodium hydride produces the α-carboethoxy anion which is transformed into the alkylated derivative **114** with alkyl halides. Cleavage of the dithiane ring using mercuric salts produces α-keto carboxylic acids **115** in good yield. Alternatively, the alkylated dithiane **114** may be transformed into substituted acetic esters **116** by reductive cleavage using nickel chloride-sodium borohydride. Although this entry into α-keto esters is a variation of the earlier sequence[35] involving dithiane **117**, it possesses distinct

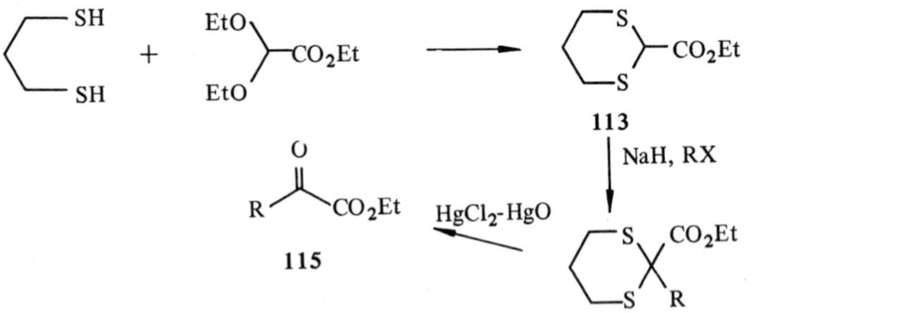

R	%
PhCH$_2$	76
n-Bu	60
sec-Bu	62

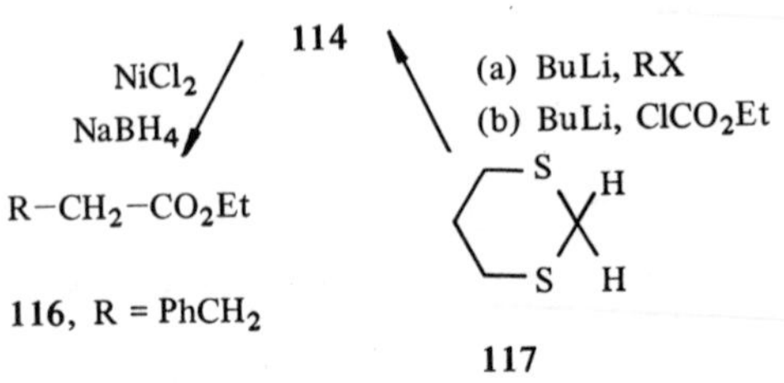

advantages. The approach originating from the parent dithiane requires two successive steps involving butyllithium, whereas only sodium hydride is required in the present example. Furthermore, **114** appears to serve as a common precursor for a wide variety of a-keto esters by appropriate choice of alkyl, allyl, and benzyl-type halides.

An intriguing application of dithiane chemistry toward the synthesis of functionalized carboxylic acids was offered by Marshall. The transformation of the dithiane **118** to the unsaturated acid **119** has already been mentioned in

connection with olefin syntheses (Chapter 3, Section IX). Further developments in this area center in the nonoxidative cleavage[36] of cyclic ketones **120**, via the a-dithiane derivative **121**,[37] to ring opened carboxylic acids **124**. The method is based on the nucleophilic attack on the a-keto dithiane leading to the carboxylic acid **122**. Ring cleavage is undoubtedly assisted by the ability of the dithiane to form a stable anion. Removal of the dithiane was accomplished using mercuric salts providing the aldehyde ester **123**. Oxidation of the latter to the dicarboxylic acid **124** completed the sequence. This novel approach to ring cleavage not only features the dithiane as a directing group for specific

C–C bond cleavage (**120** → **121**), but also leads to a useful intermediate **122** through which further synthetic manipulations should be feasible.

I. From 1,3-Oxazines

The versatile dihydro-1,3-oxazine system has also demonstrated its ability to serve as a precursor to acetic acid derivatives.[38] Previously the utility of the ketene *N,O*-acetal **126**, derived from the oxazine methiodide salt **125**, was described with respect to aldehyde preparations (Chapter 9, Section III.C). Carboxylic acids may also be synthesized by a simple modification in procedure. The highly nucleophilic nature of the ketene *N,O*-acetal was demonstrated

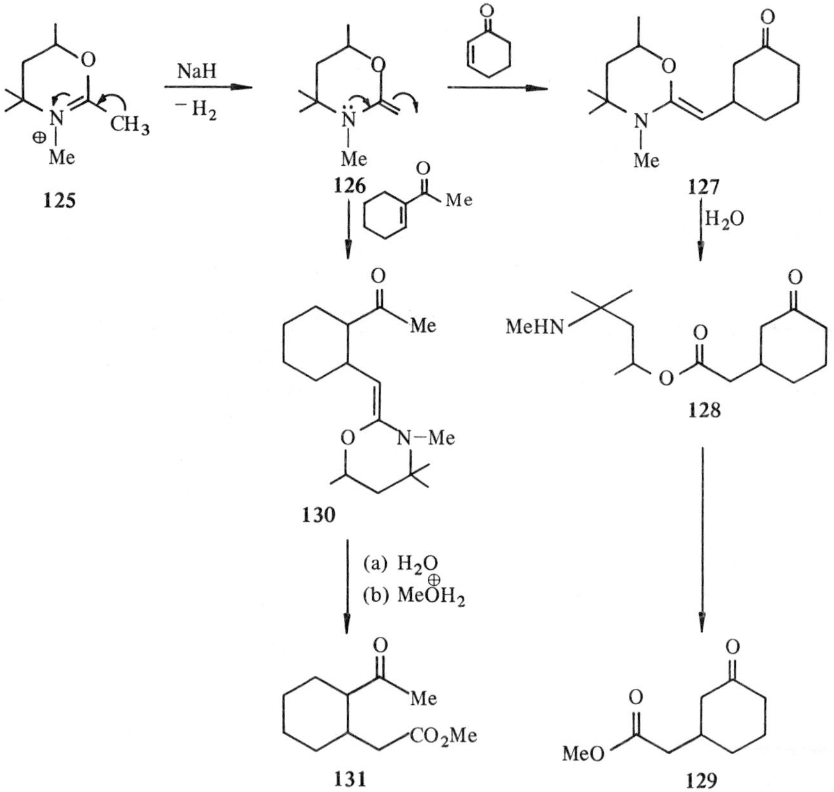

by its rapid addition to electrophilic olefins, such as 2-cyclohexenone, giving the adduct **127**. Instead of borohydride reduction, necessary for the route to aldehydes, **127** was treated with water resulting in rapid hydrolytic cleavage to the open chain amino ester **128**. The scheme was completed by transesterification with methanol generating the acetic ester **129**. Similarly, 1-acetyl cyclohexene was converted to its homologated acetic ester **131** after transesterification of the initial intermediate **130**. Additional examples leading to homologated acetic esters may be seen by the conversion of 2-cyclopentenones to **132** and progesterone to **133**.[39] Certain substitution patterns on the electrophilic olefin

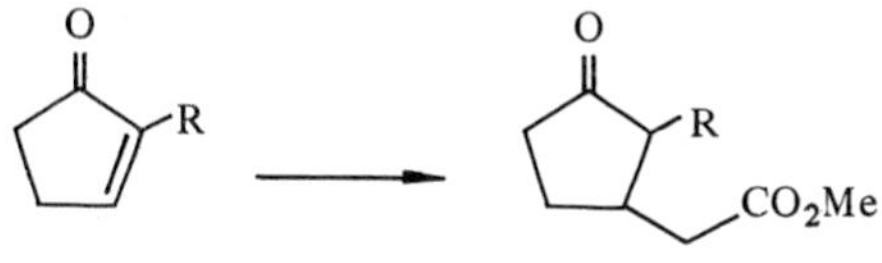

132 R = H, alkyl

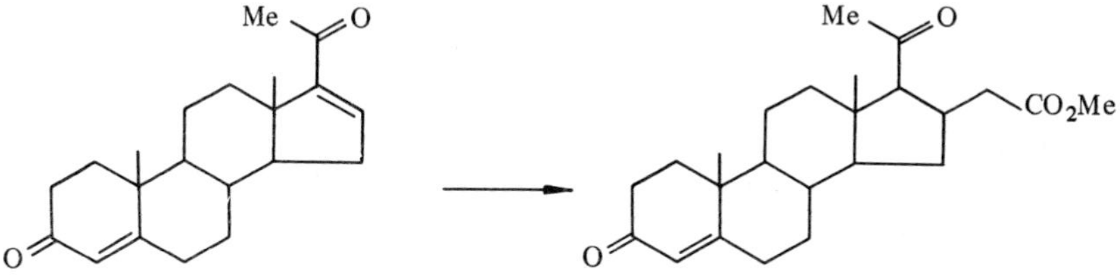

133

retard the addition of the ketene N,O-acetal. Thus in the formation of **133**, only the more accessible olefinic carbon at C-16 is attacked, whereas the C-5 carbon is inert to addition by the heterocyclic reagent. Although this acetic ester homologation could conceivably be implemented using the malonic ester synthesis, the absence of a hydrolysis-decarboxylation step provides a distinct advantage.

A stereospecific deuterium labeling technique, based on a temporarily constructed oxazine **137**, opens a route to phenylpropionic acids possessing this substituent.[40] The scheme begins with a deuterium labeled aldehyde **134**

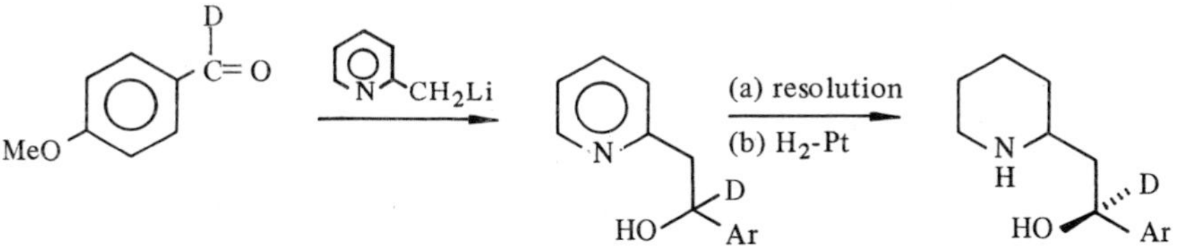

134   135   136

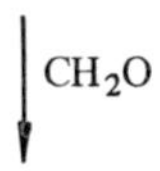

137

139   **138**   **137**

140

which is alkylated by a pyridine carbanion to the hydroxyl derivative **135**. Resolution of the latter is followed by catalytic reduction to the chiral piperidine carbinol **136**. The key feature of the entire process now emerges as the optically active piperidine carbinol is transformed into the oxazine **137**. Treatment with palladium-on-carbon results in ring opening of the optically active oxazine with inversion of configuration at the carbinol carbon producing the *N*-formyl piperidine **138**. Through a series of standard manipulations, the amino-olefin **139** is reached which gives the asymmetrically labeled phenyl propionic acid **140**. Although the method appears lengthy and tedious, further work may eliminate some of the intermediate steps. Furthermore, the interesting oxidative rupture of the oxazine **137**, in a stereospecific fashion, illustrates the continued synthetic value of this heterocycle.

II.   SYNTHESIS OF AMINO ACIDS

A.   From Thiophenes

Just as thiophenes have played a major role in the synthesis of hydrocarbons (Chapter 2, Section I.A) and simple carboxylic acid syntheses (Chapter 10, Section I.A) so have they also been widely employed to prepare amino acids. This approach to amino acids has been singularly developed by Goldfarb and his co-workers. It should also be stated that Goldfarb has pioneered much of the other syntheses of aliphatic compounds from thiophenes and has been the victim of incomplete reporting by many reviewers. This book will not attempt to recoup these omissions to Goldfarb's efforts in synthesis but will only single out to the English-speaking scientific community the excellent article published by the Russian group in 1962.[41] In addition, the reader is also directed to the review by Gronowitz[42] which pays considerable tribute to Goldfarb's efforts in thiophene chemistry.

Virtually every conceivable straight and branched chain amino acid can be prepared using the thiophene method.[41] By utilizing the appropriate 2-formyl

thiophene **141** (R = alkyl, aryl, or H) the Strecker synthesis may be applied giving **142** and, after hydrolysis to the thiophene α-amino acid, reductive cleavage generates the α-amino acid **143**. Alternatively, the thiophene may be elaborated to **144** using malonic ester followed by ammonia addition. Hydrolysis and decarboxylation of **144** is succeeded by treatment with Raney nickel producing the β-amino acid **145**. In this fashion, a large number of racemic α- and β-amino acids bearing a host of substituents have been prepared in respectable overall yields. More highly branched amino acids were prepared by initiating the synthesis with 3-formyl thiophenes **146**. Via this intermediate,

according to the previously mentioned method, the α-amino acid **147** was prepared.

The thiophene approach to amino acids is readily modified to provide almost any number of carbons between the carboxyl and amino group. This was accomplished by acylating thiophenes with ester-acid chlorides of dicarboxylic

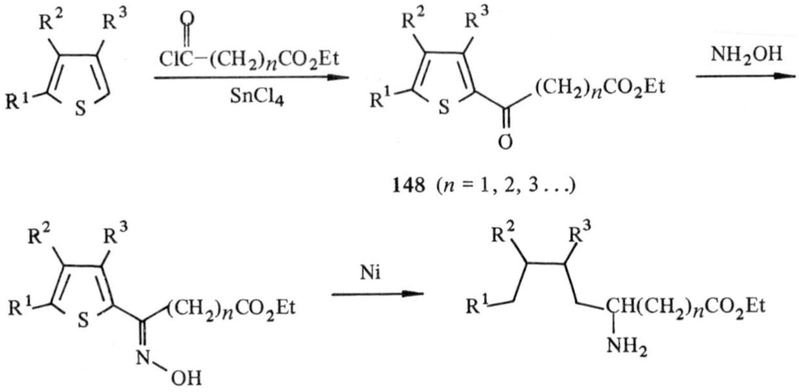

148 ($n = 1, 2, 3 \ldots$)

149   150

acids resulting in the keto thiophenes **148**. Conversion to the oximes **149** was then followed by the usual reductive cleavage giving the amino acids **150** whose carbon chain linking the amino and carboxylic groups is determined solely by the choice of the dicarboxylic acid derivative. So far, these methods are characterized by a common strategy—namely, that both functional groups (amino and carboxyl) are part of the same thiophene side chain. However, the functional groups may also be part of *different* side chains as seen from the example that follows:

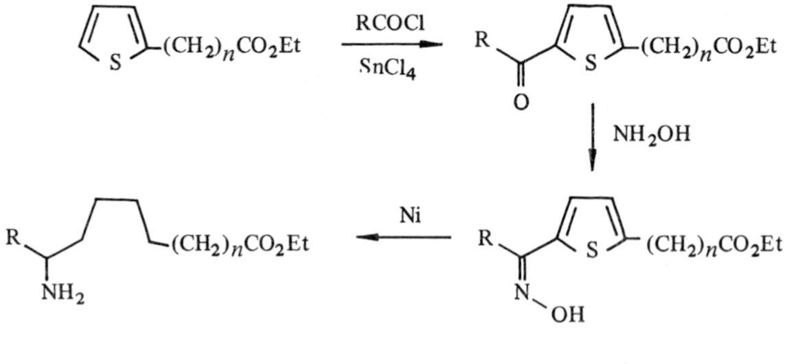

Another useful variation involves the reductive cleavage of ring substituted nitro thiophenes **151** to γ-amino acids **152** containing various alkyl substituents

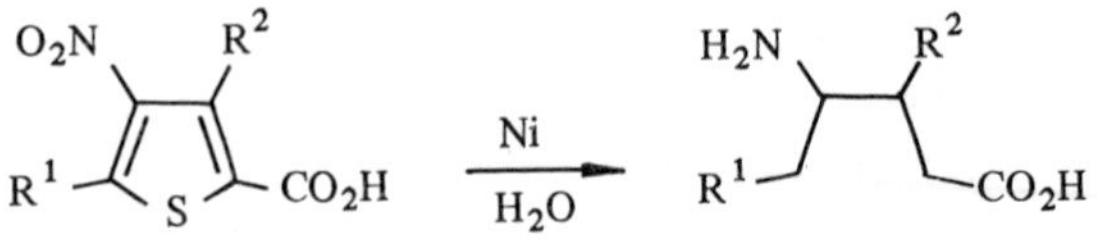

151   152

at the β- and δ-positions. In this particular sequence the Raney nickel desulfurization was performed in aqueous medium since the yields of amino acids were poor in alcoholic solvents.

Before leaving this subject, it should be mentioned that Goldfarb's technique has also led to cyclic lactones formed by a minor variation in the scheme. Thus, by cyclizing the thiophene acid chloride via an intramolecular acylation, the fused thienyl ketone 153 is obtained. Formation of the oxime 154 is succeeded

by a Beckmann rearrangement to 155. Reductive desulfurization produces the lactam 156 in good overall yields.

Without question, the thiophene approach to amino acids is truly powerful, at least with respect to variations in structural array. If there were some way to introduce the amino or alkyl moieties in an asymmetric manner or, even more exciting, to reductively desulfurize the thiophene nucleus in an asymmetric manner, the result would be optically active amino acids. The successful implementation of this idea would certainly represent a "giant step" for the organic chemist.

B. From Pyrroles

Oxidation of pyrrole with hydrogen peroxide has been studied and the process brought under control so that the usual tarry products have been minimized.[43] Thus, by treatment of pyrrole with 30% hydrogen peroxide in the presence of barium carbonate, a 30% yield of a mixture containing the pyrrolin-2-ones 157 was obtained (67% based on reacted pyrrole). The mixture was utilized[44] without purification since the subsequent step presumably leads to a common electrophilic species 158. Introduction of various aromatic substrates leads to good yields of the 5-aryl pyrrolidones **159** which, upon hydrolysis, generate the γ-aryl-γ-amino butyric acids **160**. A similar route to 4-furyl amino acids **161** was also described. The major limitation of this method appears to lie in the nature of the aromatic substrate utilized. Since the protonated pyrrolin-2-one is a weak electrophile, only highly nucleophilic aromatic systems may be introduced. In the case where indole was used as the substrate, the alkylation

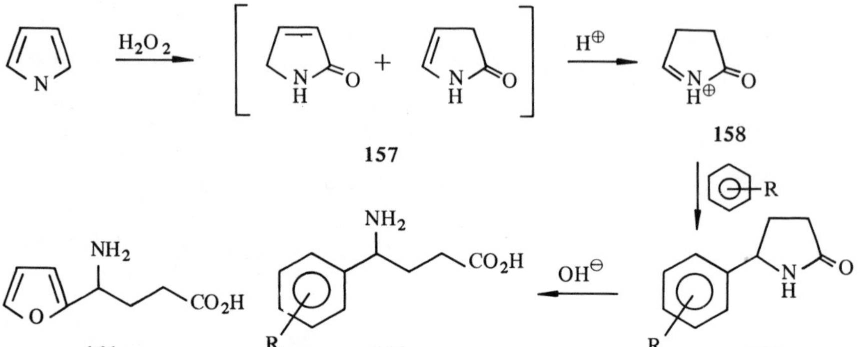

R	%
o-OH	70
p-Me$_2$N	74

proceeded in good yield to **162**; however, attempts to prepare the indole amino acid were fruitless.

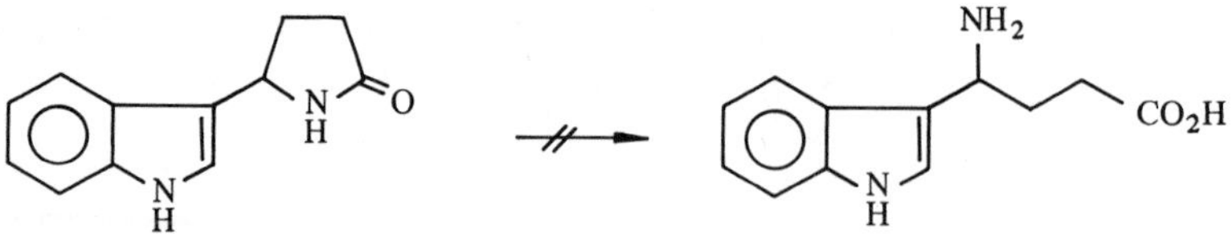

162

C.   From Oxazolones

The familiar reaction of *a*-amino acids with acid chlorides leading to their *N*-acyl derivatives **163** and ultimately, via cyclodehydration, to the oxazolone **164** has provided the basis for one of the oldest *a*-amino acid syntheses.[45,46]

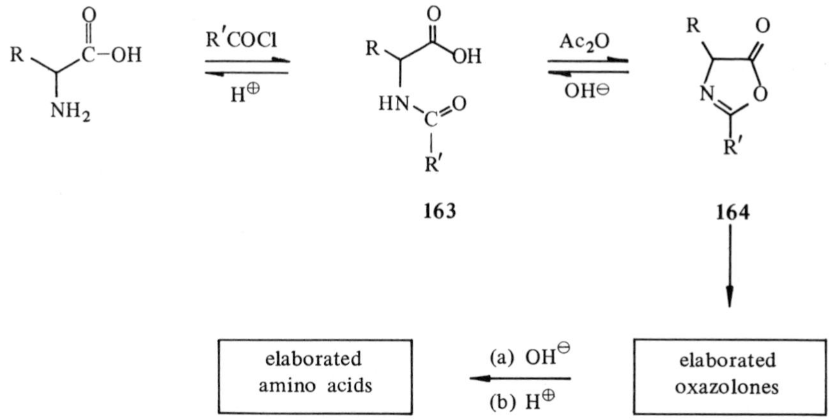

163   164

| elaborated amino acids | (a) $OH^\ominus$ ← (b) $H^\oplus$ | elaborated oxazolones |

By subjecting the oxazolone to hydrolytic conditions, the starting amino acid is returned. With this property available to the chemist, all that remains for a useful amino acid synthesis is to elaborate the oxazolone at any of several sites and thereafter hydrolyze back to the elaborated amino acid.

To illustrate this sequence, the preparation of (±)-phenylalanine **168** from glycine **165** via the temporarily constructed oxazolone **166** is presented.[47]

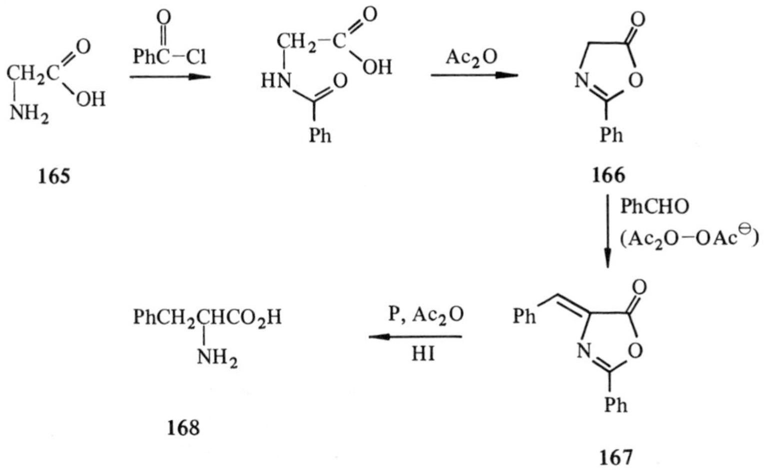

165   166

168   167

By condensing benzaldehyde with the oxazolone the benzylidene derivative **167** is formed which is reduced and cleaved using phosphorous and hydrogen iodide to the desired amino acid.[47a] On the other hand, instead of reductive cleavage of **167**, it may be subjected to a Friedel-Crafts-type alkylation involving an aromatic substrate. Treatment of **167** with a typical Lewis acid results in an electrophilic species which attacks certain aromatics present in the medium.

167 169 170

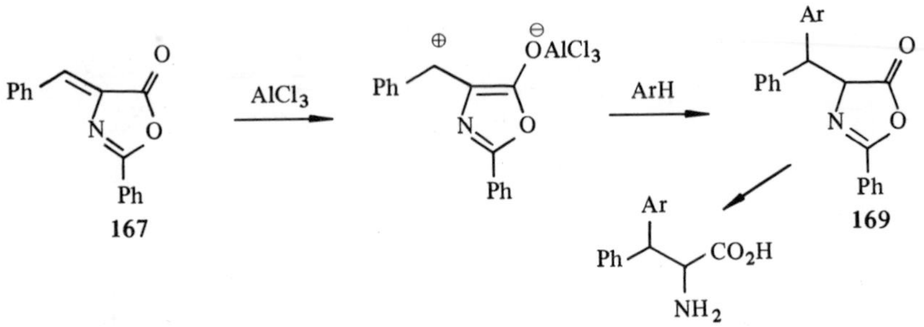

The arylated oxazolone **169** is then transformed into the β,β-diaryl-α-amino acid **170**. The yields for this process range from 60 to 90% when the aryl substrate is of the "activated type" in normal electrophilic aromatic substitutions.[48] It was observed by Horner[49] that 4-alkylidene or 4-arylidene oxazolones **171** react with *alkyl* Grignard reagents by conjugate addition to give the precursors to β-disubstituted amino acids **172**. It is, therefore, clear that the 4-alkylidene oxazolones **171** are very important precursors to amino acids. Since these derivatives are commonly obtained by a Perkin-type condensation with the appropriate carbonyl component, their access is limited when the carbonyl

171 172

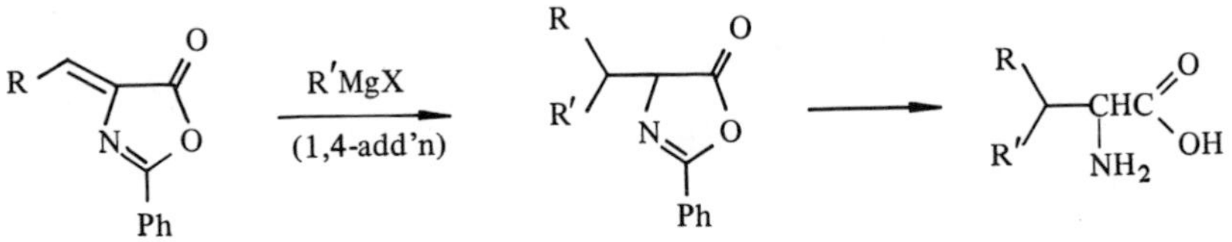

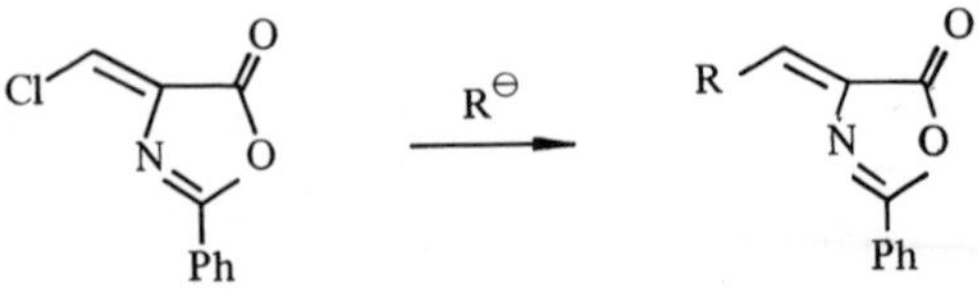

component is not available. To this end, there exists a useful synthesis of unsaturated oxazolones which relies on the chlorovinyl derivative **173**.[50] Reaction with various carbanions leads to the substituted oxazolone **171** via an

173 171

addition-elimination pathway. The anions generally employed were aryl Grignard and cadmium reagents (and presumably also organocuprates). The synthesis of 5-methoxy-tryptophan **175**, via the adduct **174**, is an example of this approach.

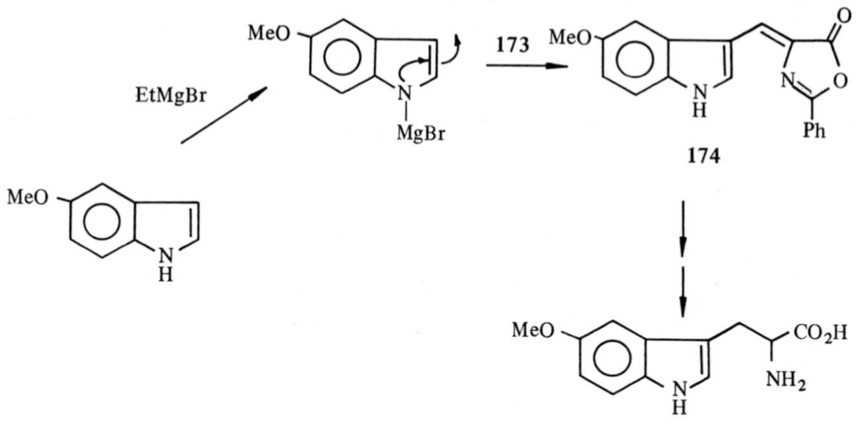

Amino acids containing additional functionality are also available through the oxazolone route. The preparation of the serine **181** indicates the versatility of

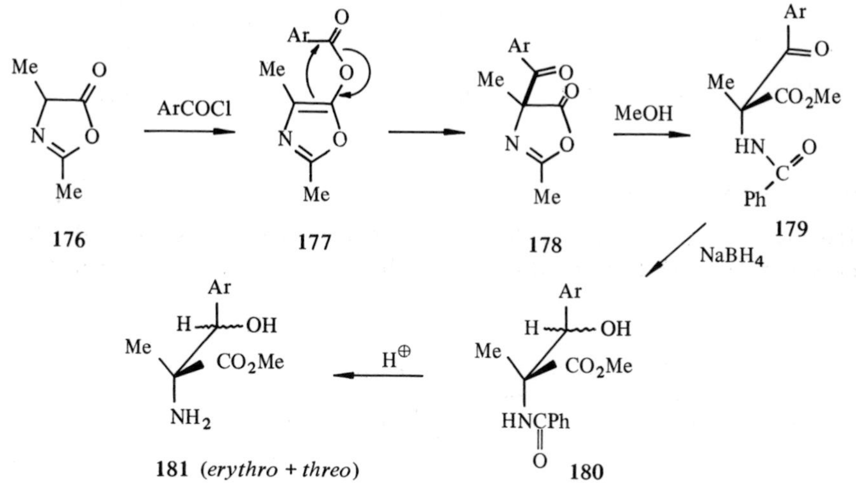

the oxazolone method.[51] Starting with the 4-methyl oxazolone **176** (derived from alanine), O-acylation to **177** is achieved by treatment with benzoyl chloride. Base catalyzed O- to C-migration generates **178** which is solvolyzed to the amino ester **179**. The keto group in **179** is reduced to the carbinol **180** and the synthesis is completed by hydrolysis to the serine. Although the method is useful, it lacks the stereospecificity of reduction in **179** and thus produces the serine as a mixture of its *threo* and *erythro* isomers. A variation on the entry into the serines invokes the same oxazolone as above, but instead of aroyl halides to introduce the aryl group, an organometallic is employed.[52] The sequence leading to serines via the 4-carboethoxy oxazolone **182** is presented below. In this case the borohydride reduction of the ketone **183** gave, after hydrolysis, predominantly the *erythro* isomer (60-66%) **181**. The review by Filler[46] should be consulted for more examples concerning the utility of oxazolones in synthesis.

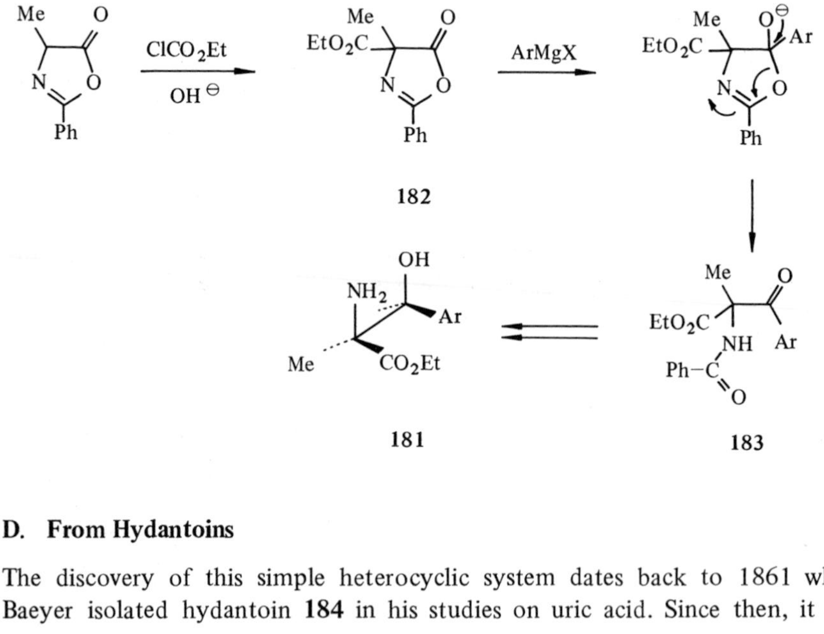

182

181

183

D.   From Hydantoins

The discovery of this simple heterocyclic system dates back to 1861 when Baeyer isolated hydantoin **184** in his studies on uric acid. Since then, it has

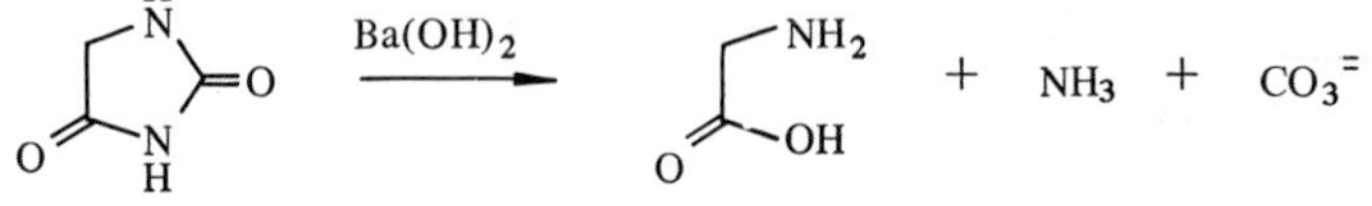

184

become an important precursor to many α-amino acids because of its lability to alkaline cleavage. Once again, the ability to elaborate hydantoins becomes compulsory if a versatile amino acid synthesis is to be realized. Many investigations have been performed in a manner analogous to that of the previously discussed oxazolones to produce a-amino acids. These have been reviewed in detail.[53]

In recent years there have been some important new synthetic approaches to the elaboration of hydantoins and ultimately to their derived amino acids. Finkbeiner[54] reported that the reaction of methylmagnesium carbonate (MMC) with hydantoins leads to a magnesium chelate 186 which is capable of being alkylated[55] with a variety of alkyl halides to 187. The yields of alkylated hydantoins range from fair to excellent; destruction of the hydantoin with alkali affords the a-amino acids 188. This use of a simple heterocycle provides a convenient route to a number of amino acids from readily available materials.

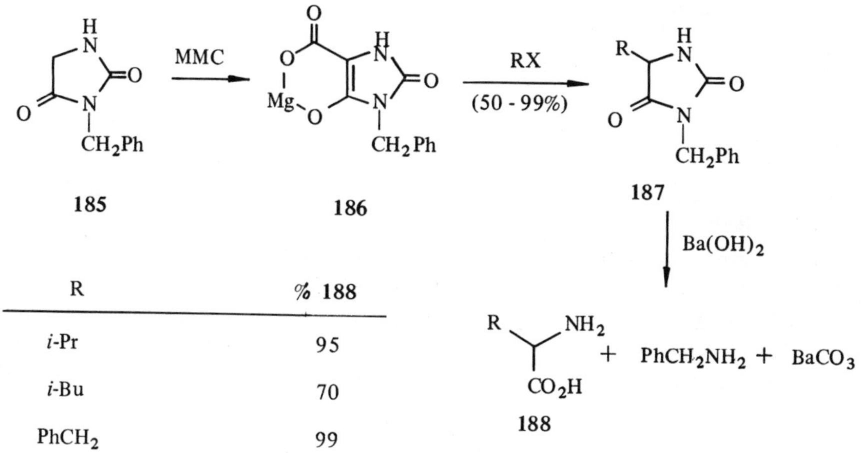

R	% 188
i-Pr	95
i-Bu	70
PhCH$_2$	99
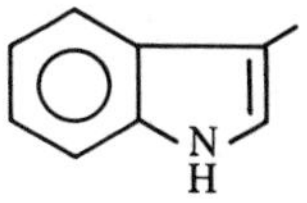 indolyl	97
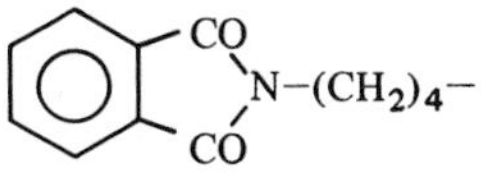 phthalimido-(CH$_2$)$_4$-	85
PhCH$_2$SCH$_2$-	84

The magnesium chelate 186 is also capable of double alkylation and this behavior was capitalized on in order to obtain proline 189 and its homologs. Thus the magnesium carbonate chelate of carbonyl compounds 190 displays

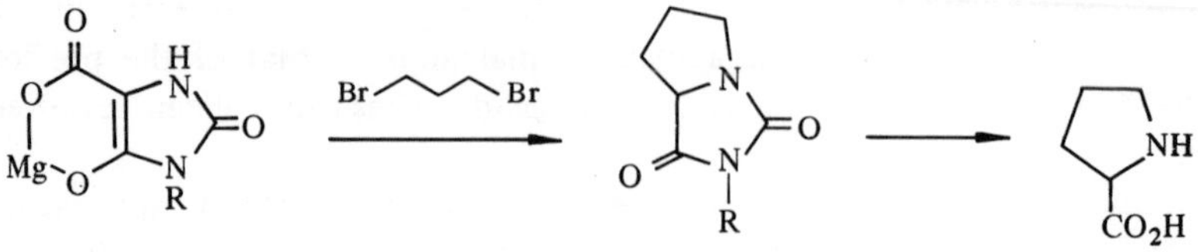

186   189

highly versatile synthetic features which presumably involve C-alkylation
followed by loss of carbon dioxide from the adduct **191** during the quenching

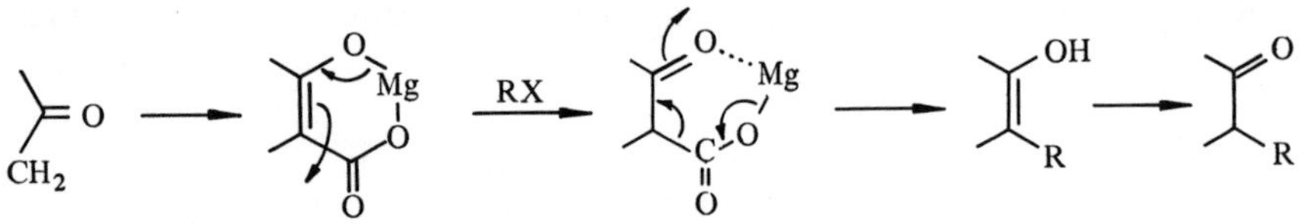

190   191

step. Anions derived from nitroparaffins may also be carboxylated using MMC
as well as other nucleophilic species.[56]

The 5-methoxyhydantoin **192** has been shown by Ben-Ishai to be a synthetically useful system for introduction of alkyl[57] and cycloalkyl[58] substituents
generating **194** and **194a**, respectively. The reactions are acid catalyzed and
supposedly involve the highly electrophilic intermediate **193**. The products
derived from alkene addition were obtained, in some instances, as mixtures of

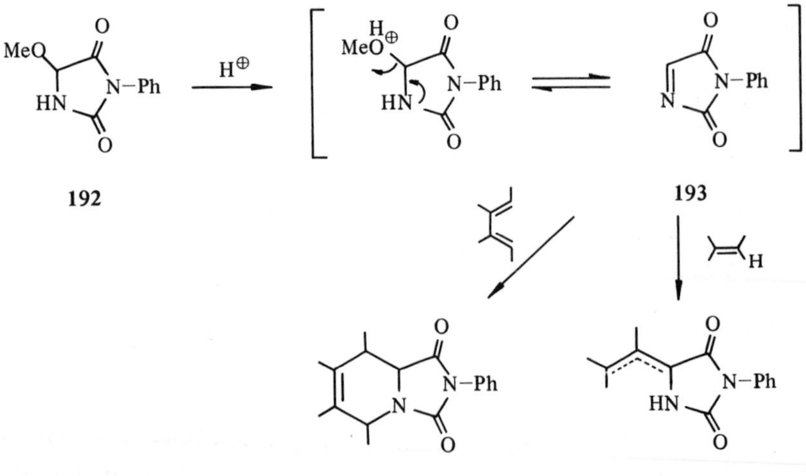

192   193

194a   194

double bond isomers **194**. Although 1,1-diphenyl propene gave only **195**,
1-phenyl propene gave both **196** and **197**. It is not surprising that olefinic

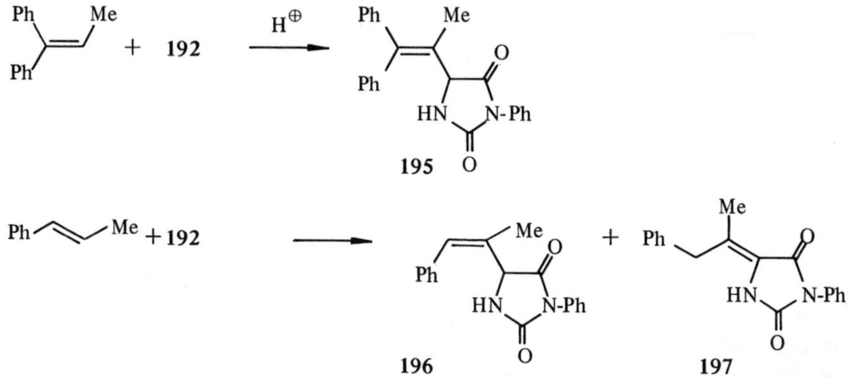

isomers are obtained in this condensation since the conditions allow the products to be formed under thermodynamic control. The 5-methoxy group is
sufficiently labile in the hydantoin **192** that it also provides an *in situ* source of
the reactive "dienophile" **193** and cycloaddition with a host of dienes results.
That the cycloaddition of dienes to **192** also occurs thermally (without an
acid catalyst) indicates that the reaction may, in some cases, be a true Diels-
Alder condensation. A variety of dienes was examined with or without acid
catalysts and the products are shown. The dotted lines indicate the correspond-

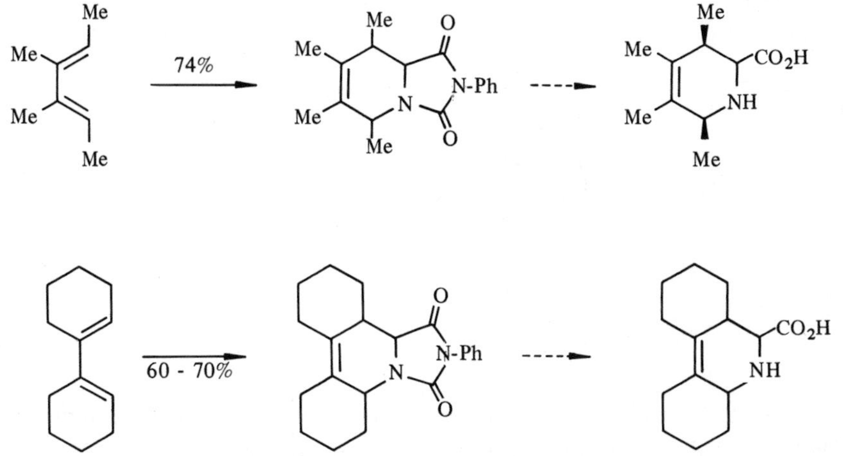

(two isomers)

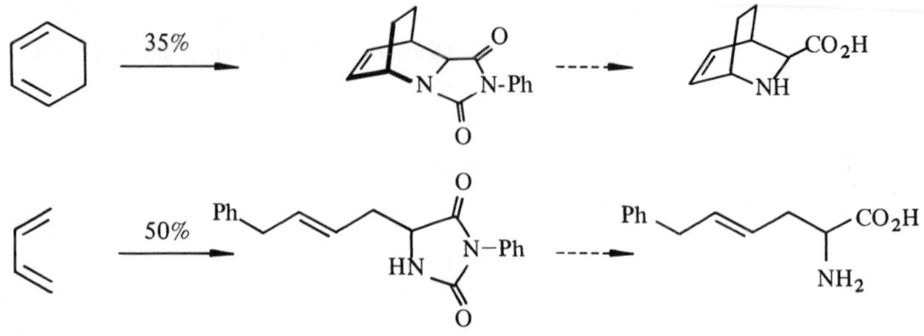

198

ing amino acids which would result upon alkaline hydrolysis. When butadiene was employed in the reaction, only a minor amount of the normal cycloadduct was obtained, the major product being the 5-(4-phenyl-2-butenyl) derivative **198**. This interesting product was stated to have arisen from participation of the benzene solvent. Arylation of hydantoins was also shown to be a feasible process.[59] Thus the 5-butoxyhydantoin **199**, when treated with aromatic

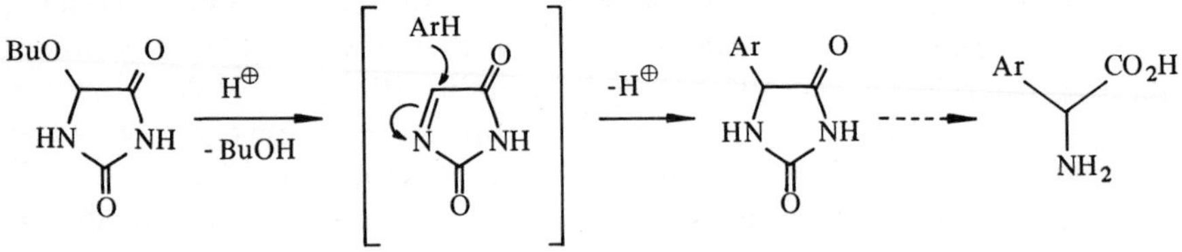

ArH	% **201**
Benzene	93
Toluene	53
Chlorobenzene	93
Acetanilide	51
Anisole	63
p-Xylene	70
Naphthalene	67
Phenanthrene	74

compounds, underwent arylation to **201**. The highly electrophilic intermediate **200** is probably responsible for this result. In this fashion a number of aryl substituted hydantoins which, in effect, represent α-aryl-α-amino acids **202**, were produced.

The 5-alkoxyhydantoins **192** and **199** were prepared for this study in 43 and 48% yield, using the following routes:

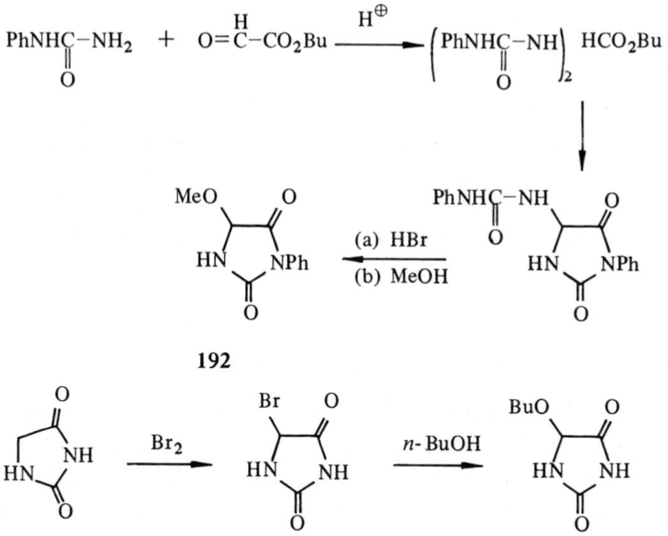

$$\text{PhNH}\underset{\underset{O}{\|}}{C}\text{-NH}_2 \quad + \quad O=\underset{\underset{}{\overset{H}{C}}}{C}\text{-CO}_2\text{Bu} \quad \xrightarrow{H^{\oplus}} \quad \left(\text{PhNH}\underset{\underset{O}{\|}}{C}\text{-NH}\right)_2 \quad \text{HCO}_2\text{Bu}$$

192

199

E. From Oxazolines

An approach to α-amino acids **207** using the oxazoline system was reported by Schöllkopf[60] in his continued studies on α-carbanions of isonitriles. Formation of the anion of α-carbethoxy isonitriles **203** and treatment with aromatic aldehydes leads to the adduct **204** which, upon quenching, generates the 2-oxazoline **205**. This sequence is mainly an extension of earlier work already discussed in a previous chapter (Chapter 3, Section VII). However, the novel feature in this study arises from the facile reductive cleavage of the oxazoline to α-(N-formyl)

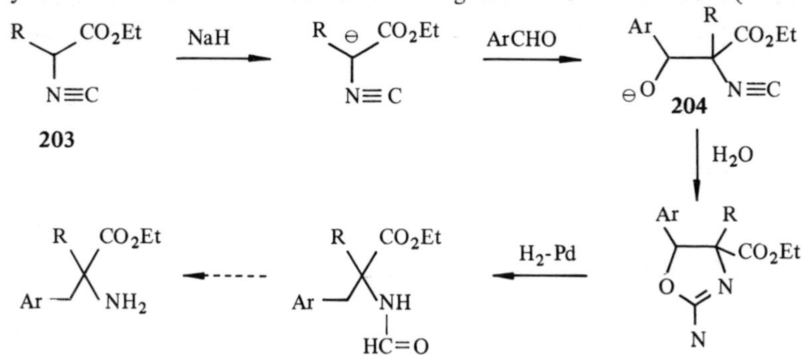

203

204

205

206

207

Ar	R	% 206
Ph	H	99
p-MeOPh	H	95
Ph	Me	99

amino esters **206**, a ready precursor to the corresponding amino esters **207**. It is noteworthy that this reductive cleavage to the *N*-formyl derivatives is similar, if not identical, to that mentioned earlier[40] (**137** → **138**). In both studies, hydrogenolysis occurred at the site possessing a benzylic carbon and, in the case of **138**, *ring cleavage took place stereospecifically.* It would be of interest to evaluate the present oxazoline route to amino acids with regard to this behavior.

F. From 1,4-Oxazinones

Modern concepts in organic chemistry have now been focused on the synthesis of optically active molecules. If the wonders of nature are truly to be capitalized upon, it is important for the scientist to raise his level of sophistication in synthesis to that which is found in natural processes. A synthetic approach to optically pure amino acids from racemic and achiral precursors has long been the goal of organic chemists and indeed some significant progress toward this end has been realized. The fact that heterocyclic compounds have played a major role in this effort further underlines their value in synthetic programs.

The preparation of aspartic acid **211** in very high optical purity has been efficiently achieved[61] starting with optically active *erythro*-1,2-diphenyl ethanolamine **208**. By condensation with acetylene dicarboxylic ester, the requisite heterocycle, **209**, was obtained in excellent yield. Catalytic hydrogenation proceeded quantitatively in a highly biased manner generating the oxazinone **210** which, upon hydrolysis, led to R(+) aspartic acid whose optical purity was 98%. Unfortunately, the optically active reagent was sacrificed in the process and this presents a distinct disadvantage for the technique. Nevertheless, its inherent concept prevails and further efforts should provide an alternative to

rotate

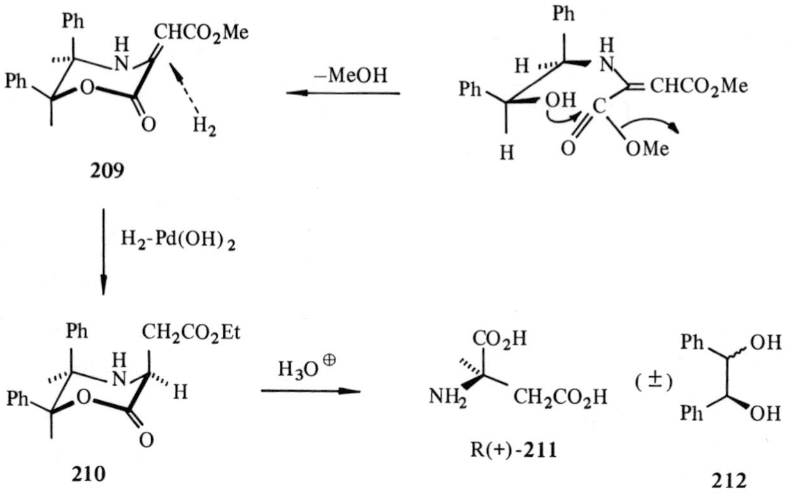

209

$$\xrightarrow{-MeOH}$$

H$_2$-Pd(OH)$_2$

210

$$\xrightarrow{H_3O^{\oplus}}$$

R(+)-211

(±)

212

the hydrolysis technique to preserve the chiral amino alcohol. The employment of a rigid heterocycle to direct asymmetric addition of hydrogen (or other incoming groups) will undoubtedly be an avenue for future successes in this area.

G. From Indolines

Corey[62,63] has developed an asymmetric synthesis of α-amino acids which shows considerable promise. In addition to producing amino acids of exceedingly high optical purity (92-99%) the chiral centers of the reagent needed to perform the synthesis were also recovered unscathed. The chiral indoline reagent **213** was condensed with pyruvic esters to form the expected hydrazone

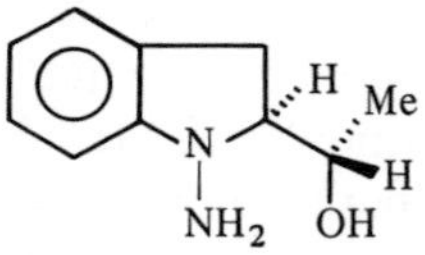

213 (S$_N$S$_O$)

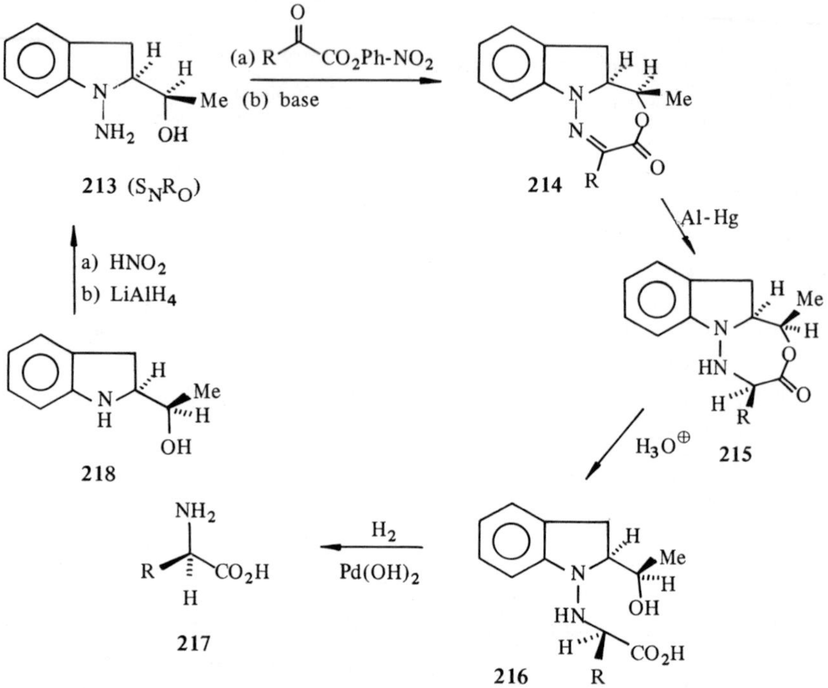

R	% O.P.
Me	96
Et	97
i-Pr	97
i-Bu	99

derivative which underwent lactone formation to **214** upon treatment with base. Reduction of the C=N link took place *only* with aluminum amalgam, since both catalytic and hydride techniques failed. The reduced product **215**, containing the newly formed chiral center, was subjected first to hydrolysis giving **216** and then to hydrogenolysis providing the α-amino acid **217** and the indoline carbinol **218**. The optical purity of the amino acids from the (S,R)-indoline **213** ranged from 96-99% while the same sequence starting from the (S,S)-indoline **213** gave amino acids whose optical purity was 92-96%. There are obviously some subtle steric interactions present in the (S,S)-indolene which account for the lower optical yield. Since the mechanism of aluminum-amalgam reductions is still an open question, the complete comprehension of this process must await further study. Nevertheless, the virtually pure amino acids obtained make this approach a major advance in synthetic methodology. Readily isolated from the reaction mixture was the indoline **218** which was nitrosated and subsequently reduced to the starting reagent **213**. The multistep synthesis, including a resolution necessary to obtain the indolines **213**, indicates that more efficient routes are still required. These are presented in Schemes 1 and 2.

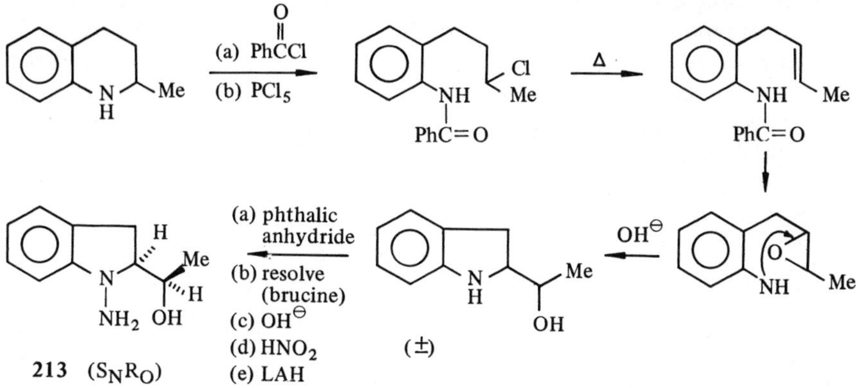

Scheme 1

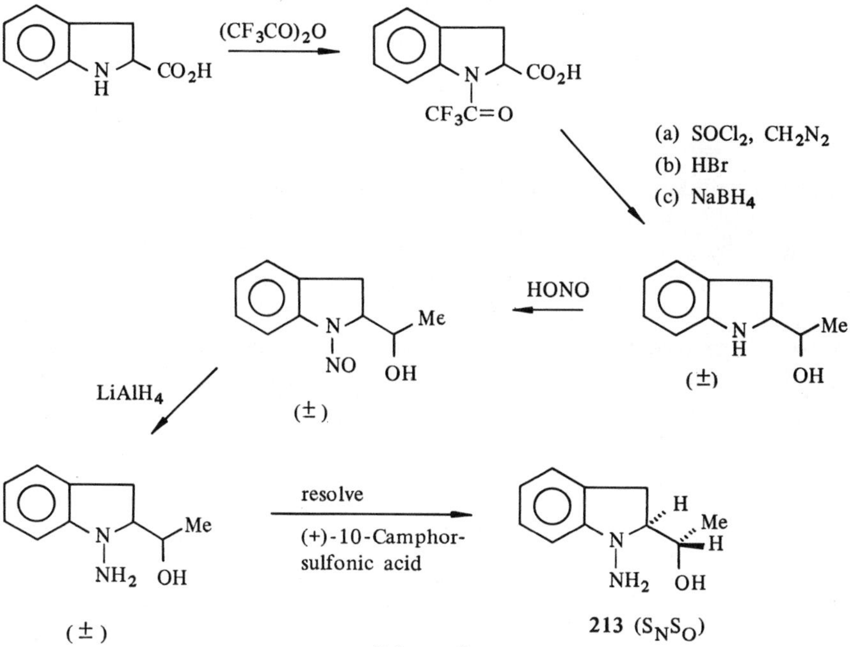

Scheme 2

Corey also showed that by specific structural modifications it was possible to isolate either chiral antipode (S,R)-**213** and (S,S)-**213** and these may in turn be used to predictably prepare either enantiomer of amino acids.

H. From Nicotinamides

A novel, although limited, approach to ω-amino acids **222** bearing an α-cyano group was reported by Quin.[64] This rather interesting process was based on the easy accessibility of the tetrahydronicotinamide **220** from partial hydrogenation[65] of nicotinamide **219**. By simply treating a methanolic solution of **220** with hydroxylamine at room temperature, the isoxazolin-5-one **221** was isolated in 48% yield. It was observed that the latter was quite labile to heat

and upon refluxing a solution of **221** in ethanol, the amino acid **222** was formed in good yield. On the other hand, the amino acid could be isolated in a single operation by subjecting a solution of **220** and hydroxylamine to reflux. The availability of the tetrahydronicotinamide **220** is significant in light of the many synthetic uses that systems of this type have been associated with. Wenkert[66] has commented on the value of β-acyl enamines **223** in alkaloid synthesis but a discussion of this subject is outside the scope of this monograph.

X = R, OEt, H
R = Alkyl

223

Whether or not **223** could be utilized as a masked ω-amino pentanoyl group (e.g., **220**) in complex syntheses remains to be seen.

I. From Azetidinones

Hydrolysis of the four-membered heterocycle, azetidinone **224**, leads to β-amino acids. Most of the research efforts over the years have concerned themselves

with the reverse reaction, namely azetidinone formation from β-amino acids. This was almost exclusively prompted by the presence of the azetidinone moiety in penicillins and related antibiotics.[67] However, an efficient synthesis of azetidinones would also serve as a route to β-amino acids provided the method originated from sources other than the amino acid. Such an approach was described by Graf[68,69] who found that olefins underwent cycloaddition to chlorosulfonyl isocyanate (CSI) **225** producing the N-chlorosulfonyl azetidinones **227** which, after treatment with thiophenol gave the azetidinones **227**.

Thus a channel to β-amino acids is cleared starting with alkenes. Moriconi[70] described the stereospecific *cis* addition of **CSI** to geometrically pure olefins. The adducts, upon removal of the chlorosulfonyl group and hydrolysis, led to geometrically pure β-amino acids. For example, **CSI** addition to *cis*-β-methylstyrene gave the *cis*-azetidinone **228** while addition to *trans*-β-methylstyrene gave the *trans*-azetidinone **230**. Removal of the chlorosulfonyl groups[71] and hydrolytic cleavage delivered the *erythro*-(**229**) and *threo*-(**231**) β-aminopropionic acids, respectively. The reactions are simple and the yields are high making this method potentially quite useful in amino acid syntheses.

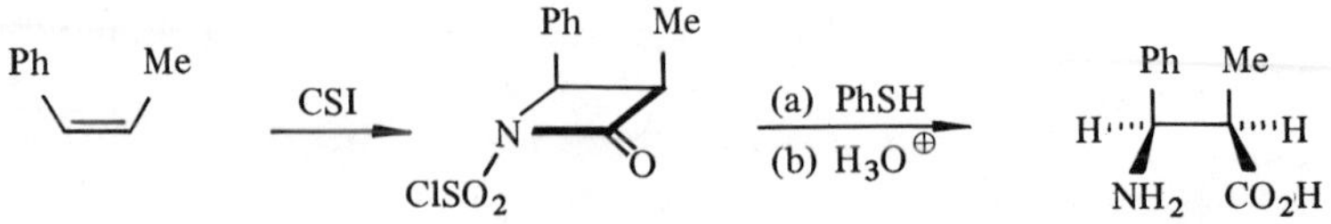

III.   AMIDES AND PEPTIDES

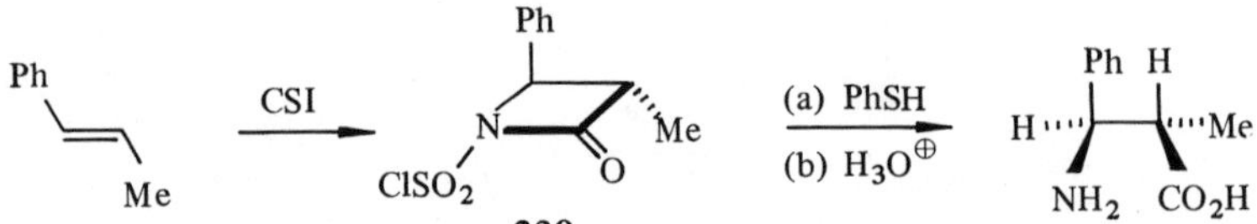

A.   From Oxaziranes

Since their discovery in 1952 by Krimm[72] and independently by Emmons[73] and Horner and Jurgens,[74] these simple and interesting heterocycles 232 have demonstrated their high energy content in a number of varied reactions. A review by Schmitz[75] describes their formation and properties. Of interest

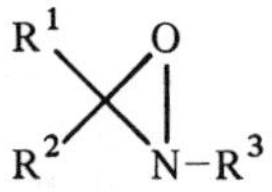

$$(R^1 = R^2 = R^3 = \text{alkyl, aryl})$$

in this work is the facile oxidative ring opening which leads to amides and shown by Emmons[76] to have preparative value. For example, peracid oxidation

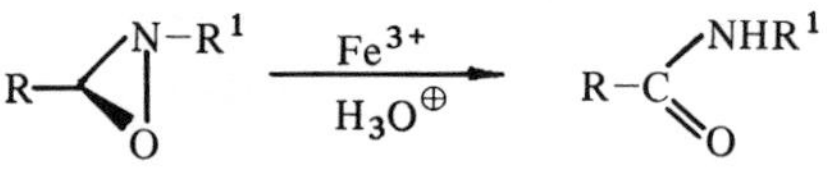

of Schiff bases 235 in dichloromethane generates the oxazirane 236 which, after treatment with aqueous ferrous ammonium sulfate, is further oxidized to N-t-butylbenzamide 237 in 98% yield. The conversion of an aldehyde to its N-alkyl amide via 236 is a useful synthetic process. Other examples include the conversion of oxazirane 238, derived from formaldehyde, to the formamide 239. When the oxazirane is geminally substituted (e.g., 240), the amido

$$PhCHO + H_2N-\hspace{-4pt}\overset{|}{\underset{|}{-}}\hspace{-4pt}\longrightarrow PhCH=N-\hspace{-4pt}\overset{|}{\underset{|}{-}}\hspace{-4pt}\xrightarrow{CH_3CO_3H} Ph-\hspace{-4pt}\underset{O}{\overset{N-\hspace{-4pt}\overset{|}{\underset{|}{-}}}{<}}$$

235 236

$$Fe^{3+} \downarrow \begin{array}{l} H_3O^{\oplus} \\ (98\%) \end{array}$$

238   239   237

240   241   dimerization   242

radical **241** is postulated as an intermediate enroute to the C_{12}-carboxylic acid diamide **242**. The mechanism for the conversion of oxaziranes to amides seems generally accepted to involve a one-electron transfer from ferrous ion to

$$Fe^{2+} + R-\underset{O}{\overset{N-R'}{<}} \xrightarrow{H^{\oplus}} Fe^{+3} + R-\underset{O\cdot}{\overset{H}{<}}\overset{NR'}{} \xrightarrow[R-\underset{O}{\overset{NR'}{<}}]{} \underset{O}{\overset{}{R}}CNHR'$$

243   244

$$+$$

$$etc. \longleftarrow R\underset{O\cdot}{\overset{}{C}}-NHR'$$

the oxazirane ring forming the radical **243**. Attack by a second molecule of oxazirane leads to the amide **244** and a new radical which propogates the chain reaction. Because of the presence of the oxy-radical **243**, the nature of the adjacent R group may alter the course of the reaction which, in such instances, may lead to other products. Thus, in the case of **240**, alkyl cleavage results, albeit in an advantageous manner. Thermolysis also causes rearrangement of oxaziranes and in some cases, very useful ones. Pyrolysis of *N-isobutyl-3,3-*

pentamethylene oxazirane **245** gave *N*-isobutyl caprolactam in 83% yield.

245

B. From Imidazoles

As already discussed in Chapter 9, Section I.E, the ability to transform a carboxylic acid to its acyl imidazole **247** via Staab's reagent **246** has provided the chemist with a convenient active ester. This was immediately accepted by the peptide community as a potentially valuable coupling reagent for amino acids. Anderson[77] has demonstrated that a number of fairly complex

246   **247**

$R'NH_2$ (R = peptide residue)

RC—NHR'

248

peptides **248** could be efficiently formed with a minimum degree of racemization and thus the acyl imidazole may take its place among the most reliable of peptide coupling reagents.

C. From Isoxazolium Salts

Woodward[78] showed that carboxylic acids (i.e., **250**) react rapidly and smoothly with *N*-ethyl isoxazolium salts **249** under very mild conditions to yield activated enol esters **251**. The latter undergo a facile reaction with amino groups (i.e., **252**) affording the coupled peptide **254** in high yield with less than 2% racemization (in the case of optically active substances). The accompanying product **253**, being a sulfonic acid, is easily separated from the peptide by aqueous washing. In this manner a number of polypeptides were prepared and attested to the value of the reagent, now termed "Woodward's Reagent K." Since the original disclosure of this method, there have been many variations on the

249 250 (Z = Cbz) 251 252

254 253

structure of **249** in order to enhance its reactivity, convenience in handling, and accessibility. The initial step in the coupling process has been postulated to proceed via the keto-ketenimine **256** formed by ring opening of the ylid **255**. The ketenimine has been directly observed[79] and its existence subsequently confirmed by isolation.[79a] Attack by the carboxylate species proceeds across

255 256 257

251

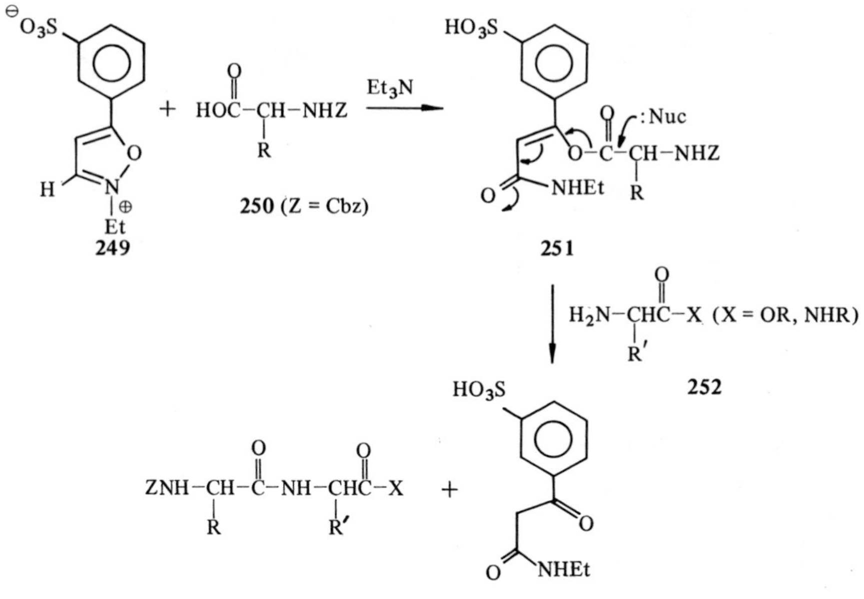

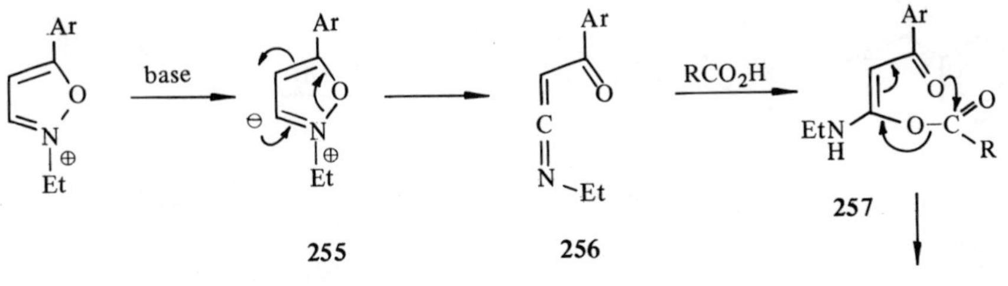

the reactive cumulative system producing **257** which transacylates to the active ester **251**. The use of the benzisoxazolium salt **258** was also reported[80,81] to yield active esters **259**, presumably through the same general pathway. Peptide coupling occurs in good yield (70-90%) leading to protected polypeptides (e.g., **260**). The benzoisoxazolium reagent **258** is obtained[81] from salicylaldehyde and amino sulfonic acid followed by quaternization with triethyloxonium fluoroborate. Synthesis of the monocyclic isoxazolium salts 249 is accomplished[78] from acetophenone and ethyl formate followed by oxime formation. Dehydration to the isoxazole was then performed. Chlorosulfonation of the aryl

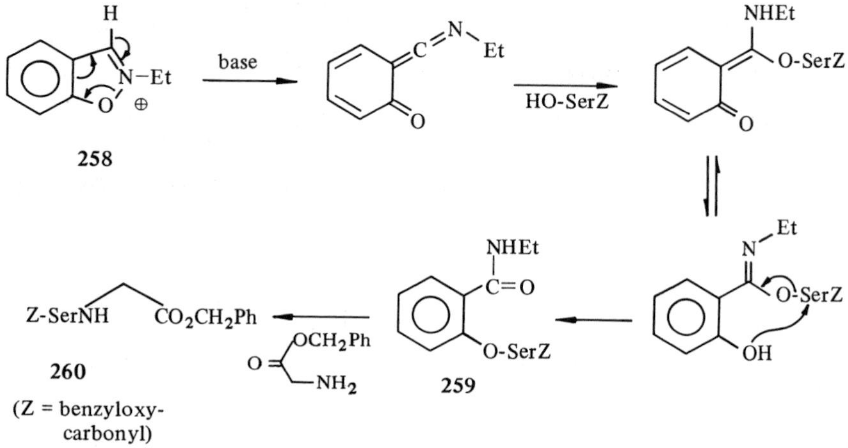

substituent gave a mixture (*meta* and *para*) of sulfonyl chlorides which were transformed into the methyl esters. The latter, when heated neat, generated the zwitterionic isoxazolium salts in what Woodward termed a "bootstrap" reaction, the sulfonate ester acting both as an alkylating agent and the species alkylated. The *meta*-sulfonic acid derivative could be separated from the *para*-derivative by crystallization from acid solution.

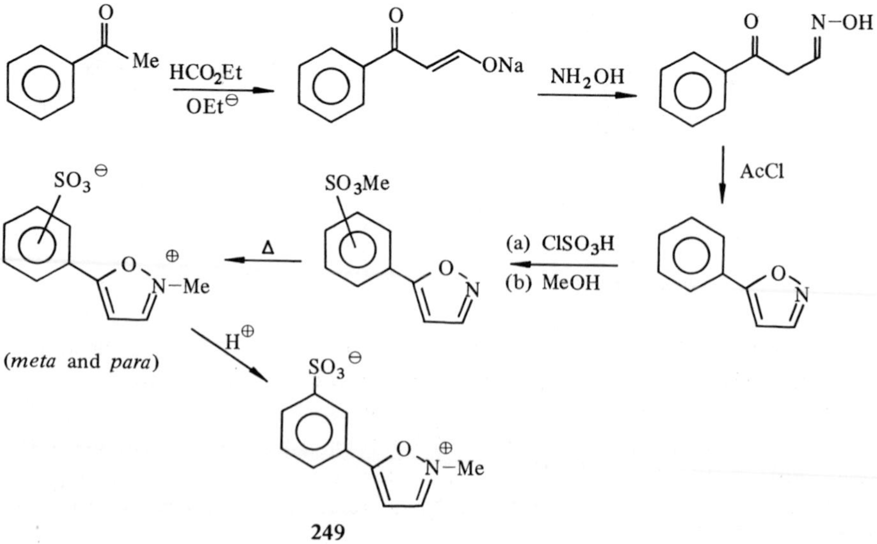

D.   From Oxazoles

Earlier in this chapter (Section I.D), the reaction of 4,5-polymethyleneoxazoles with singlet oxygen was mentioned as a route to ω-cyano carboxylic acids. With respect to the synthesis of amide derivatives, the oxazole has also exhibited

its versatile chemistry. Wasserman[82] described this behavior for 4,5-disubstituted oxazoles which are not part of a cyclic system. Photooxidation of various oxazoles **261** produced the triamides **262** in 55-85% yields. The process may be

261 (R = Ph, Me, H)   **262**

implemented with several different substituents, thus giving mixed triamides.[82a] The mechanism has been elucidated recently[83] using O^{18} labeling and proceeds according to the equations set forth below. Since there are very few methods to cleanly prepare triamides, the route opened up by oxazoles should be considered

preparatively useful. The potential of triamides as acylating agents should also be examined.

E.   From 2-Ethoxypyridine 1-Oxide

Paquette[84] described a new class of highly activated esters **265** derived from the pyridine 1-oxide **263**, via O-acylation and subsequent halide displacement **(264)**. The ester **265** was found to react smoothly at room temperature with an amine component (e.g., glycine methyl ester) producing the dipeptide **266** in 80% yield. When the process was repeated with optically active derivatives, the coupling took place without any significant racemization. Although the method demonstrates the ability of a heterocycle to serve, once again, as a useful

reagent, it suffers from the fact that acid chlorides (themselves activated carbonyls) are required to reach 265. Taylor and McKillop,[85] armed with thallium chemistry, enhanced the efficiency of the formation of the active ester 265 by reacting the thallium salt of 267 with protected amino acid halides. In this fashion yields were somewhat higher and conditions for the formation generally milder. Of greater significance was their preparation of active esters 265 directly from the carboxylic acid—the more coveted route to peptide coupling. By treating the 1-hydroxy pyridone 267 with thionyl chloride, the presumed chlorosulfite 268 intermediate was subjected, without isolation, to reaction with the thallium(I) carboxylate and then the amine component. In this manner dipeptides were formed, without racemization, in yields of 30-50%.

F. From 2,2'-Dipyridyldisulfide

A useful and practical peptide synthesis was described by Mukaiyama[86] which results in high yields and high optical purity. The method calls upon the oxidized dimer of 2-mercaptopyridine, namely 2,2'-dipyridyl disulfide 269. A mixture of the latter, triphenyl phosphine, the acid, and the amine components

269 **270**

271 R^1 = alkyl, aryl
$R^2 = CH_2CO_2Et$

were simply allowed to react in methylene chloride at room temperature. After washing and evaporation of the solvent, the peptide **271** was isolated in 91% yield and 96% optical purity. This remarkably simple procedure was repeated with a number of examples and gave yields of 70-93% with optical purities of 95-97%. The pentavalent phosphorous species **270** is believed to be responsible for the coupling reaction by allowing the amine to be acylated. The concomitant formation of triphenylphosphine oxide and pyridine-2-thione provides the necessary driving force.

G. From Saccharin

When saccharin **272** is heated with phosphorus pentachloride, the chloro derivative **273** is obtained. This heterocycle has been successfully employed[87] as a peptide coupling agent and possesses many of the virtues required for this process. Besides allowing coupling of carboxylic acids and amines without prior activation, it may also be recovered and recycled for further use. The

272 **273** **274**

276 **275**

carboxyl component reacts in most cases with **273** at 0° in dichloromethane to produce the active ester **274**. The yields of the latter vary from moderate to excellent. Addition of the amine component (which may be used with an unprotected carboxyl group) leads to efficient acylation with minimal racemization and ultimately to the peptide **275**. The highly delocalized saccharin anion **276** represents an excellent leaving group during the acylation and may be removed from the peptide product simply by washing.

H.   From Quinolines

In a search for new central nervous system depressants, Belleau discovered the highly active[88] dihydroquinoline derivative **277**. However, in addition to its potent pharmocological activity, this easily accessible[89] heterocycle also exhibited chemical properties that are of significant synthetic value. The key feature of **277** is its ability to react with OH and SH functions present in

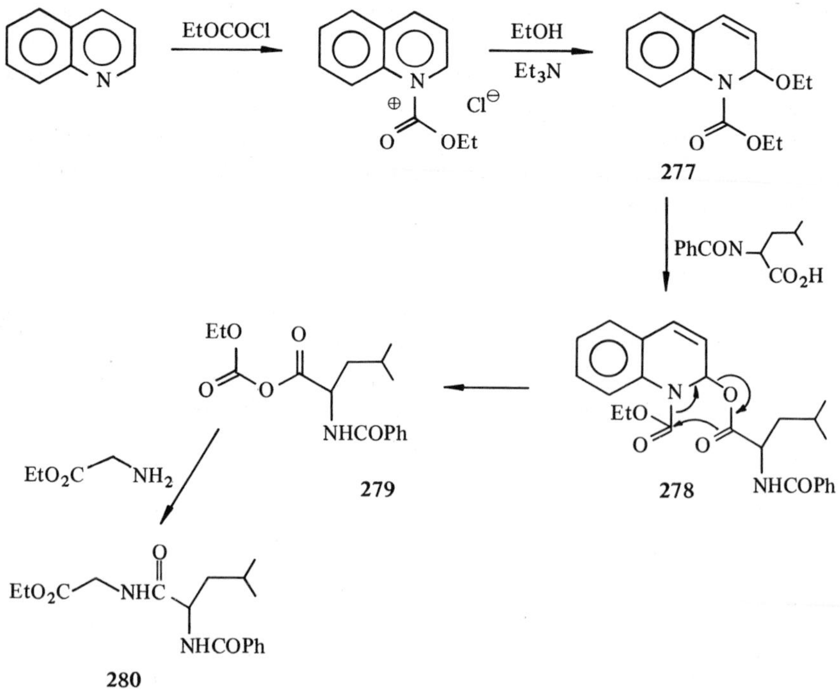

carboxylic acids and thiols to form the transient adduct **278**. By rapid intramolecular rearrangement, the mixed anhydride is efficiently produced which reacts further with an amine component present in the medium. In this fashion, a high yield (95%) of Bz-Leu-Gly-OEt **280** results and renders this technique one of the more powerful routes to peptides.[90] Of further note is the fact that the dihydroquinoline reagent 277 does not react with amines and, therefore, all the components of the peptide synthesis may be added at once. Racemization of optically active amino acid derivatives during this process has been reported to be less than 0.2%.[91] Recently the isobutyl derivative 281 was prepared and found to also possess excellent coupling capabilities.[92]

281

IV. NITRILES

A. From Isoxazoles

A practical technique for introducing a cyano group α- to a carbonyl involves the temporary construction of an isoxazole. Meyer[93] prepared the 2-cyano-cyclohexanone 286 by initially treating 2,2-dimethylcyclohexanone with ethyl formate under Claisen conditions producing the 2-hydroxymethylene derivative 282. Direct treatment with hydroxylamine gave an isoxazole 287 which proved to be synthetically useless because of the absence of a proton at the 3-position. By converting 282 to its isopropyl enol ether 283 and then preparing the transient oxime 284, the required isoxazole 285 was obtained. Base catalyzed ring opening by removal of the proton at C-3 led to the desired cyano-ketone. Similarly, Kuehne[94] prepared the cyano ketones 288 and 289 by calling upon the isoxazole to perform this useful task. In effect, the isoxazole represents

282

283

287

248

287

284

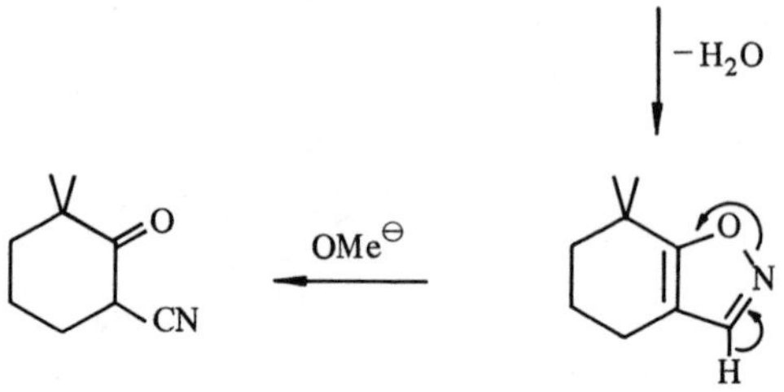

$-H_2O$

286

$OMe^{\ominus}$

285

288

Me

289

an electrophilic cyano group ($\oplus$CN) which may be envisioned to "cyanate" enolate ions.

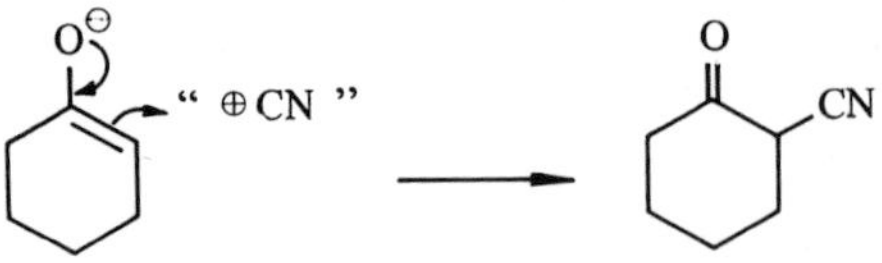

"$\oplus$ CN"

B. From 1,2,5-Oxadiozoles

Boulton[95] and Moffat[96] independently reported the decomposition of 1,2,5-oxadiazoles to nitriles. Although the former author's interest was aimed at furazan syntheses, the latter recognized the general synthetic potential and reported some cursory experiments. Since it was already known[97] that N-oxides of furazans 290 could be deoxygenated to the furazan (1,2,5-oxadiazole) by heating with trialkylphosphites at 160-170°, Boulton attempted this reaction and obtained only 1,8-dicyanonaphthalene. Similarly, treatment of the camphor heterocycle **291** under these conditions gave, as the only product, 1,2,2-trimethyl-1,3-dicyanocyclopentane. Moffat, on the other hand, heated simple disubstituted oxadiazoles **292** at 270° and observed, in the several cases studied, preparatively useful yields of the nitriles. The deoxygenated heterocycle **293** was undoubtedly formed first since oxygen extrusion by phosphorous compounds occurs, as mentioned above, at 160-170°.

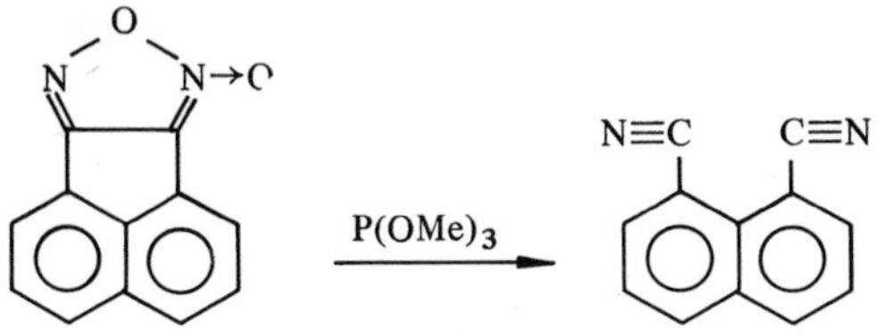

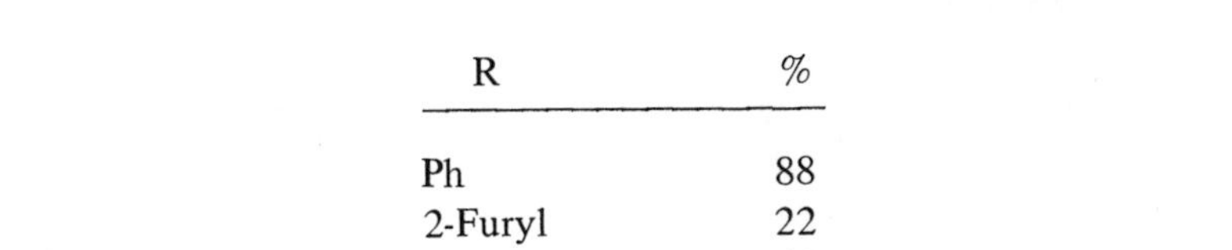

R	%
Ph	88
2-Furyl	22
Et	65

If the method of oxadiazole construction is taken into account, namely the dehydration of dioximes derived from α-diketones (Scheme 3), a synthetically

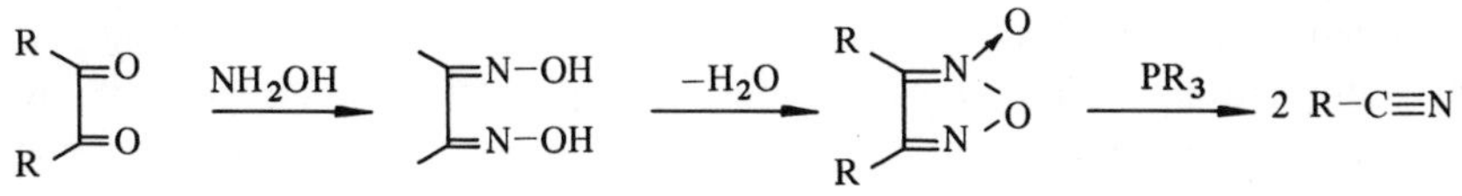

Scheme 3

useful route to nitriles begins to take form. From this author's vantage point regarding elaboration of heterocyclic compounds prior to their destruction, it becomes rapidly evident that starting from the readily available and simplest diketone, biacetyl, one may have in hand a generally useful synthesis of substituted acetonitriles (Scheme 4).

Ⓔ = alkyl halides, carbonyl compounds, epoxides, etc.

Scheme 4

The report[98] that the 1,2,5-oxadiazole **294** gave an 82% yield of benzonitrile upon irradiation is of further interest in this regard.

294, R = Ph
295, R = Me

296

The 3,4-dimethyl derivative **295** under similar circumstances produced aceto-nitrile, in somewhat lower yield (41%) along with the trapped adduct **296** originating from acetonitrile oxide and cyclopentene. The latter was used as a cosolvent during the irradiation. When the sulfur analogs, 1,2,5-thiadiazoles **297** and **298** were subjected to light, benzonitrile and acetonitrile were

297, R = Ph
298, R = Me

generated, respectively. In this series, elemental sulfur was formed. Armed with this information, it should be feasible to employ the thiadiazole **298** as a precursor to nitriles (according to Scheme 4) rather than the oxygen hetero-cycle. Because of the greater ease of sulfur extrusion over oxygen with phos-phorous nucleophiles, the final destruction of the elaborated heterocycle derived from **298** could conceivably take place under milder conditions.

C. From 1,3-Dithiol-2-ones

Dicyanoacetylene **301**, whose interesting structural features have been the subject of considerable study, has recently been prepared in preparative quantities by a method involving the dithiolone **300**.[99] The procedure relies on the reaction of sodium cyanide and carbon disulfide to produce the dimercapto salt of maleonitrile **299** and further reaction with phosgene leading to the

heterocyclic precursor **300**. Pyrolysis, under reduced pressure, affords dicyanoacetylene **301** in 59-76% yield. The overall conversion from sodium cyanide is 43% without purification of intermediates **299** and **300**.

D. From Tetrazolo[1,5-b]pyridazines

A route to cyclopropenyl nitriles **306** has been opened by irradiation of the titled heterocycle **302** in dichloromethane for 6-7 hours.[100] The yields of isolated cyclopropenyl nitriles in this one-step process ranged from 20 to 25%— not an inefficient conversion for reaching this interesting nitrile. The pathway leading from starting material to product has been envisioned as proceeding through intermediates **303, 304,** and **305**. Since the latter intermediate may be expected to tautomerize to the cyano pyrazole **307** and this was indeed isolated (0.1% yield), the suggested mechanism is probably correct.

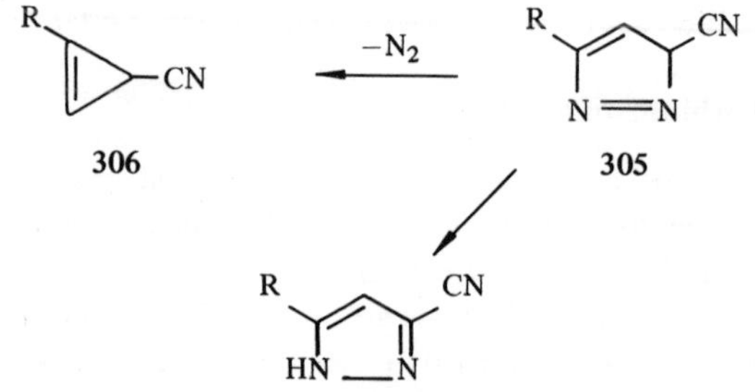

306 ← $-N_2$ ← **305**

307 (trace)

E. From Pyridine-*N*-oxide

Addition of Grignard reagents to pyridine *N*-oxide was reported[101] to lead to 1,2-dihydropyridines **313** which were then converted to 2-substituted pyridines **312** by acetic anhydride. Kellogg has reinvestigated this reaction[102] and found that the Grignard-pyridine *N*-oxide adduct was not **313**, but the open chain unsaturated oxime **310**. A series of aryl Grignard reagents all gave the oxime in 10-45% yield. This unusual ring opening process, postulated to proceed initially to the adduct **308**, followed by rearrangement to **309** provides a simple and synthetically useful route to conjugated nitriles **311**. The oxime **310** was

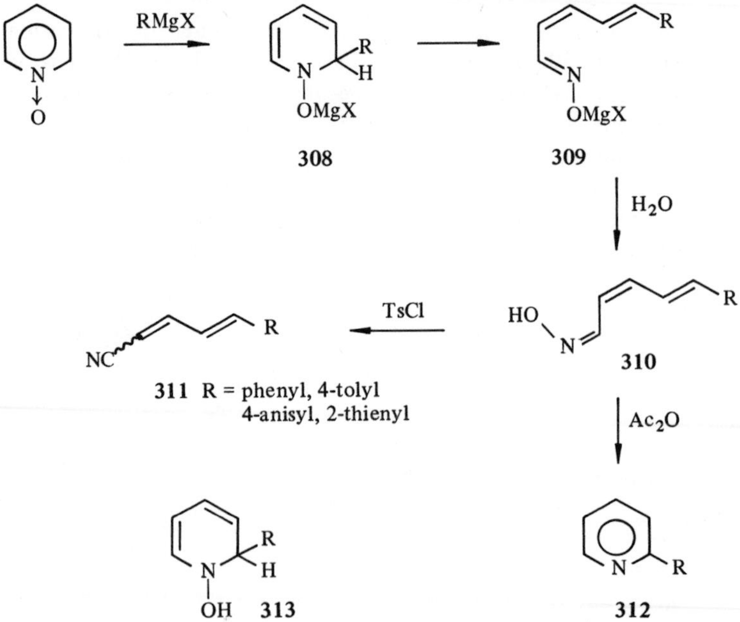

308   **309**

H_2O

311 R = phenyl, 4-tolyl
4-anisyl, 2-thienyl

TsCl

310

Ac_2O

313   **312**

formed stereospecifically and its configuration was established by spectroscopic means. When treated with acetic anhydride, the oxime underwent cyclodehydration to the 2-arylpyridine, **312**, the product observed in the initial[100] study on this reaction. Production of the unsaturated nitriles **311** was readily accomplished by dehydrating the isolated oximes with toluenesulfonyl chloride in pyridine in yields of 67-85%. This behavior of pyridine N-oxide depicts still another example in the growing number of discoveries concerning "invisible" reactions of heterocycles. That is, the "apparent" direct introduction of substituents on a heterocycle which in reality proceeds via a ring opening-ring closing pathway.[103]

F. From Benzotriazoles

Pyrolysis of benzotriazoles **314** has been extensively studied by Crow[104] and found to be a unique route to 1-cyano cyclopentadienes **315** (80-100%). Although the procedure requires temperatures of 500-800°, the apparatus is not

unduly difficult to utilize[105] and may possess some preparative value. Evidence that the reaction proceeds through the diradical **316** has been offered by the isolation of aniline.

G. From 2-Oxazolines

A general synthetic method for transforming ketones into nitriles **321** of one additional carbon relies on the transient formation of an oxazoline **318**.[106] By addition of a ketone to the anion of toluenesulfonylmethyl cyanide (TosMIC), the adduct **317** is obtained which has already been described in earlier sections of this chapter (Sections I.C and II.E). The intermediate oxazoline **318**, a derivative of which has been isolated and characterized, proceeds

$$\text{R}_2\text{C=O} \xrightarrow[\text{OEt}^{\ominus}]{\text{TsCH}_2\text{NC}} \left[\begin{array}{c} \text{317} \\ \text{318} \end{array} \right]$$

317

318

$$\text{R}_2\text{C-C≡N} \longleftarrow \text{R}_2\text{C=C=N-CHO} \longleftarrow \text{319}$$

321   **320**   **319**

to ring open to **319** and after elimination of the tosyl group furnishes the N-formyl ketenimine **320**. In the protic solvent (ethanol) and the presence of base, the latter decomposes to the nitrile **321**. The pathway from ketone to nitrile has been documented by C^{14} labeling techniques which showed that the methylene carbon of TosMIC eventually becomes the cyanide carbon in the product.

This method, carried out in a single operation, gives yields of nitriles in the 75-85% range. In the case of a hindered ketone, t-butylmethylketone, the yield of nitrile drops to 36%. A series of nitriles were prepared using this route and they are presented below.

Ketone	Nitrile	%
Adamantanone	2-Cyanoadamantane	85
Cycloheptanone	Cycloheptyl cyanide	80
Cyclohexanone	Cyclohexyl cyanide	80
4-Heptanone	4-Cyanoheptane	75
t-Butylmethyl ketone	2-Cyano-3,3-dimethylbutane	36
Acetophenone	1-Phenylpropionitrile	80
p-Bromoacetophenone	1-(p-Bromophenyl)propionitrile	79

It is interesting to compare this reaction route to that reported by Schöllkopf[18] leading to homologated carboxylic acids from ketones (p. 252). In the earlier process, aqueous quenching of the nitrile precursor **319** affords the acid, whereas omitting this step provides the nitriles.

REFERENCES

1. H. Wynberg and A. Logothetis, *J. Am. Chem. Soc.*, **78**, 1958 (1956).
2. S. Hansen, *Acta Chem. Scand.*, **8**, 695 (1954).
3. N. P. Buu-Hoi, M. Sy, and N. D. Xuong, *Bull. Soc. Chim., France*, 1583 (1955).
4. S. Gronowitz, "Chemistry of Thiophenes," in *Advances in Heterocyclic Chemistry*, **1**, A. R. Katritzky, ed., Academic Press, New York, 1963, p. 1.
5. A. Hassner, M. J. Haddadin, and P. Catsoulacos, *J. Org. Chem.*, **31**, 1363 (1966).
6. J. W. Cornforth, *Heterocyclic Compounds*, Vol. 5, R. C. Elderfield, ed., Wiley, New York, 1957, p. 386.
7. H. L. Wehrmeister, *J. Org. Chem.*, **27**, 4418 (1962).
8. A. I. Meyers and D. L. Temple, *J. Am. Chem. Soc.*, **92**, 6644 (1970).
8a. Unpublished results, E. D. Mihelich and A. I. Meyers.
9. M. W. Rathke and J. Deitch, *Tetrahedron Lett.*, 2953 (1971), and earlier papers cited.
10. P. E. Pfeffer, E. Kinsel, and L. S. Silbert, *J. Org. Chem.*, **37**, 1256 (1972), and earlier papers cited.
11. A. I. Meyers and D. L. Temple, Jr., *J. Am. Chem. Soc.*, **92**, 6646 (1970).
12. R. L. Shriner, *Org. React.*, **1**, 14 (1942).
13. G. W. Moersch and A. R. Burkett, *J. Org. Chem.*, **36**, 1149 (1971).
14. P. Allen and J. Ginos, *ibid.*, **28**, 2759 (1963).
15. R. Adams and F. Leffler, *J. Am. Chem. Soc.*, **59**, 2252 (1937).
16. A. I. Meyers and D. Haidukewych, *Tetrahedron Lett.*, 3031 (1972).
17. Dr. D. Haidukewych, unpublished results.
18. U. Schöllkopf and R. Schroder, *Angew. Chem. Int. Ed.*, **11**, 311 (1972).
19. A. I. Meyers and E. W. Collington, *J. Am. Chem. Soc.*, **92**, 6676 (1970); A. I. Meyers and H. W. Adickes, *Tetrahedron Lett.*, 5151 (1969).
20. F. E. Ziegler and P. A. Wender, *J. Am. Chem. Soc.*, **93**, 4318 (1971).
21. H. H. Wasserman and E. Druckery, *ibid.*, **90**, 2440 (1968); H. H. Wasserman, E. Druckery, and G. R. Lenz, *ibid.*, **95**, 0000 (1973).
22. H. Bredereck and R. Gompper, *Chem. Ber.*, **87**, 726 (1954).
23. W. Steglich and P. Gruber, *Angew. Chem. Int. Ed.*, **10**, 655 (1971).
24. L. A. Carpino, *J. Am. Chem. Soc.*, **80**, 599 (1958).
25. L. A. Carpino, *ibid.*, **80**, 601 (1958).
26. L. A. Carpino, P. H. Terry, and S. D. Thatte, *J. Org. Chem.*, **31**, 2867 (1966).
27. L. A. Carpino and E. G. S. Rundberg, *ibid.*, **34**, 1717 (1969).
28. E. C. Taylor, R. L. Robey, and A. McKillop, *Angew. Chem. Int. Ed.*, **11**, 48 (1972).
29. E. C. Taylor, R. L. Robey, and A. McKillop, *J. Org. Chem.*, **37**, 2798 (1972).
30. W. Adam and R. Rucktäschel, *J. Am. Chem. Soc.*, **93**, 557 (1971).
31. E. J. Corey and G. Märkl, *Tetrahedron Lett.*, 3201 (1967).

32. E. D. Bergmann and I. Shahak, *J. Chem. Soc., C,* 1005 (1966).
33. E. L. Eliel and A. A. Hartmann, *J. Org. Chem.,* **37,** 505 (1972).
34. R. B. Moffett, *Org. Syn. Coll. Vol. IV,* 427 (1963).
35. D. Seebach, *Synthesis,* **1,** 17 (1969).
36. J. A. Marshall and H. Roebke, *Tetrahedron Lett.,* 1555 (1970).
37. J. A. Marshall and H. Roebke, *J. Org. Chem.,* **34,** 4188 (1969).
38. A. I. Meyers and N. Nazarenko, *ibid.,* **38,** 175 (1973).
39. A. I. Meyers and T. Kowar, unpublished results.
40. C. Fuganti, D. Ghiringhelli, and P. Grasselli, *Chem. Commun.,* 1152 (1972).
41. Ya. L. Goldfarb, B. P. Fabrichnyi, and I. F. Shalavina, *Tetrahedron,* **18,** 21 (1962).
42. See ref. 4, p. 113.
43. V. Bocchi, L. Chierici, G. P. Gardini, and R. Mondelli, *Tetrahedron,* **26,** 4073 (1970).
44. V. Bocchi and G. P. Gardini, *Org. Prep. Proced.,* **1,** 271 (1969).
45. J. W. Cornforth, "Oxazolones," in *Heterocyclic Compounds,* R. C. Elderfield, ed., Vol. 5, Wiley, New York, 1957, p. 336.
46. R. Filler, "Recent Advances in Oxazolone Chemistry," in *Advances in Heterocyclic Chemistry,* Vol. 4, A. R. Katritzky, ed., Academic Press, New York, 1965, p. 75.
47. H. B. Gillespie and H. R. Snyder, *Org. Syn. Coll. Vol. II,* 489 (1943).
47a. For improved procedures, see A. Badshah, N. H. Khan, and A. R. Kidwai, *J. Org. Chem.,* **37,** 2916 (1972).
48. R. Filler and Y. S. Rao, *J. Heterocycl. Chem.,* **1,** 153 (1964).
49. L. Horner and H. Schwahn, *Annals,* **591,** 99 (1955).
50. H. Behringer and H. Taul, *Berichte,* **90,** 1398 (1957).
51. S. H. Pines and M. Sletzinger, *Tetrahedron Lett.,* 727 (1969).
52. S. H. Pines, S. Karady, and M. Sletzinger, *J. Org. Chem.,* **33,** 1758 (1968).
53. E. Ware, *Chem. Rev.,* **46,** 403 (1950); E. S. Schipper and A. R. Day, in *Heterocyclic Compounds,* Vol. 5, R. C. Elderfield, ed., Wiley, New York, 1957.
54. H. Finkbeiner, *J. Org. Chem.,* **30,** 3414 (1965).
55. M. Stiles, *J. Am. Chem. Soc.,* **81,** 2598 (1959).
56. H. L. Finkbeiner and M. Stiles, *ibid.,* **85,** 616 (1963).
57. D. Ben-Ishai and G. Ben-Et, *Chem. Commun.,* 1399 (1969).
58. E. Goldstein and D. Ben-Ishai, *Tetrahedron Lett.,* 2631 (1969).
59. G. Ben-Et and D. Ben-Ishai, *Chem. Commun.,* 376 (1969).
60. U. Schöllkopf and D. Hoppe, *Angew. Chem. Int. Ed.,* **9,** 459 (1970).
61. J. P. Vigneron, H. Kagan, and A. Horeau, *Tetrahedron Lett.,* 5681 (1968).
62. E. J. Corey, R. J. McCaully, and H. S. Sachdev, *J. Am. Chem. Soc.,* **92,** 2476 (1970).
63. E. J. Corey, H. S. Sachdev, J. Z. Gougoutas and W. Saenger, *ibid.,* **92,** 2488 (1970).
64. L. D. Quin and D. O. Pinion, *J. Org. Chem.,* **35,** 3130 (1970).
65. P. M. Quan and L. D. Quin, *ibid.,* **31,** 2487 (1966).
66. E. Wenkert, *Acc. Chem. Res.,* **1,** 78 (1968).
67. M. S. Manhas and A. K. Bose, *Beta-Lactams,* A. K. Bose, ed., Wiley-Interscience, New York, 1971.
68. R. Graf, *Berichte,* **89,** 1071 (1956); *Angew. Chem. Int. Ed.,* **7,** 172 (1968).

69. R. Graf, *Annals,* **661,** 111 (1963); see *Org. Syn.,* **46,** 23, 51 (1966).

70. E. J. Moriconi and J. F. Kelly, *Tetrahedron Lett.,* 1435 (1968).

71. T. Durst and M. J. O'Sullivan, *J. Org. Chem.,* **35,** 2043 (1970).

72. H. Krimm and K. Hamann, Ger. Patent 952,895 (1952); H. Krimm, K. Hamann, and K. Bauer, U.S. Patent 2,686,739 (1953).

73. W. D. Emmons, *J. Am. Chem. Soc.,* **78,** 6208 (1956).

74. L. Horner and E. Jurgens, *Berichte,* **90,** 3285 (1957).

75. E. Schmitz, *Advances in Heterocyclic Chemistry,* Vol. 2, A. R. Katritzky, ed., Academic Press, New York, 1963, p. 85; *Angew. Chem. Int. Ed.,* **3,** 333 (1964).

76. W. D. Emmons, *J. Am. Chem. Soc.,* **79,** 5739 (1957).

77. R. Paul and G. W. Anderson, *J. Org. Chem.,* **27,** 2094 (1962); *J. Am. Chem. Soc.,* **82,** 4596 (1960).

78. R. B. Woodward, R. A. Olofson, and H. Mayer, *Tetrahedron, Supplement No. 8,* Part I, **22,** 321 (1966).

79. R. B. Woodward and R. A. Olofson, *ibid., Supplement No. 7,* 415 (1966).

79a. R. B. Woodward and D. J. Woodman, *J. Am. Chem. Soc.,* **88,** 3169 (1966).

80. S. Rajappa and A. S. Akerkar, *Chem. Commun.,* 826 (1966).

81. D. S. Kemp and R. B. Woodward, *Tetrahedron,* **21,** 3019 (1965).

82. H. H. Wasserman and M. B. Floyd, *Tetrahedron, Supplement No. 7,* **22,** 441 (1966).

82a. Thiazoles have also been reported to photolytically cleave to (monothio)-*tris*-amides; T. Matsuura and I. Saito, *Bull. Chem. Soc. (Japan),* **42,** 2973 (1969); H. H. Wasserman and G. R. Lenz, private communication.

83. H. H. Wasserman, F. J. Vinick, and Y. C. Chang, *J. Am. Chem. Soc.,* **94,** 7182 (1972).

84. L. A. Paquette, *ibid.,* **87,** 5186 (1965).

85. E. C. Taylor, F. Kienzle, and A. McKillop, *J. Org. Chem.,* **35,** 1672 (1970).

86. T. Mukaiyama, R. Matsueda, and M. Suzuki, *Tetrahedron Lett.,* 1901 (1970).

87. F. Micheel and M. Lorenz, *ibid.,* 2119 (1963).

88. B. Belleau, R. Martel, G. Lacasse, M. Menard, N. L. Weinberg, and Y. G. Perron, *J. Am. Chem. Soc.,* **90,** 823 (1968).

89. B. Belleau, in *Reagents for Organic Synthesis,* M. Fieser and L. F. Fieser, Vol. 2, Wiley-Interscience, New York, 1969, p. 191.

90. B. Belleau and G. Malek, *J. Am. Chem. Soc.,* **90,** 1651 (1968).

91. N. Izumiya and M. Muraoka, *ibid.,* 2391 (1969).

92. Y. Kiso and H. Yajima, *Chem. Commun.,* 942 (1972).

93. W. L. Meyer, G. B. Clemans, and R. W. Huffman, *Tetrahedron Lett.,* 4255 (1966).

94. M. E. Kuehne, *J. Org. Chem.,* **35,** 171 (1970).

95. A. Rahman and A. J. Boulton, *Chem. Commun.,* 73 (1968).

96. S. M. Katzman and J. Moffat, *J. Org. Chem.,* **37,** 1842 (1972).

97. T. Mukaiyama, H. Nambu, M. Okamoto, *ibid.,* **27,** 3651 (1962); C. Grundmann, *Berichte,* **97,** 575 (1964).

98. T. S. Cantrell and W. S. Haller, *Chem. Commun.,* 977 (1968).

99. E. Ciganek and C. G. Krespan, *J. Org. Chem.,* **33,** 541 (1968).

100. T. Tsuchiya, H. Arai, and H. Igeta, *Chem. Commun.,* 1059 (1972).

101. T. Kato and H. Yamanaka, *J. Org. Chem.,* **30,** 910 (1965).

102. T. J. van Bergen and R. M. Kellogg, *ibid.*, **36,** 1705 (1971).

103. A. I. Meyers and G. Knaus, *J. Am. Chem. Soc.,* **95,** 3408 (1973); describes other so-called "invisible" intermediates involved in heterocyclic chemistry.

104. W. D. Crow and C. Wentrup, *Chem. Commun.,* 1026 (1968); *Tetrahedron Lett.,* 4379 (1967).

105. W. D. Crow and R. K. Solly, *Austr. J. Chem.,* **19,** 2119 (1966).

106. O. H. Oldenziel and A. M. van Leusen, *Tetrahedron Lett.,* 1357 (1973).

AUTHOR INDEX

307

SUBJECT INDEX